Universitext

Universitext

Universitext is a series of textbooks that presents material from a wide variety of mathematical disciplines at master's level and beyond. The books, often well class-tested by their author, may have an informal, personal, even experimental approach to their subject matter. Some of the most successful and established books in the series have evolved through several editions, always following the evolution of teaching curricula, into very polished texts.

Thus as research topics trickle down into graduate-level teaching, first textbooks written for new, cutting-edge courses may make their way into *Universitext*.

For further volumes:
http://www.springer.com/series/223

Wolfram Koepf

Hypergeometric Summation

An Algorithmic Approach to Summation and Special Function Identities

Second Edition

 Springer

Wolfram Koepf
University of Kassel
Kassel
Germany

Maple^{TM} is a trademark of Maplesoft, headquartered in Waterloo, Ontario, Canada (http://www.maplesoft.com/), and a member of Cybernet Systems Group, headquartered in Tokyo, Japan (http://www.cybernet.co.jp/english/).

The software of this book, as well as the Maple^{TM} sessions, can be obtained from the book's web site http://www.hypergeometric-summation.org.

ISSN 0172-5939   ISSN 2191-6675   (electronic)
ISBN 978-1-4471-6463-0   ISBN 978-1-4471-6464-7   (eBook)
DOI 10.1007/978-1-4471-6464-7
Springer London Heidelberg New York Dordrecht

Library of Congress Control Number: 2014938224

Mathematics Subject Classification: 33C20, 33D15, 33F10, 68W30, 33C45, 33D45, 11B35, 05A19, 05A30, 11B65, 13P05

1st edition: © Vieweg+Teubner Verlag, Wiesbaden, Germany 1998

Printed on acid-free paper

Springer is part of Springer Science+Business Media (www.springer.com)

Preface

The first edition of this book appeared in 1998 and was published by Vieweg (Braunschweig/Wiesbaden). Several years later, the book was sold out and no longer available. Some time ago, I discussed this situation jointly with Ulrike Schmickler-Hirzebruch from Vieweg (which is now Springer-Vieweg) and with Clemens Heine from Springer Germany, and we opted for a second edition published by Springer, this publisher being better linked to the English language market.

The book covers many algorithms for summation and integration, most of which have not changed much in the meantime and are still up-to-date. Fasenmyer's algorithm for definite summation (Chap. 4) is very old, nevertheless it is so easy to describe that it must be included for didactical reasons. Gosper's algorithm (Chap. 5) solves the problem of how to find a hypergeometric antidifference, and it is the starting point of Zeilberger's celebrated algorithm for definite summation (Chap. 7). The book also covers the differential counterpart of Zeilberger's summation algorithm (Chap. 10) as well as its integration counterparts (Chaps. 11 and 12), and Gosper's algorithm is the driving force for all these algorithms. Therefore, its description remained unchanged. The other mentioned algorithms are also still up-to-date. Therefore, the above chapters have been updated only cautiously. However, in most chapters, new developments are cited and suggestions for further reading are given. As in the first edition, in all chapters an introduction to the corresponding q-theory is given.

The situation is quite different concerning the following parts of the book. Multivariate hypergeometric summation was still unfeasible when the first edition was written. In the meantime, ideas by Wegschaider cleared the way. These newer developments are incorporated and illustrated in Chap. 4, and the corresponding `multsum`-package is introduced. Furthermore, van Hoeij's algorithm has dramatically improved the efficiency of finding hypergeometric term solutions of holonomic recurrence equations over Petkovšek's original approach. Therefore, his ideas have been incorporated in Chap. 8 so that the reader gets a clear impression of where the new efficiency comes from. Nevertheless, the presentation of Petkovšek's original algorithm has not been withdrawn since it is still interesting from a historical point of view. More decisively, the efficiency of van Hoeij's algorithm can only be understood by comparison with Petkovšek's approach. The chapter finishes with the Maple package `qFPS` which contains the

q-case of van Hoeij's algorithm. More details about operator factorization are given in Chaps. 9 and 12. Finally, there were some new developments on discrete Rodrigues formulas, by my Ph.D. student Kornelia Fischer, which have been incorporated in Chap. 13.

For the first edition I had selected Maple as the computer algebra system in which the algorithms were programmed and demonstrated. Moreover, these (and some more) algorithms were incorporated in the packages hsum (and qsum for the q-case). This selection has proven successful, and since the other packages mentioned (multsum and qFPS) are also written in Maple, Maple is still the best system to keep the book self-contained.

On the web resource www.hypergeometric-summation.org/ all the Maple packages

- hsum.mpl (programs for hypergeometric summation)
- qsum.mpl (programs for q-hypergeometric summation)
- multsum.mpl (programs for multiple hypergeometric summation)
- qFPS.mpl (contains the q-Petkovšek-van-Hoeij algorithm)

and further materials such as the book's Maple sessions are available. These packages are regularly updated to work with newer versions of Maple.

I would like to thank Mama Foupouagnigni, Jürgen Gerhard, Dieter Schmersau[†] and Glenn P. Tesler who had read the first edition very carefully and identified several errors that I could correct. Special thanks go to Mark van Hoeij for his warm hospitality when I visited him in November–December 2013 at Florida State University (FSU) in Tallahassee. He gave me important assistance, especially concerning Chap. 9, about his brilliant algorithm. Also special thanks go to Torsten Sprenger who updated the hsum and qsum packages, contributed the multsum (Chap. 4) and qFPS packages (Chap. 9) and incorporated the FormalPower-Series package, which is mentioned in Chaps. 10 and 13, into Maple. Finally I am very grateful to Martin Muldoon who smoothed out the English of the manuscript again.

The finalization of this project was made possible by a sabbatical from the University of Kassel, and by the Alexander von Humboldt Foundation who financed my stay in the USA by awarding an alumni research scholarship. I am very grateful for this invaluable support.

Last but not least, I thank Ulrike Schmickler-Hirzebruch from Vieweg, Clemens Heine from Springer Germany, and Lynn Brandon from Springer London, for their good collaboration and for making this second edition possible.

January 15, 2014   Wolfram Koepf

Preface to the First Edition

The current book is the result of a lecture course that I gave at the Free University, Berlin, during the spring semester 1995. This course was influenced by the remarkable book *Concrete Mathematics* by Graham, Knuth, and Patashnik, and by the interesting lecture notes *Identities and Their Computer Proofs* by Herbert Wilf [Wilf93]. In the meantime, these notes have appeared together with other material in the book $A = B$ by Petkovšek, Wilf, and Zeilberger [PWZ96].

In contrast to the books just mentioned, it is my objective to present the material by giving more detailed advice on implementation. Furthermore, I wished to cover not only material about recurrence equations but also about differential equations, not only about sums but also about integrals, and finally not only the hypergeometric case but also its q-analogue.

In the current book, up-to-date algorithmic techniques for summation are described in detail, and worked out using Maple programs. With Maple release V.4 and higher, some of these tools are available through Maple's `sum` command and `sumtools` package, by an implementation that I incorporated in the Maple library prior to my lecture course. In this book, readers are invited to implement the algorithms step by step. This will give them a detailed insight into the structure of the algorithms under consideration, and will enable them to solve quite involved problems.

The book covers Gosper's algorithm for indefinite hypergeometric summation and Zeilberger's algorithm for definite hypergeometric summation, as well as the WZ method and extensions of these algorithms. Petkovšek's decision procedure for hypergeometric term solutions of holonomic recurrence equations completes the picture on the summation topic.

By an analogous technique, differential equations, derivative rules, and similar identities for sums can be generated, and a chapter on this topic is included. An equivalent theory of hyperexponential integration, both indefinite and definite, which was given by Almkvist and Zeilberger, completes the book.

The combination of all results considered gives work with orthogonal polynomials and (hypergeometric type) special functions a solid algorithmic foundation. Hence, many examples from this very active field are given.

Although multiple sums are briefly mentioned in Chapter 4, I have not covered the algorithmic theory of multisums, integral sums, etc., which was developed by Wilf and Zeilberger. Instead, by many examples I show how the one-dimensional theory can be applied successfully to double sums and integral sums, in particular to sums and integrals involving orthogonal polynomials.

The book contains many worked examples of the algorithms considered, and Maple implementations of them are presented. Furthermore, a lot of exercises encourage the readers to do their own implementations in Maple, and to study the topics included in detail. Exercises that demand Maple implementations are marked by a diamond ($\diamond$).

In all chapters, an introduction to the corresponding q-theory is given, whereas in the hypergeometric case, the algorithms are rigorously presented and detailed proofs of the statements are given, in the q-case we state only the results, indicate their proofs, present Maple implementations, and give examples and exercises.

A basic knowledge of a programming language such as Pascal or C should be sufficient to understand the Maple programs and to solve the corresponding exercises since all major Maple procedures that are used are briefly described. On the other hand, a deeper familiarity with Maple might help the reader to understand the code in more detail.

I could have presented the algorithms in pseudo code, without giving preference to a particular computer algebra system. On the other hand, an implementation in an existing and widely distributed computer algebra system makes the algorithms ready for execution, and therefore fills them with life. As a result, every student can execute all the examples no matter how complicated they may be.

Hence I had to decide on one of the major systems. Of the most important general purpose systems, Axiom [JS93], Macsyma [Macsyma], Maple [Char-et-al91]–[Char-et-al92], Mathematica [Wolfram96] and REDUCE [Hearn95], undoubtedly Maple and Mathematica have the largest audiences, since they are accessible at most universities and research institutions.

I wished to write my code as near as possible to the mathematical description of the corresponding algorithms, and since the latter depend heavily on the fast symbolic solution of (sometimes very complicated) systems of linear equations, the poor performance of Mathematica's `Solve` command for linear systems (see [PS95]) supported my decision to choose Maple. Furthermore, Maple is much friendlier with respect to user information (e.g., the `infolevel` routine).

Readers who use one of these systems can access some of the algorithms considered:

Axiom	The `sum` command contains an implementation of Gosper's algorithm.
Macsyma	The `sum` command contains an implementation of Gosper's algorithm written by Gosper.

Maple   Maple's sum command contains an implementation of Gosper's algorithm, completely rewritten by the author for Maple V.4. There are implementations of Zeilberger ([Zeilberger91b], [PWZ96]), and Koornwinder [Koornwinder93] of Zeilberger's algorithm; Almkvist and Zeilberger [AZ91] implemented the continuous version. Maple V.4's sumtools package was written by the author [Koepf96] and contains an implementation of Zeilberger's algorithm. In the present book, structured implementations of Gosper's algorithm, Zeilberger's algorithm, Petkovšek's algorithm and their q-analogues are developed. Salvy and Zimmermann's Generating Functions package gfun [SZ94] and Chyzak's Mgfun package [Chyzak94] are also strongly connected with the algorithms developed in the current book.

Mathematica   Implementations of Gosper's and Zeilberger's algorithms were done by Paule and Schorn [PS95], and Petkovšek implemented his algorithm and the corresponding q-version ([Petkovšek92], [PWZ96], and [APP98]). Also Paule and Riese [PR97] implemented the q-analog of Zeilberger's algorithm. A package on multidimensional summation is due to Wegschaider [Wegschaider97].[1]

REDUCE   Gosper's and Zeilberger's algorithms are accessible by an implementation of Koepf and Stölting [Koepf95b]; Böing and Koepf [BK97] implemented the q-analogs of Gosper's and Zeilberger's algorithm.

The Maple programs for the current book are discussed in detail in the text. Some of the implementations are even printed in the book. The programs are collected in the package hsum and can be obtained from the URL http://www. hypergeometric-summation.org/. Worksheets containing the examples given in the text, as well as Maple solutions of the exercises are available at the same URL. The corresponding q-analogs of Gosper's, Zeilberger's and Petkovšek's algorithms are implemented in the package qsum [BK99], written by Harald Böing, and can be obtained from the same site.

The present book is designed for use in the framework of a seminar. In seminars at German universities, every participating student is asked to present a lecture about a certain topic. The arrangement of the book makes the division into lectures easy. Each chapter covers a certain subtopic which can be presented by one or two students. Obviously, the book is also suitable for a lecture course in this area since it was written in connection with such a course presented by the author. Special notational conventions used in the book are defined at their first occurrence, and are listed in the *List of Symbols*.

---

[1] On the web see http://www.math.upenn.edu/~wilf/progs.html and http://www.risc.jku.at/research/combinat/software/.

I would like to thank Peter Deuflhard, who introduced me to the study of this topic, for his support and encouragement. Furthermore, I thank Martin Grötschel, without whose support the final version would not have been possible. Thanks go to Herbert Melenk for his advice on Gröbner bases, and for his excellent REDUCE implementation [MA94]. Due to his severe bicycling accident, the paper [MK95] is still unfinished. Also, I am very grateful for the warm hospitality of the ETH Zürich, where I visited to install my code in the Maple library, and especially to Mike Monagan, who headed the installation. Furthermore, thanks go to Tom Koornwinder for his implementation `zeilb` which was the starting point of my Maple implementations, and to Harald Böing who did some extensions of the implementations of this book that are covered in the `hsum` package as well as the q-implementation under my supervision. A few of the exercises have been collected by Torsten Thiele, and Lisa Temme corrected some of my English language mistakes. I am very grateful to Martin Muldoon who smoothed out the English of the final manuscript and to Harald Böing for the final proofreading.

Last but not least, I thank Ulrike Schmickler-Hirzebruch from Vieweg as well as the editor of the current book series Martin Aigner for their good collaboration and for making this project happen.

April 15, 1998   Wolfram Koepf

Contents

Introduction

Although the first steps towards an algorithmic treatment of summation go back to Celine Fasenmyer in the 1940s, these methods have not been used widely because of the lack of tools, such as fast computers and computer algebra systems, for the necessary calculations.

Perhaps the first algorithm which probably would not have been found without the use of a computer algebra system (as the developer states), is Gosper's algorithm for *indefinite hypergeometric summation,* which was discovered in 1978.

In the past two decades computer algebra systems like Maple have achieved recognition because computers with large amounts of memory are now available cheaply for everybody. In several papers appearing in the early nineties, Doron Zeilberger went back to the ideas of Celine Fasenmyer and used Gosper's algorithm in a non-obvious way to find a very efficient algorithm for *definite hypergeometric summation.*

Zeilberger's paradigm is to generate a recurrence equation with respect to some discrete variable for a sum under consideration. In the current text this method is extended in an obvious way and the generation of differential equations with respect to non-discrete variables is investigated. Following work of Almkvist and Zeilberger, it is further shown how recurrence and differential equations for *definite integrals* are established.

In particular, the combination of all these results gives work with orthogonal polynomials and special functions a solid algorithmic foundation.

I would like to explain these ideas in more detail in connection with the *Legendre polynomials* $P_n(x)$, that form a polynomial system which is orthogonal with respect to the scalar product

$$\langle f(x), g(x) \rangle = \int_{-1}^{1} f(x)\, g(x)\, dx.$$

Note that by using the Gram-Schmidt orthogonalization process applied to the monomial list $(1, x, x^2, \ldots)$ the orthogonal polynomials related to the given scalar product are well-defined up to constant factors.

In different books on orthogonal polynomials one finds quite different characterizations for the Legendre polynomials $P_n(x)$. Any of the representations

$$P_n(x) = \sum_{k=0}^{n} \binom{n}{k} \binom{-n-1}{k} \left(\frac{1-x}{2}\right)^k , \tag{1}$$

$$P_n(x) = \frac{1}{2^n} \sum_{k=0}^{n} \binom{n}{k}^2 (x-1)^{n-k} (x+1)^k \tag{2}$$

or

$$P_n(x) = \frac{1}{2^n} \sum_{k=0}^{\lfloor n/2 \rfloor} (-1)^k \binom{n}{k} \binom{2n-2k}{n} x^{n-2k} \tag{3}$$

as a sum might be used to define $P_n(x)$, or these polynomials might be defined by means of the *Rodrigues formula*

$$P_n(x) = \frac{(-1)^n}{2^n \, n!} \frac{d^n}{dx^n} (1-x^2)^n . \tag{4}$$

Another possibility is to introduce the Legendre polynomials via one of the *generating functions*

$$\sum_{n=0}^{\infty} P_n(x) \, z^n = \frac{1}{\sqrt{1-2xz+z^2}} , \tag{5}$$

and

$$\sum_{n=0}^{\infty} \frac{1}{n!} P_n(x) \, z^n = e^{xz} J_0(z\sqrt{1-x^2}) , \tag{6}$$

the latter with the aid of the *Bessel function.*

Still other possibilities are to define the Legendre polynomials as a solution of the *recurrence equation*

$$(n+2) P_{n+2}(x) - (2n+3)x P_{n+1}(x) + (n+1) P_n(x) = 0 \tag{7}$$

having the initial values

$$P_0(x) = 1, \qquad P_1(x) = x,$$

or as a solution of the *differential equation*

$$(1-x^2) P_n''(x) - 2x P_n'(x) + n(n+1) P_n(x) = 0 \tag{8}$$

satisfying

$$P_n(1) = 1, \qquad P'_n(1) = \frac{n(n+1)}{2}.$$

Both the recurrence and differential equations are *holonomic*, i.e. they are homogeneous and linear, and have polynomial coefficients.[2]

These different representations of $P_n(x)$ give a remarkable set of identities which are proved by an amazing variety of methods in books on orthogonal polynomials. This situation raises the question of which of these representations might be the most natural.

No satisfying answer can be given to the question of which of the three sum representations (1)–(3) might be preferable, and there are even further different representations. Rodrigues type representations are specific to the *classical orthogonal polynomials* and many interesting function families cannot be described by such formulas. Furthermore, generating functions cannot always be represented by elementary functions; see (6). Hence, the recurrence equation (7) and the differential equation (8) are most natural since both turn out to constitute a *normal form*: When common factors of their coefficients are canceled, then they are the uniquely determined holonomic equations of lowest order valid for $P_n(x)$.

The crucial point is that the algorithms that we describe provide a method for finding these recurrence and differential equations for functions given by sums like (1)–(3), by Rodrigues type formulas like (4) or by generating functions like (5)–(6). In particular, this serves as an algorithmic method to prove that the above representations of the Legendre polynomials all constitute the same family of functions.

Next, I would like to emphasize another aspect. Even though the sum representations (1)–(3) cannot be sorted by preference, they can be *categorized*. The underlying idea for this categorization is the concept of *hypergeometric functions*, in terms of which (1)–(3) read as

$$P_n(x) = {}_2F_1\left(\begin{matrix} -n, n+1 \\ 1 \end{matrix} \middle| \frac{1-x}{2} \right),$$

$$P_n(x) = \left(\frac{x-1}{2} \right)^n {}_2F_1\left(\begin{matrix} -n, -n \\ 1 \end{matrix} \middle| \frac{x+1}{x-1} \right)$$

and

$$P_n(x) = \binom{2n}{n} \frac{x^n}{2^n} {}_2F_1\left(\begin{matrix} -n/2, -n/2 + 1/2 \\ -n + 1/2 \end{matrix} \middle| \frac{1}{x^2} \right). \tag{9}$$

Hence, this procedure places us in the world of *hypergeometric transformations*.

---

[2] Note that since there is a strong analogy between a differential equation and a recurrence equation, we try to avoid the notions *recursion*, *recurrence relation* etc. throughout the book; these are not relations, but equations.

It turns out that there is a deep connection between this categorization and the algorithms that we describe. Indeed, the notion of hypergeometric functions is central to the presentation in this book, and hence we restrict ourselves to the consideration of hypergeometric sums.

As a final comment, I mention that the evaluation $P_n(1) = 1$ is quite simple if one of the representations (1)–(2) is at hand, and is quite involved in terms of (3). In this case, the evaluation $P_n(1) = 1$ is equivalent to the identity

$$\sum_{k=0}^{\lfloor n/2 \rfloor} (-1)^k \binom{n}{k} \binom{2n - 2k}{n} = 2^n . \tag{10}$$

Hence we are led to a *summation identity* that is a by-product of the method used. This is a typical connection between orthogonal polynomials and summation, and one of the reasons why the algorithmic examination of such summation identities is one of the primary concerns of the present text. According to (9), the identity (10) is a particular case of the *Chu-Vandermonde identity* (Chap. 2).

I see no need to give a rigorous introduction to orthogonal polynomials and special functions. Many good introductions exist, for example [AAR99, Chihara78, Gautschi04, Ismail09, KLS10],[3] [NU88, OLBC10, Rainville60, Szegö39] and [Tricomi55]. The level of activity in the field of orthogonal polynomials and special functions is illustrated by the fact that five of the above monographs have appeared since the first edition of my book.

The purpose of my presentation is to give an algorithmic viewpoint on the topic, and to present implementations of efficient algorithms in computer algebra.

References

AAR99. Andrews, G., Askey, R.A., Roy, R.: Special Functions. Encyclopedia of Mathematics and its Applications. Cambridge University Press, Cambridge (1999)

Chihara78. Chihara, T.S.: An Introduction to Orthogonal Polynomials. Gordon and Breach Publishers, New York (1978) (reprinted by Dover Publ., Dover, 2011)

Gautschi04. Gautschi, W.: Orthogonal Polynomials: Computation and Approximation. Oxford University Press, New York (2004)

Ismail09. Ismail, M.E.H.: Classical and Quantum Orthogonal Polynomials in one Variable. Cambridge University Press, Cambridge (2005)

KLS10. Koekoek, R., Lesky, P., Swarttouw, R.F.: Hypergeometric Orthogonal Polynomials and their q-Analogues. Springer Monographs in Mathematics. Springer, Berlin (2010)

KS98a. Koekoek, R., Swarttouw, R.F.: The Askey-scheme of hypergeometric orthogonal polynomials and its q-analogue. Report 98–17. Faculty of Technical Mathematics and Informatics, Delft University of Technology (1998). http://aw.twi.tudelft.nl/koekoek/askey.html

---

[3] Which contains in particular an extension of the widely distributed internet resource [KS98a].

NU88. Nikiforov, A.F., Uvarov, V.B.: Special Functions of Mathematical Physics (Translated from the Russian by R.P. Boas). Birkhäuser, Basel (1988)

OLBC10. Olver, F.W.J., Lozier, D.W., Boisvert, R.F., Clark, C.W.: NIST Handbook of Mathematical Functions. Cambridge University Press, New York (2010). http://dlmf.nist.gov

Rainville60. Rainville, E.D.: Special Functions. The MacMillan Co., New York (1960)

Szegö39. Szegö, G.: Orthogonal Polynomials, Vol. 23. American Mathematical Society College Publishers, New York (1939) (Fourth Edition and extended version, 1975)

Tricomi55. Tricomi, F.G.: Vorlesungen über Orthogonalreihen. Grundlehren der Mathematischen Wissenschaften 76, Springer, Berlin (1955)

Chapter 1
The Gamma Function

Apart from the elementary transcendental functions such as the exponential and trigonometric functions and their inverses, the Gamma function is probably the most important transcendental function. It was defined by Euler to interpolate the factorials at noninteger arguments.

Following Euler, we define

$$\Gamma(z) := \int_0^\infty t^{z-1} e^{-t}\, dt,$$

and call it the *Gamma function*.

This improper integral exists for complex $z \in \mathbb{C}$ with $\operatorname{Re} z > 0$ (or, if you prefer only to think of real variables, for real $z > 0$). Using integration by parts, we get the fundamental functional equation

$$\Gamma(z+1) = \int_0^\infty t^z e^{-t}\, dt = -t^z e^{-t}\Big|_{t=0}^{t=\infty} + z \int_0^\infty t^{z-1} e^{-t}\, dt = z\,\Gamma(z). \qquad (1.1)$$

From the initial value

$$\Gamma(1) = \int_0^\infty e^{-t}\, dt = -e^{-t}\Big|_0^\infty = 1,$$

it follows further by induction that

$$\Gamma(k+1) = k! \qquad (1.2)$$

for $k \in \mathbb{N}_{\geq 0} := \{0, 1, 2, \ldots\}$. Therefore the Γ-function interpolates the factorial function continuously, and we may *define* the factorial function by (1.2), for

W. Koepf, *Hypergeometric Summation*, Universitext,
DOI: 10.1007/978-1-4471-6464-7_1, © Springer-Verlag London 2014

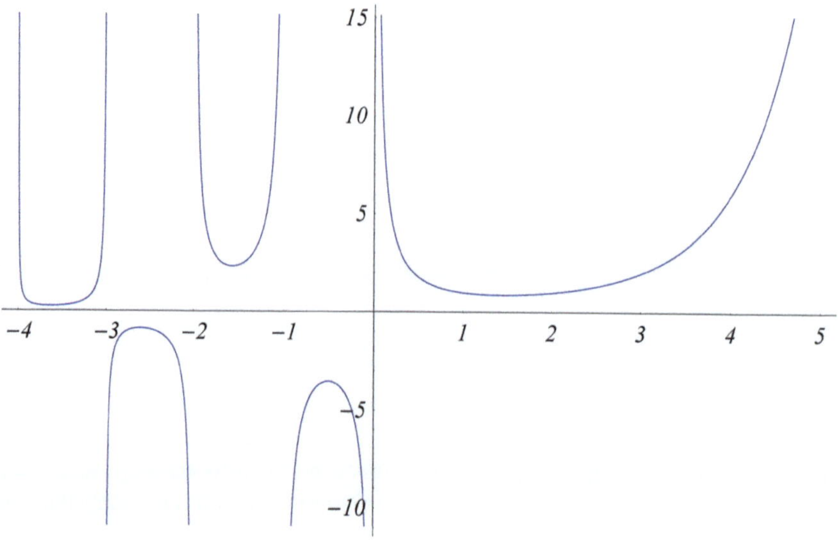

Fig. 1.1 The Gamma function on the real axis

noninteger values $k \in \mathbb{C}$. In this book we will use the Gamma and factorial functions interchangeably, related by (1.2).[1]

For points $z \notin \mathbb{Z} := \{\ldots, -2, -1, 0, 1, 2, \ldots\}$ with nonpositive real part one reads the fundamental functional equation (1.1) from right to left to obtain

$$\Gamma(z) = \frac{\Gamma(z+1)}{z},$$

and *defines* the Γ-function by a recursive application of this rule for $z \in \mathbb{C}\backslash\mathbb{Z}$ with nonpositive real part (in particular for $z \in \mathbb{R}\backslash\mathbb{Z}$ with $z < 0$) (Figs. 1.1 and 1.2).

The resulting function is differentiable in the whole complex plane (proved by standard differentiation under the integral sign) except at the nonpositive integers where it has poles of order 1. By continuity, we may set

$$\frac{1}{\Gamma(-k)} = 0 \qquad (k \in \mathbb{N}_{\geq 0}), \tag{1.3}$$

and by our general interpretation this reads as $\frac{1}{k!} = 0$ for $k = -1, -2, \ldots$ In function-theoretic terms this means that the function $1\backslash\Gamma$ is an entire function, i.e., it is analytic in the entire complex plane with zeros exactly at the negative integers and the origin, the poles of Γ.

By induction, we get from (1.1) for $k \in \mathbb{N} := \{1, 2, 3, \ldots\}$

---

[1] Computer algebra systems like Maple and Mathematica share this policy.

$$\Gamma(z+k) = z\,(z+1)\cdots(z+k-1)\,\Gamma(z) = (z)_k\,\Gamma(z)\,. \tag{1.4}$$

The *shifted factorial*

$$(z)_k := \prod_{j=0}^{k-1}(z+j) = \frac{\Gamma(z+k)}{\Gamma(z)} \quad (k \in \mathbb{N})\,, \tag{1.5}$$

which occurs in (1.4), is also called the *Pochhammer symbol*. It will occur frequently in this book. For $k = 0$ the Pochhammer symbol is defined as $(z)_0 := 1$ and for arbitrary $k \in \mathbb{C}$ it can be defined by the right term of (1.5). In this book, however, we will use the Pochhammer symbol only for integer values of k.

From the fundamental identities (1.1) and (1.4), we get the following limit relation at the poles $-k$ ($k \in \mathbb{N}_{\geq 0}$) of the Γ-function

$$\lim_{z\to-k}(z+k)\Gamma(z) = \lim_{z\to-k}\frac{(z+k)\Gamma(z+k)}{(z)_k} = \lim_{z\to-k}\frac{\Gamma(z+k+1)}{(z)_k}$$

$$= \frac{1}{(-k)_k} = \frac{(-1)^k}{k!}\,. \tag{1.6}$$

This computation may be interpreted as the residue computation

$$\operatorname*{Res}_{z=-k}\Gamma(z) = \frac{(-1)^k}{k!}\,.$$

Note that the identity

$$(z)_k = z\,(z+1)\cdots(z+k-2)\,(z+k-1) \tag{1.7}$$
$$= (-1)^k\,(1-z-k)\,(2-z-k)\cdots(-z-1)\,(-z) = (-1)^k\,(1-z-k)_k$$

reads, in terms of Γ-functions, as

$$\frac{\Gamma(z+k)}{\Gamma(z)} = (-1)^k\,\frac{\Gamma(1-z)}{\Gamma(1-z-k)}$$

or equivalently

$$\Gamma(z)\,\Gamma(1-z) = (-1)^k\,\Gamma(z+k)\,\Gamma(1-(z+k))\,. \tag{1.8}$$

It turns out that in general

$$\Gamma(z)\,\Gamma(1-z) = \frac{\pi}{\sin(\pi z)} \quad \text{or equivalently} \quad \frac{\sin(\pi z)}{\pi} = \frac{1}{\Gamma(z)}\cdot\frac{1}{\Gamma(1-z)}\,, \tag{1.9}$$

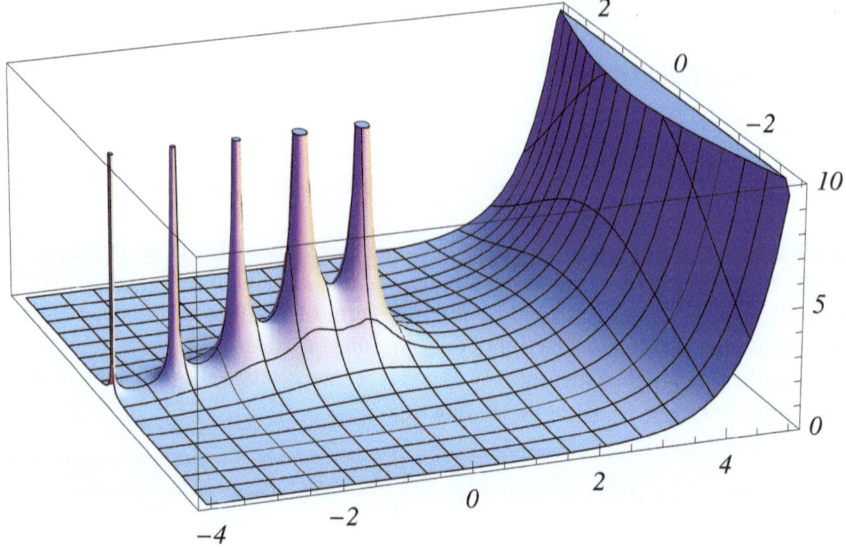

Fig. 1.2 The function $|\Gamma(z)|$ for complex z

which puts (1.8) in a more general setting. Equation (1.9) is called the *reflection formula* of the Γ-function, a proof of which is outlined in Exercise 1.7. Note that a function theoretic interpretation of (1.9) is essentially given by the fact that the functions on both the left and the right hand sides of (1.9) have the same set of zeros over $\mathbb{C}$. This set consists of precisely the integers, all zeros having order one and identical local behavior. That's why both functions have the same *Weierstrass product representation*. Without proof we note the product representations of the sine function

$$\sin(\pi z) = \pi z \prod_{k=1}^{\infty} \left(1 - \frac{z^2}{k^2}\right)$$

whose zeros obviously are precisely the integers, and of the Gamma function

$$\Gamma(z) = \lim_{n \to \infty} \frac{n!\, n^z}{(z)_{n+1}} = \frac{e^{-\gamma z}}{z} \prod_{k=1}^{\infty} \left(1 + \frac{z}{k}\right)^{-1} e^{\frac{z}{k}} \tag{1.10}$$

where

$$\gamma := \lim_{n \to \infty} \left(\sum_{k=1}^{n} \frac{1}{k} - \ln n\right) \approx 0.57721\,56649\,01532\,86060\,65120\,90082$$

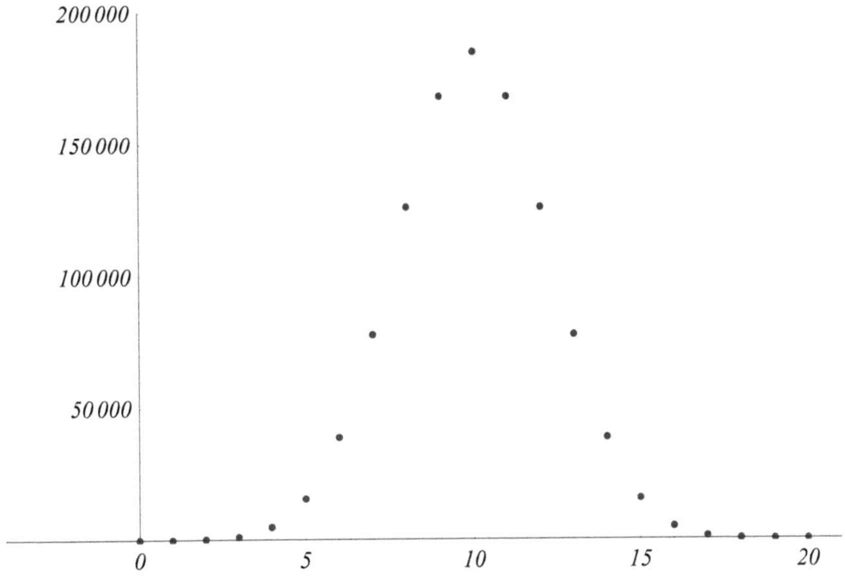

Fig. 1.3 The binomial coefficients $\binom{20}{k}$ for $k = 0, \ldots, 20$

denotes the *Euler-Mascheroni constant*. Of course both representations can be combined to obtain the reflection formula (1.9).

Furthermore, by (1.4) the binomial coefficients can be written in terms of the Γ-function as

$$\binom{z}{k} = \frac{z\,(z-1)\cdots(z-k+1)}{k!} = \frac{\Gamma(z+1)}{k!\,\Gamma(z-k+1)} \tag{1.11}$$

for arbitrary $z \in \mathbb{C}$, $z+1 \neq 0, -1, \ldots$, and $z-k+1 \neq 0, -1, \ldots$. In particular, by (1.3), for $k, n \in \mathbb{N}_{\geq 0}$, we have

$$\binom{n}{k} = 0 \quad \text{for } k < 0 \text{ and } k > n$$

in agreement with the elementary definition of the binomial coefficient as a product (Fig. 1.3).

Note, further, the following relation between the Pochhammer symbol and the binomial coefficient,

$$\binom{z}{k} = (-1)^k \binom{k-z-1}{k} = \frac{(-1)^k}{k!}(-z)_k.$$

Now we would like to consider the *Beta function*, defined by

$$B(z, w) := \int_0^1 t^{z-1} (1 - t)^{w-1} \, dt.$$

This improper integral exists for $\operatorname{Re} z > 0$, $\operatorname{Re} w > 0$. We shall show now that the Beta function can be expressed in terms of the Γ-function. The substitution $t = 1 - x$ shows immediately that

$$B(z, w) = B(w, z) . \tag{1.12}$$

Next, we use the substitution $t = \sin^2 \varphi$ to obtain the trigonometric representation

$$B(z, w) = 2 \int_0^{\pi/2} \sin^{2z-1} \varphi \cdot \cos^{2w-1} \varphi \, d\varphi . \tag{1.13}$$

Now we consider the product

$$\Gamma(z) \, \Gamma(w) = \int_0^\infty t^{z-1} e^{-t} dt \cdot \int_0^\infty u^{w-1} e^{-u} du$$

and use the substitutions $t = x^2$ and $u = y^2$ to obtain

$$\Gamma(z) \, \Gamma(w) = 4 \int_0^\infty e^{-x^2} x^{2z-1} dx \int_0^\infty e^{-y^2} y^{2w-1} dy$$

$$= 4 \int_0^\infty \int_0^\infty e^{-x^2 - y^2} x^{2z-1} y^{2w-1} dx dy. \tag{1.14}$$

Applying polar coordinates $x = r \cos \varphi$, $y = r \sin \varphi$ to this double integral, we get

$$\Gamma(z) \, \Gamma(w) = 4 \int_0^{\pi/2} \int_0^\infty e^{-r^2} r^{2z+2w-2} \cos^{2z-1} \varphi \cdot \sin^{2w-1} \varphi \cdot r \, dr \, d\varphi$$

$$= 2 \int_0^\infty e^{-r^2} r^{2z+2w-1} dr \cdot 2 \int_0^{\pi/2} \cos^{2z-1} \varphi \cdot \sin^{2w-1} \varphi \, d\varphi$$

$$= \Gamma(z + w) \, B(w, z) = \Gamma(z + w) \, B(z, w)$$

where the substitution $r = \sqrt{t}$, and Eqs. (1.13) and (1.12) are utilized.
   Therefore we have deduced

Theorem 1.1 *For* $\operatorname{Re} z > 0$ *and* $\operatorname{Re} w > 0$ *the identity*

$$B(z, w) = \int_0^1 t^{z-1} (1-t)^{w-1} \, dt = 2 \int_0^{\pi/2} \sin^{2z-1} \varphi \cdot \cos^{2w-1} \varphi \, d\varphi = \frac{\Gamma(z)\Gamma(w)}{\Gamma(z+w)}$$

is valid. $\qquad\qquad\square$

Session 1.2 Maple knows the Γ-function. We define

```
>   term :=GAMMA(z)*GAMMA(1-z);
```
$$term := \Gamma(z)\,\Gamma(1-z)$$

Maple uses the reflection formula:

```
>   simplify(term);
```
$$\frac{\pi}{\sin(\pi z)}$$

We can convert between Γ, factorial, and binomial terms.

```
>   convert(n!,GAMMA);
```
$$\Gamma(n+1)$$

```
>   res :=convert(binomial(n,k),factorial);
```
$$res := \frac{n!}{k!\,(n-k)!}$$

```
>   convert(res,binomial);
```
$$\binom{n}{k}$$

```
>   res :=expand(GAMMA(n+4));
```
$$\Gamma(n)\,n^4 + 6\,\Gamma(n)\,n^3 + 11\,\Gamma(n)\,n^2 + 6\,\Gamma(n)\,n$$

```
>   collect(res,GAMMA,factor);
```
$$\Gamma(n)\,n\,(n+1)\,(n+2)\,(n+3)$$

```
>   simplify(GAMMA(n+4)/n!);
```
$$(n+3)\,(n+2)\,(n+1)$$

```
>   expand(binomial(n+2,k-1));
```
$$\frac{(n+2)\,(n+1)\,k\binom{n}{k}}{(n+3-k)\,(n+2-k)\,(n+1-k)}$$

```
>   simplify(pochhammer(z,k)*GAMMA(z)/GAMMA(z+k));
```
$$1$$

Maple can simplify the integrals

```
>   int(t^(z-1)*exp(-t),t=0..infinity);
```

$$\Gamma(z)$$

```
>   int(t^(z-1)*(1-t)^(w-1),t=0..1);
```

$$\frac{\Gamma(w)\,\Gamma(z)}{\Gamma(z+w)}$$

and, if we make the correct assumption on z and w, Maple discovers the Beta function

```
>   assume(z,positive);
>   assume(w,positive);
>   int(t^(z-1)*(1-t)^(w-1),t=0..1);
```

$$B(w,z)$$

Maple also evaluates the trigonometric integral

```
>   int(sin(phi)^(2*z-1)*cos(phi)^(2*w-1),phi=0..Pi/2);
```

$$\frac{1}{2}\frac{\Gamma(w)\,\Gamma(z)}{\Gamma(z+w)}$$

By (1.2), we know the exact values of the Γ-function at the positive integers. To calculate the exact value of the Γ-function at the point $1/2$, we use the substitution $t = x^2$ in our definition and utilize the well-known probability integral, which gives us

$$\Gamma(1/2) = \int_0^\infty t^{-1/2}\,e^{-t}\,dt = 2\int_0^\infty e^{-x^2}\,dx = \sqrt{\pi}\,. \tag{1.15}$$

Note that this result is also easily deduced from (1.14) (for $z = w$), or from the reflection formula (1.9), and yields a closed form for all half integer values using the basic functional equation (1.1).

Further Reading

For further reading on the Gamma function we recommend the book [AAR99], Chap. 1.

Exercises

Exercise 1.1 *Use the value $\Gamma(1/2)$ to determine formulas for $\Gamma(1/2 + k)$ and $\Gamma(k/2)$ in terms of factorials with integer arguments, assuming $k \in \mathbb{Z}$.*

Exercise 1.2 *Show that for $k \in \mathbb{N}_{\geq 0}$*

$$\lim_{z \to k} \frac{\Gamma(1-z)}{\Gamma(1-2z)} = (-1)^k \frac{(2k)!}{k!}\,.$$

Hint: *Use* (1.6).

Exercise 1.3 *Use* (1.10) *to prove the limits*

(a) $\lim\limits_{n\to\infty} \frac{\Gamma(n+1)\,n^z}{\Gamma(n+z+1)} = 1,$

(b) $\lim\limits_{n\to\infty} \frac{\Gamma(n+a)}{\Gamma(n+b)}\, n^{b-a} = 1.$

Exercise 1.4 *Show that for* $z \in \mathbb{C}$, $k \in \mathbb{N}_{\geq 0}$

(a) $(2z)_{2k} = 4^k\,(z)_k\,(z+1/2)_k,$
(b) $(3z)_{3k} = 27^k\,(z)_k\,(z+1/3)_k\,(z+2/3)_k.$

Deduce a similar rule for $(mz)_{mk}$ *where* $m \in \mathbb{N}$, $m \geq 4$.

Exercise 1.5 *Prove the following rules for the Pochhammer symbol*

(a) $(z)_{n+k} = (z)_n\,(z+n)_k,$
(b) $(1/2)_k = \frac{(2k)!}{4^k\,k!}.$

Exercise 1.6 *Give a definition of the Pochhammer symbol* $(a)_k$ *for negative* k *which is consistent with* (1.4).

Exercise 1.7 *(Reflection Formula) Use Theorem 1.1 to prove* (1.9). *Hint: The identity*
$$\Gamma(z)\,\Gamma(1-z) = B(z, 1-z),$$

and the substitution $t = x/(1+x)$ *yields an integral representation that can be evaluated by the residue theorem.*

Exercise 1.8 *For* $\mathrm{Re}\,z > 0$, $\mathrm{Re}\,w > 0$ *write*

$$\int_0^{\pi/2} \sin^{2z-1}\varphi \cdot \cos^{2w-1}\varphi\, d\varphi$$

in terms of the Γ-*function.*

Exercise 1.9 *Use Theorem 1.1 to represent the integrals* $(n, m \in \mathbb{R})$

$$\int_0^{\pi/2} \cos^n t\, dt\,, \qquad \int_0^{\pi/2} \sin^m t\, dt\,, \qquad \text{and} \qquad \int_0^{\pi/2} \cos^n t\, \sin^m t\, dt$$

in terms of the Γ-*function. For* $n, m \in \mathbb{N}_{\geq 0}$, *express the result in terms of factorials with integer arguments.*

Exercise 1.10 *Calculate for $m, n \geq 0$ the definite integral*

$$\int_0^1 x^m \ln^n x \, dx.$$

Exercise 1.11 *Evaluate*

$$\int_{-1}^1 (1+t)^{z-1} (1-t)^{w-1} \, dt.$$

Reference

AAR99. Andrews, G., Askey, R.A., Roy, R.: Special Functions. Encyclopedia of Mathematics and its Applications. Cambridge University Press, Cambridge (1999)

Chapter 2
Hypergeometric Identities

In this chapter we deal with hypergeometric identities. These are identities like

$$\sum_{k=0}^{n} \binom{n}{k} = 2^n, \tag{2.1}$$

$$\sum_{k=0}^{n} (-1)^k \binom{n}{k} = 0 \quad (n \neq 0), \tag{2.2}$$

$$\sum_{k=0}^{n} \binom{n}{k}^2 = \frac{(2n)!}{n!^2} = \binom{2n}{n}, \tag{2.3}$$

$$\sum_{k=0}^{n} (-1)^k \binom{n}{k}^2 = \begin{cases} 0 & \text{if } n \text{ is odd} \\ \dfrac{(-1)^{n/2}\, n!}{(n/2)!^2} & \text{otherwise} \end{cases}, \tag{2.4}$$

$$\sum_{k=0}^{n} (-1)^k \binom{n}{k}^3 = \begin{cases} 0 & \text{if } n \text{ is odd} \\ \dfrac{(-1)^{n/2}\, (3n/2)!}{(n/2)!^3} & \text{otherwise} \end{cases}, \tag{2.5}$$

or

$$\sum_{k=-n}^{n} (-1)^k \binom{n+b}{n+k} \binom{n+c}{c+k} \binom{b+c}{b+k} = \frac{\Gamma(b+c+n+1)}{n!\,\Gamma(b+1)\,\Gamma(c+1)}, \tag{2.6}$$

involving sums of a special type. We will meet the above identities—and many more—in one form or another at several places later on in this book. For the moment, we will not prove any of these identities. However, all of them will be proved by several methods later.

We would like to mention that these kinds of identities can often be interpreted combinatorially. Assume S is a set with n elements. The left-hand side of (2.1) counts the number of subsets of S with k elements and sums these. The right-hand side counts the total number of subsets of S. As soon as we have this combinatorial interpretation and proofs using combinatorial arguments for both sides, we have

W. Koepf, *Hypergeometric Summation*, Universitext,
DOI: 10.1007/978-1-4471-6464-7_2, © Springer-Verlag London 2014

proved (2.1) combinatorially. On the other hand, in many cases we have the opposite situation: By combinatorial considerations, a sum of the above type occurs, but we are lacking a (combinatorial) method to evaluate this sum directly. One may ask whether the sum under consideration can be rewritten in a simpler form.

We will not deal with combinatorial interpretations of identities in this book. Instead, we will introduce several methods to find simpler form representations for sums of the above type.

What do the above sums have in common? They all are *definite sums* of the type

$$F = \sum_{k=-\infty}^{\infty} a_k \tag{2.7}$$

the sum to be taken over all integers k. This is so since (for any $n \in \mathbb{N}_{\geq 0}$) all summands vanish outside a finite set. We say that a_k has *finite support* in this situation. In most of the above cases this is the interval $k = 0, \ldots, n$.

A sum of type (2.7) is called a *hypergeometric series* if the term ratio a_{k+1}/a_k represents a rational function of k. In this case we call the summand a_k a *hypergeometric term*.[1]

Note that the summands of the above identities (2.1)–(2.6) not only represent hypergeometric terms with respect to the summation variable k, but form hypergeometric terms with respect to all variables (k, n, a, b, c) involved.

Without giving a formal definition, we call an equation a *hypergeometric identity* if it represents a hypergeometric series (2.7) by hypergeometric terms like the right-hand sides of (2.1)–(2.6).[2] If the sum is written in terms of products of binomial coefficients, we will frequently also speak of a *binomial sum identity*. Binomial sum identities are hypergeometric ones if the arguments of the binomial coefficients occurring are *integer-linear* in the summation variable k, i.e., they are of the form $\alpha k + \beta$ with $\alpha \in \mathbb{Z}$, $\beta \in \mathbb{K}$, $\mathbb{K}$ denoting any field of characteristic zero, e.g. $\mathbb{K} = \mathbb{Q}$, $\mathbb{R}$ or $\mathbb{C}$. For simplicity, in the current text, we generally assume $\mathbb{K} = \mathbb{Q}$ or a transcendental extension of $\mathbb{Q}$ with a finite number of variables adjoined, i.e. $\mathbb{K} = \mathbb{Q}(x_1, x_2, \ldots, x_m)$,[3] and it is implicitly understood that the variables $x_1, x_2, \ldots, x_m$ are independent of the other variables occurring in the given context.

We will further frequently deal with the case of *rational-linear* input with arguments of the form $\alpha k + \beta$ with $\alpha \in \mathbb{Q}$.

Assume now a hypergeometric series (2.7) is given. In this chapter we begin by considering how to find a representation of F in terms of the *generalized hypergeometric function* ${}_pF_q$ given by

---

[1] A hypergeometric term is always the *summand*, not the sum!

[2] The right-hand sides a_n form m-fold hypergeometric terms. These are generalizations of hypergeometric terms satisfying a recurrence equation of the type $a_{n+m} = R(n)\, a_n$ for some $m \in \mathbb{N}$ with rational $R(n)$.

[3] $\mathbb{Q}(x_1, x_2, \ldots, x_m)$ denotes the field of rational functions in the variables $x_1, x_2, \ldots, x_m$ over $\mathbb{Q}$.

$$
{}_pF_q \left(\left. \begin{matrix} \alpha_1, \alpha_2, \ldots, \alpha_p \\ \beta_1, \beta_2, \ldots, \beta_q \end{matrix} \right| x \right) := \sum_{k=0}^{\infty} A_k\, x^k = \sum_{k=0}^{\infty} \frac{(\alpha_1)_k \cdot (\alpha_2)_k \cdots (\alpha_p)_k}{(\beta_1)_k \cdot (\beta_2)_k \cdots (\beta_q)_k} \frac{x^k}{k!}. \tag{2.8}
$$

This is the appropriate thing to do since we shall see soon that the term ratio a_{k+1}/a_k of the summand $a_k := A_k x^k$ of ${}_pF_q$ is a general rational function in k in factored form.

The numbers α_k are called the *upper* and β_k the *lower parameters* of ${}_pF_q$. Note that ${}_pF_q(x)$ is well-defined if no lower parameter is a negative integer or zero and it constitutes a convergent series if $p \le q$, or if $p = q + 1$ and $|x| < 1$.

We will, however, deal almost exclusively with the case where ${}_pF_q(x)$ is a polynomial so that convergence is not an issue. This situation occurs if one of the upper parameters is a negative integer. Throughout the present book, the letter n will denote a nonnegative integer and $-n$, $-2n$, or $-n - 1$, etc. might denote upper parameters. In such a case, ${}_pF_q(x)$ is a polynomial in x of degree (at most) n, $2n$, or $n + 1$, respectively.

Since by the definition of the shifted factorial

$$
\frac{(\alpha)_{k+1}}{(\alpha)_k} = k + \alpha,
$$

the summand $a_k = A_k x^k$ of the generalized hypergeometric function has the rational term ratio

$$
\frac{a_{k+1}}{a_k} = \frac{A_{k+1} x^{k+1}}{A_k x^k} = \frac{(k + \alpha_1) \cdot (k + \alpha_2) \ldots (k + \alpha_p)}{(k + \beta_1) \cdot (k + \beta_2) \ldots (k + \beta_q)} \frac{x}{k + 1} \qquad (k \in \mathbb{N}_{\ge 0}), \tag{2.9}
$$

i.e., the first order recurrence equation

$$
(k + \beta_1) \cdot (k + \beta_2) \cdots (k + \beta_q) \cdot (k + 1)\, A_{k+1} - (k + \alpha_1) \cdot (k + \alpha_2) \cdots (k + \alpha_p)\, A_k = 0 \tag{2.10}
$$

is valid for A_k.

Note that the extra factor $(k + 1)$ in the denominator of (2.9) which does not occur in the list of lower parameters guarantees that ${}_pF_q(x)$, which is a *power series*, corresponds to a *bilateral sum* (2.7), i.e., for arbitrary $A_0 \ne 0$, the statement $A_{-1} = 0$ can be deduced from (2.10), so that all coefficients A_k with negative k vanish.[4] This argument applies whenever none of the upper parameters is a positive integer, whereas in the latter case the lower bound $k = 0$ of ${}_pF_q(x)$ is not the *natural* one, i.e., the summand a_k is not identically zero for negative k, and therefore ${}_pF_q(x)$ cannot be considered as a bilateral sum.

The generalized hypergeometric series generalizes the exponential and geometric series: For

---

[4] This fact is also expressed by the $k!$-term in the denominator of the right-hand sum (2.8).

$$e^x = \sum_{k=0}^{\infty} \frac{x^k}{k!},$$

we have $a_k = x^k/k!$, and therefore $a_{k+1}/a_k = x/(k+1)$ so that $e^x = {}_0F_0(x)$. For

$$\frac{1}{1-x} = \sum_{k=0}^{\infty} x^k,$$

we have $a_k = x^k$ ($k \in \mathbb{N}_{\geq 0}$), hence $a_{k+1}/a_k = x$ ($k \in \mathbb{N}_{\geq 0}$). Note that this term ratio is *not* valid for $k = -1$; however, after multiplication by $(k+1)$, the recurrence equation $(k+1)a_{k+1} - (k+1)xa_k = 0$ is valid for all $k \in \mathbb{Z}$, and we have for $|x| < 1$

$$\frac{1}{1-x} = {}_1F_0 \left(\begin{matrix} 1 \\ - \end{matrix} \middle| x \right).$$

In particular, since the upper parameter is a positive integer, this is not a bilateral sum of type (2.7).

Note that the function ${}_2F_1(x)$ (whose radius of convergence is 1) was introduced by Gauss and is therefore called *Gauss' hypergeometric function*. On the other hand, the series ${}_1F_1(x)$ converges for all $x \in \mathbb{C}$ and is called *Kummer's confluent hypergeometric function*.

If a hypergeometric series (2.7) is given and if $a_0 \neq 0$, then it is easy to represent F in terms of a generalized hypergeometric function if we are able to find polynomials u_k and v_k such that

$$\frac{a_{k+1}}{a_k} = \frac{u_k}{v_k} \tag{2.11}$$

and if we assume a complete factorization in linear factors of u_k and v_k, by comparison with (2.9).

We introduce some notation. $\mathbb{K}(k)$ denotes the field of rational functions in the variable k over $\mathbb{K}$, and $\mathbb{K}[k]$ denotes the ring of polynomials in k over $\mathbb{K}$. Similarly, $\mathbb{K}(n, k)$ and $\mathbb{K}[n, k]$ are the field of rational functions and the ring of polynomials in two variables, respectively.

Session 2.1 As we saw above, given a_k, it is crucial to find polynomials $u_k, v_k \in \mathbb{Q}[k]$ such that (2.11) is valid. How can we find these with Maple? We saw in Session 1.2 that Maple's expand command expands binomial coefficients and factorial and Γ function terms. Therefore, we have for example

```
>    expand(binomial(n+2,k-1)/binomial(n-1,k+2));
```

$$\frac{(n+2)(n+1)kn(k+2)(k+1)}{(n+3-k)(n+2-k)(n+1-k)(n-k-2)(n-k-1)(n-k)}$$

However, we must say that this is not a safe procedure. This is shown in the following example that we will consider in more detail in Example 2.5.

```
>    summand:=binomial(n,k)/2^n:
>    term:=subs(n=n+1,summand)-summand:
>    expr:=expand(subs(k=k+1,term)/term);
```

$$
expr := -1/2 \binom{n}{k} n \left(1/2 \frac{\binom{n}{k} n}{(n+1-k)\,2^n} + 1/2 \frac{\binom{n}{k}}{(n+1-k)\,2^n} - \frac{\binom{n}{k}}{2^n} \right)^{-1} (k+1)^{-1} \left(2^n\right)^{-1}
$$

$$
+ 1/2 \binom{n}{k} \left(1/2 \frac{\binom{n}{k} n}{(n+1-k)\,2^n} + 1/2 \frac{\binom{n}{k}}{(n+1-k)\,2^n} - \frac{\binom{n}{k}}{2^n} \right)^{-1} (k+1)^{-1} \left(2^n\right)^{-1}
$$

$$
+ \binom{n}{k} k \left(1/2 \frac{\binom{n}{k} n}{(n+1-k)\,2^n} + 1/2 \frac{\binom{n}{k}}{(n+1-k)\,2^n} - \frac{\binom{n}{k}}{2^n} \right)^{-1} (k+1)^{-1} \left(2^n\right)^{-1}
$$

```
>    normal(expr);
```

$$
-\frac{(-n-1+k)(2k+1-n)}{(-n-1+2k)(k+1)}
$$

We see that in this example, the situation can be resolved by a further application of
`normal` or `simplify`.[5] On the other hand, this procedure is not at all efficient in
cases like

```
>    normal(expand(GAMMA(k+10000)/GAMMA(k+9999)));
```

$$
k + 9999
$$

Issuing this command gives you time to have lunch before you receive the *trivial*
result. The reason is that both numerator and denominator are expanded indepen-
dently as multiples of $\Gamma(k)$. In the worst case you receive the error message

```
    Error, (in expand/GAMMA) object too large
```

depending on the memory situation on your computer! We will now present a better
method for the given purpose which is implemented in the `hsum` package[6]

```
>    read "hsum.mpl";
```

which gives an instant answer to the much more complicated question

---

[5] Note that `simplify` can easily handle the next question. However, `simplify` does not always
simplify towards a rational function, even if the result is rational. Moreover, `simplify` is a
combination of so many algorithms so that it is not even possible to describe its full mechanism. It
is better to use simplification commands that have a clear description.

[6] The current updated version is `hsum17.mpl`. In future Maple sessions we will always assume
that the `hsum` package is loaded by the `read` command.

```
>   simpcomb(GAMMA(k+1000000)/GAMMA(k+999999));
```
$$k + 999999$$

The following algorithm, which is almost trivial but decisive, describes how u_k and v_k can be identified (at least) for input of a special type. We will later see that the same algorithm applies for input of a more general type.

Algorithm 2.2 (`simpcomb`)
The following algorithm decides the rationality of term ratios a_{k+1}/a_k:

1. Input: a_{k+1}/a_k, where $a_k \neq 0$ is a ratio of products of rational functions, powers, factorials, Γ function terms, binomial coefficients and Pochhammer symbols that are rational-linear in their arguments.[7]
2. (`togamma`)
    Build a_{k+1}/a_k, and convert all occurrences of factorials, binomial coefficients, and Pochhammer symbols to Γ function terms according to (1.2), (1.5), and (1.11), avoiding negative arguments. The case of binomial coefficients is done by the rules

$$\binom{a}{k} \rightarrow \begin{cases} (-1)^k \frac{\Gamma(k-a)}{\Gamma(k+1)\,\Gamma(-a)} & \text{if } a \in \mathbb{Z},\ a < 0 \\ \qquad 0 & \text{if } a - k \in \mathbb{Z},\ a - k < 0 \ . \\ \frac{\Gamma(a+1)}{\Gamma(k+1)\,\Gamma(a-k+1)} & \text{otherwise} \end{cases}$$

3. (`simplify_gamma`)
    Rewrite the preceding expression recursively according to the rule (see (1.4))

$$\Gamma(a + j) = a(a + 1) \cdots (a + j - 1) \cdot \Gamma(a)$$

    whenever the arguments a and $a + j$ of two representing Γ function terms have positive integer difference j. Reduce the final fraction canceling common Γ terms.
4. (`simplify_power`)
    Rewrite the preceding expression recursively according to the rule

$$b^{a+j} = b^j\, b^a$$

    whenever the arguments a and $a+j$ of two representing power terms have positive integer difference j. Reduce the final fraction canceling common power terms.
5. The expression a_{k+1}/a_k is rational if and only if the resulting expression u_k/v_k in step 4 is rational, i.e., $u_k, v_k \in \mathbb{Q}[k]$.
6. Output: (u_k, v_k).

---

[7] If the input terms have integer-linear arguments in k, then the ratio a_{k+1}/a_k is clearly a rational function; if the input terms are rational-linear in k, then this is not automatically the case, and the algorithm detects this.

Proof Note that this result follows immediately from the given form of a_k (as a ratio) and therefore of the expression a_{k+1}/a_k. The given form guarantees that common Γ and power terms in the numerator and denominator cancel in steps (3) and (4) if a_{k+1}/a_k is rational.

Note that, for integer-linear input, it is clear, by the use of the given rewrite rules, that all Γ and power terms cancel and polynomials $u_k, v_k \in \mathbb{Q}[k]$ are constructed.

Example 2.3 The sine function has the power series representation

$$\sin x = \sum_{k=0}^{\infty} \frac{(-1)^k x^{2k+1}}{(2k+1)!}.$$

To find its hypergeometric counterpart, we start with $a_k = \frac{(-1)^k x^{2k+1}}{(2k+1)!}$ and use Algorithm 2.2. We get the term ratio

$$\begin{aligned}
\frac{a_{k+1}}{a_k} &= \frac{(2k+1)!}{(-1)^k x^{2k+1}} \cdot \frac{(-1)^{k+1} x^{2k+3}}{(2k+3)!} \\
&= \frac{(2k+1)!}{(-1)^k x^{2k+1}} \cdot \frac{-(-1)^k x^2 x^{2k+1}}{(2k+3)(2k+2)(2k+1)!} \\
&= -\frac{x^2}{(2k+3)(2k+2)} = \frac{1}{\left(k+\frac{3}{2}\right)(k+1)} \left(-\frac{x^2}{4}\right).
\end{aligned}$$

Since $a_0 = x$, this leads finally to the hypergeometric representation

$$\sin x = \sum_{k=0}^{\infty} \frac{(-1)^k x^{2k+1}}{(2k+1)!} = x \cdot {}_0F_1\left(\begin{array}{c} - \\ \frac{3}{2} \end{array} \middle| -\frac{x^2}{4}\right)$$

by Algorithm 2.2.

Example 2.4 As another example, the rationality of a_{k+1}/a_k for

$$a_k = \frac{\Gamma(2k)}{4^k \, \Gamma(k) \, \Gamma(k+1/2)}$$

is recognized using the given procedure by the stepwise transformations

$$\begin{aligned}
\frac{a_{k+1}}{a_k} &= \frac{\Gamma(2k+2)}{4^{k+1}\Gamma(k+1)\Gamma(k+3/2)} \Big/ \frac{\Gamma(2k)}{4^k\Gamma(k)\Gamma(k+1/2)} \\
&= \frac{(2k)(2k+1)\Gamma(2k)4^k \, \Gamma(k) \, \Gamma(k+1/2)}{\Gamma(2k) \, 4 \cdot 4^k \, k\Gamma(k)(k+1/2)\Gamma(k+1/2)} = \frac{(2k)(2k+1)}{4k(k+1/2)} = 1.
\end{aligned}$$

From the resulting information, it follows by induction (or easier, by the hypergeometric coefficient formula (2.8)) that for $k \in \mathbb{N}$

$$\frac{\Gamma(2k)}{4^k\,\Gamma(k)\,\Gamma(k+1/2)} = a_k = a_1 = \frac{\Gamma(2)}{4\,\Gamma(1)\Gamma(3/2)} = \frac{1}{2\,\Gamma(1/2)} = \frac{1}{2\sqrt{\pi}}, \quad (2.12)$$

using (1.15). Note that (2.12), which is called the *duplication formula* of the Γ function, is valid for all $k \in \mathbb{C}$, k not the half of a negative integer or zero, a fact which, however, cannot be proved by the present method.

Algorithm 2.2 also applies to

$$b_k = \Gamma(2k) - \alpha\, 4^k\, \Gamma(k)\, \Gamma(k+1/2)$$

and the same procedure leads to

$$\frac{b_{k+1}}{b_k} = 2k\,(2k+1)$$

(check!) which is true whenever $\alpha \neq \frac{1}{2\sqrt{\pi}}$. If $\alpha = \frac{1}{2\sqrt{\pi}}$, however, by the above computation, $b_k \equiv 0$ and therefore b_{k+1}/b_k is not properly defined.

Note that the occurrence of another variable, α, had the side effect that our calculation was not valid for a particular value of α. This is a typical situation since we work with rational arithmetic and must make sure that no denominator which might appear in any intermediate calculation is ever equal to zero.

Example 2.5   Next, we consider the expression ($n \in \mathbb{N}_{\geq 0}$)

$$a_k := \frac{1}{2^{n+1}}\binom{n+1}{k} - \frac{1}{2^n}\binom{n}{k}. \qquad (2.13)$$

Note that a_k does not have the form required in the above algorithm since it is not just a ratio but a sum of ratios. On the other hand, it is easily seen that for any sum $a_k = \alpha_k + \beta_k$ of expressions α_k and β_k for which $\alpha_k/\beta_k \in \mathbb{K}(k)$ is a rational function,[8] the same algorithm applies. This is obviously the case for the two summands of a_k.

We obtain by the method described

$$\frac{a_{k+1}}{a_k} = \frac{\frac{1}{2^{n+1}}\binom{n+1}{k+1} - \frac{1}{2^n}\binom{n}{k+1}}{\frac{1}{2^{n+1}}\binom{n+1}{k} - \frac{1}{2^n}\binom{n}{k}}$$

$$= \frac{\dfrac{\Gamma(n+2)}{\Gamma(k+2)\Gamma(n-k+1)} - 2\dfrac{\Gamma(n+1)}{\Gamma(k+2)\Gamma(n-k)}}{\dfrac{\Gamma(n+2)}{\Gamma(k+1)\Gamma(n-k+2)} - 2\dfrac{\Gamma(n+1)}{\Gamma(k+1)\Gamma(n-k+1)}}$$

---

[8] In such a case, α_k and β_k are called *similar* hypergeometric terms, see p. 94.

$$
\begin{aligned}
&= \frac{\dfrac{(n+1)\Gamma(n+1)}{(k+1)\Gamma(k+1)(n-k)\Gamma(n-k)} - 2\dfrac{\Gamma(n+1)}{(k+1)\Gamma(k+1)\Gamma(n-k)}}{\dfrac{(n+1)\Gamma(n+1)}{\Gamma(k+1)(n-k)(n-k+1)\Gamma(n-k)} - 2\dfrac{\Gamma(n+1)}{\Gamma(k+1)(n-k)\Gamma(n-k)}} \\[2em]
&= \frac{\dfrac{n+1}{(k+1)(n-k)} - \dfrac{2}{k+1}}{\dfrac{n+1}{(n-k)(n-k+1)} - \dfrac{2}{n-k}} = \frac{n-k+1}{k+1} \cdot \frac{n+1-2(n-k)}{n+1-2(n-k+1)} \\[2em]
&= -\frac{(k-n-1)\,(k-n/2+1/2)}{(k-n/2-1/2)\,(k+1)}.
\end{aligned}
\tag{2.14}
$$

If we are now interested in

$$
F = \sum_{k=-\infty}^{\infty} a_k,
\tag{2.15}
$$

then, according to (2.9), from the final factored form of (2.14) we can read off the list of upper parameters $(-n-1, -n/2+1/2)$, the lower parameter $(-n/2-1/2)$, and $x = -1$; and by

$$
a_0 = \frac{1}{2^{n+1}} \binom{n+1}{0} - \frac{1}{2^n} \binom{n}{0} = -\frac{1}{2^{n+1}},
$$

we see (by induction, or by the hypergeometric coefficient formula (2.8)) that

$$
a_k = \frac{(-n-1)_k\,(-n/2+1/2)_k}{(-n/2-1/2)_k\,k!} (-1)^k\, a_0 = -\frac{(-n-1)_k\,(-n/2+1/2)_k}{(-n/2-1/2)_k\,k!} (-1)^k\, \frac{1}{2^{n+1}}
$$

and therefore

$$
\sum_{k=-\infty}^{\infty} \left(\frac{1}{2^{n+1}} \binom{n+1}{k} - \frac{1}{2^n} \binom{n}{k} \right) = -\frac{1}{2^{n+1}}\, {}_2F_1 \left(\begin{matrix} -n-1, -n/2+1/2 \\ -n/2-1/2 \end{matrix} \,\middle|\, -1 \right).
\tag{2.16}
$$

Note that the upper parameters of (2.16) show in particular that the sum F given by (2.15) for $n \in \mathbb{N}_{\geq 0}$ is finite with summands $k = 0, \ldots, n+1$ for even n and with summands $k = 0, \ldots, \frac{n-1}{2}$ for odd n. Furthermore we note that according to identity (2.1) $F \equiv 0$ (check!).

Let us consider the even case first. If $n = 2m \in \mathbb{N}_{\geq 0}$ is even, then by (2.16) we get for $m \in \mathbb{N}$

$$
{}_2F_1 \left(\begin{matrix} -2m-1, -m+1/2 \\ -m-1/2 \end{matrix} \,\middle|\, -1 \right) \equiv 0.
$$

On the other hand, for odd n, we divided by zero in (2.14) so that this deduction is not valid. In this case the hypergeometric series (2.16) has a negative integer lower parameter $-n/2 - 1/2$. Nevertheless, this is a valid hypergeometric series since the sum is from $k = 0, \ldots, \frac{n-1}{2}$ only. However, it turns out that its sum is *not* equal to 0 in this case.

Example 2.6 (Dixon's Identity) Identity (2.6) $(n \in \mathbb{N}_{\geq 0})$ is called *Dixon's identity*. We will now give a hypergeometric version. Therefore, for

$$a_k := (-1)^k \binom{n+b}{n+k} \binom{n+c}{c+k} \binom{b+c}{b+k},$$

we calculate by Algorithm 2.2

$$\frac{a_{k+1}}{a_k} = \frac{(k-n)\,(k-b)\,(k-c)}{(k+n+1)\,(k+b+1)\,(k+c+1)} \tag{2.17}$$

(check!), and from

$$a_0 = \binom{n+b}{n} \binom{n+c}{c} \binom{b+c}{b},$$

we are led to the hypergeometric representation

$$\binom{n+b}{n} \binom{n+c}{c} \binom{b+c}{b} \, {}_4F_3 \left(\begin{matrix} -n\,,\,-b\,,\,-c\,,\,1 \\ n+1\,,\,b+1\,,\,c+1 \end{matrix} \Bigg| \, 1 \right),$$

where we had to add the number 1 to the list of upper parameters since the denominator of (2.17) did not contain a factor $(k+1)$.

But, be careful! Did you realize that this hypergeometric function corresponds to the sum of Dixon's term for $k = 0, \ldots, \infty$ rather than for $k = -\infty, \ldots, \infty$? In a later example, we will see that in some instances this might be exactly what we want.

In our case, however, to get rid of this problem, and to deduce a ${}_3F_2$ rather than a ${}_4F_3$ representation, we realize that one of the lower parameters, $n+1$, is assumed to be an integer. In such a situation we must apply a suitable shift. Since the summation is over all $k \in \mathbb{Z}$, a shift of the summation index by an integer does not change the value of the sum. This is the nice thing when working with bilateral sums: their value is invariant with respect to shifts of the summation variable. Therefore, in our example, we shift the summation index by $-n$, i.e., we consider $b_k = a_{k-n}$ with

$$\sum_{k=-n}^{n} a_k = \sum_{k=-\infty}^{\infty} a_k = \sum_{k=-\infty}^{\infty} b_k = \sum_{k=0}^{2n} b_k$$

and we get from (2.17)

$$\frac{b_{k+1}}{b_k} = \frac{a_{k-n+1}}{a_{k-n}} = \frac{(k-2n)\,(k-n-b)\,(k-n-c)}{(k+1)\,(k-n+b+1)\,(k-n+c+1)}.$$

By this procedure, we generated a $(k+1)$-term in the denominator and since

$$b_0 = a_{-n} = (-1)^n \binom{n+c}{c-n} \binom{b+c}{b-n},$$

we have, finally, the hypergeometric representation

$$F = (-1)^n \binom{n+c}{c-n} \binom{b+c}{b-n} \, {}_3F_2 \left(\begin{array}{c} -2n\,,\,-n-b\,,\,-n-c \\ -n+b+1\,,\,-n+c+1 \end{array} \middle| \, 1 \right)$$

for the Dixon sum.

Note that we see from this hypergeometric representation and from the method of its discovery that the left-hand side is a sum in the range $k = -n, \ldots, n$. At first glance this might not have been obvious. $\triangle$

Next, we would like to give some more examples that show how one takes care of possible shifts.

Example 2.7 Let us consider $a_k = k \binom{n}{k}$. Then

$$\frac{a_{k+1}}{a_k} = \frac{n-k}{k}.$$

We see that this cannot be the term ratio of a hypergeometric representation since the denominator has a zero root. This corresponds to the fact that $a_0 = 0$, and any hypergeometric representation has $a_0 \neq 0$. By a suitable shift, however, we can overcome this difficulty and, as an important observation, the term ratio given shows us which shift will be successful! Since the denominator root is zero, we shift by one to eliminate it and to construct a $(k+1)$-term. For $b_k := a_{k+1}$, we get

$$\frac{b_{k+1}}{b_k} = \frac{a_{k+2}}{a_{k+1}} = -\frac{k+1-n}{k+1}$$

so that, from $b_0 = a_1 = n$, it follows that

$$\sum_{k=0}^{n} k \binom{n}{k} = n \cdot {}_1F_0 \left(\begin{array}{c} -n+1 \\ - \end{array} \middle| -1 \right).$$

Next, we consider the similar expression $a_k = \frac{1}{k} \binom{n}{k}$. Here we are interested in $\sum_{k=1}^{\infty} a_k$ rather than the bilateral sum. Let's see what can be done nevertheless. We have

$$\frac{a_{k+1}}{a_k} = -\frac{k(k-n)}{(k+1)^2}.$$ (2.18)

Since the numerator has a zero root, we must shift by one again. In the present case, there is no chance to keep a $(k+1)$-term in the denominator and therefore we have to increase the number of upper parameters by adding one to them.

The final result is

$$\sum_{k=1}^{n} \frac{1}{k}\binom{n}{k} = n \cdot {}_3F_2\left(\begin{array}{c} -n+1,1,1 \\ 2,2 \end{array}\middle| -1\right).$$

In this example, the extra factor $(k+1)$ that we put in both numerator and denominator of (2.18) helped us a lot since this step made the sum finite to the left—and that was exactly what we needed.

Both examples given show that for quite similar input, the orders p and q of the corresponding hypergeometric representations can be quite different. $\triangle$

Now we are prepared to state and prove the main result of this chapter. We state it only for bilateral sums and mention that a similar algorithm can be given for sums $k = k_0, \ldots, \infty$.

Algorithm 2.8 (*Conversion of Sums into Hypergeometric Notation*) The following algorithm converts hypergeometric sums into hypergeometric notation:

1. Input: the summand a_k, given as ratio of products of rational functions, powers, factorials, Γ function terms, binomial coefficients, and Pochhammer symbols that are rational-linear in their arguments, or a sum or difference of such terms like expression (2.13) in Example 2.5.
2. Calculate a_{k+1}/a_k and apply Algorithm 2.2 to it generating $u_k, v_k \in \mathbb{Q}[k]$ such that
$$\frac{a_{k+1}}{a_k} = \frac{u_k}{v_k}.$$

   If Algorithm 2.2 decides that a_{k+1}/a_k is not rational then return: "No hypergeometric representation exists."
3. Factorize u_k, v_k over the rationals.[9] If there are nonlinear factors, then return: "No rational factorization found"; exit. (For factors of degree ≤ 4, one may use symbolic complex solutions, though.) If the last step was successful, however, then we have a representation

$$u_k = A\,(k+\alpha_1)\,(k+\alpha_2)\cdots(k+\alpha_p) \text{ and } v_k = B\,(k+\beta_1)\,(k+\beta_2)\cdots(k+\beta_{q+1}).$$

4. If any of the parameters $\beta_1, \ldots, \beta_{q+1}$ is an integer, then calculate the minimal such value[10] m and shift the summation variable by $-m+1$, i.e. shift all upper

---

[9] Rational factorization will be considered in more detail on p. 83.

[10] If parameters are involved, this might be undecidable, compare e.g. the Dixon case!

and lower parameters by $-m + 1$. Denote the new upper and lower parameters by $(\alpha_1, \ldots, \alpha_p)$ and $(\beta_1, \ldots, \beta_{q+1})$ again.

5. If none of the shifted lower parameters equals one, then return:
   "The bilateral sum does not have a hypergeometric representation."
6. Calculate the initial value $b_0 = a_K$, where $K: = -m + 1$ is the total shift that occurred in step 4 if applicable, else $K: = 0$; set $\texttt{upper} := \alpha_1, \ldots, \alpha_p$, and $\texttt{lower} : = \beta_1, \ldots, \beta_q$ assuming $\beta_{q+1} = 1$; set $x: = A/B$.
7. Output: the hypergeometric function $b_0 \cdot \texttt{hypergeom}(\texttt{upper}, \texttt{lower}, \texttt{x})$.

Proof Obviously, in step 2, Algorithm 2.2 decides whether or not a_{k+1}/a_k is rational. If not, then no hypergeometric representation exists by the definition of a hypergeometric series, whereas in the affirmative case $u_k, v_k \in \mathbb{Q}[k]$ are constructed. Note that this step undoubtedly succeeds if the Γ-arguments occurring are integer-linear w.r.t. k, although this is not a necessary condition.

If the factorization in step 3 fails, no hypergeometric representation with rational parameters exists. If, on the other hand, a factorization is found, then it obviously defines a hypergeometric representation if none of the corresponding lower parameters is a negative integer. The shift in step 4—if applicable—guarantees that all negative integer lower parameters disappear (and at the same time that one of the lower parameters equals 1). This shift corresponds to a shift of the summation variable k and does not change the value of the series.

Finally, if the remaining list of lower parameters does not contain the value 1, then the bilateral sum cannot be represented by a one-sided infinite hypergeometric representation.

If the shift is K, then we work with $b_k = a_{k+K}$, and $\sum b_k = \sum a_k$, so that the initial value is given by $b_0 = a_K$.

A Maple implementation of the algorithm is given in Session 3.6. If the algorithm fails because a_{k+1}/a_k turns out not to be rational, then it may still be possible to find a number $l \in \mathbb{N}$ such that a_{k+l}/a_k is rational; compare Algorithm 8.4. In this case the series under consideration can be written as a sum of l generalized hypergeometric functions.

If other variables are involved then the shifts of steps (4) and (5) might depend on the particular values of these variables. As in Example 2.6, the knowledge that any of the variables occurring is an integer might influence this decision. We give some final examples of an application of Algorithm 2.8.

Example 2.9 (*Legendre Polynomials*) Let us give the *Legendre polynomials* by the series

$$P_n(x) := \sum_{k=-\infty}^{\infty} \binom{n}{k} \binom{-n-1}{k} \left(\frac{1-x}{2}\right)^k,$$

which, of course, is a hypergeometric one. For

$$a_k := \binom{n}{k} \binom{-n-1}{k} \left(\frac{1-x}{2}\right)^k,$$

we get by Algorithm 2.2

$$\frac{a_{k+1}}{a_k} = \frac{(k-n)\,(k+n+1)}{(k+1)^2} \cdot \frac{1-x}{2},$$

and therefore we have

$$P_n(x) = {}_2F_1\left(\begin{array}{c}-n,\,n+1\\1\end{array}\middle|\,\frac{1-x}{2}\right). \tag{2.19}$$

This shows in particular that $P_n(x)$ is a polynomial of degree n with respect to x.

Next we consider the family

$$F = P_{n+1}(x) - P_n(x) = \sum_{k=-\infty}^{\infty} a_k$$

of consecutive differences of Legendre polynomials. Note that F defines a polynomial of degree $n+1$. Does F constitute a hypergeometric series? By (2.19), it is the *difference* of two hypergeometric functions, but our question is different. Algorithm 2.8 helps us to find the answer.

An application of Algorithm 2.2 gives

$$\frac{a_{k+1}}{a_k} = \frac{(k+n+1)\,(k-n-1)}{k\,(k+1)} \cdot \frac{1-x}{2}.$$

We see that a shift by one is necessary to obtain a hypergeometric representation. For $b_k := a_{k+1}$, we have

$$\frac{b_{k+1}}{b_k} = \frac{(k+n+2)\,(k-n)}{(k+1)\,(k+2)} \cdot \frac{1-x}{2},$$

so that with

$$b_0 = a_1 = -(n+1)\,(1-x),$$

there follows

$$P_{n+1}(x) - P_n(x) = -(n+1)\,(1-x) \cdot {}_2F_1\left(\begin{array}{c}-n,\,n+2\\2\end{array}\middle|\,\frac{1-x}{2}\right).$$

Example 2.10 (*Non-Natural Bounds*) In this example, we would like to present a further method which is of value if the upper bound of a hypergeometric sum is not a natural one, i.e., the summand is not identically zero outside the summation region (so that we don't have a bilateral sum). In this case a direct application of Algorithm 2.8 is not possible. We saw how the use of an extra 1 as upper parameter can be used if the left bound is not a natural one for the sum under consideration. But what if the right bound is not natural? Here a change of variable, essentially of the type $k \to -k$, helps (hence reversing the order of summation)!

Let us consider the example

$$F = \sum_{k=0}^{m} a_k := \sum_{k=0}^{m} (-1)^k \binom{n}{k},$$

for arbitrary $m \leqq n$. Here the lower bound is a natural one, but the upper bound is not. We change the summation variable, set $b_k = a_{m-k}$ and get

$$F = \sum_{k=0}^{m} b_k = \sum_{k=0}^{m} (-1)^{m-k} \binom{n}{m-k},$$

for which we find

$$\frac{b_{k+1}}{b_k} = \frac{k-m}{k+n-m+1},$$

so that by $b_0 = a_m = (-1)^m \binom{n}{m}$, we have

$$F = (-1)^m \binom{n}{m} {}_2F_1 \left(\begin{array}{c} -m, 1 \\ n-m+1 \end{array} \middle| 1 \right).$$

Note that the extra upper parameter 1 made the lower bound a natural one, and the upper bound m was natural from the beginning! We will investigate this example further in later chapters.

Session 2.11 Maple can discover some hypergeometric identities:

```
>    sum(binomial(n,k),k=0..n);
```
$$2^n$$

However, in many cases the output is different from ours and more complicated.[11]

```
>    res := sum(binomial(n,k)^2,k=0..n);
```
$$res := \frac{4^n \, \Gamma(n+1/2)}{\sqrt{\pi} \, \Gamma(n+1)}$$
```
>    res2:=convert(res,binomial);
```
$$res2 := 4^n \binom{n-1/2}{-1/2}$$
```
>    simplify(res2) assuming n::integer;
```

---

[11] The next two sums presented are not really "nice" and need an application of the Gamma duplication formula for conversion towards simple forms. Even Maple's assume facility does not simplify appropriately.

$$4^n \binom{n-1/2}{-1/2}$$

```
>   sum((-1)^k*binomial(n,k)^2,k=0..n);
```

$$\frac{\sqrt{\pi}\, 2^n}{\Gamma\,(1+1/2\,n)\,\Gamma\,(1/2-1/2\,n)}$$

The latter output is equivalent to (2.3), see Exercise 2.7.
   Dixon's sum is also simplified:

```
>   sum((-1)^k*binomial(n+b,n+k)*binomial(n+c,c+k)*
>   binomial(b+c,b+k),k=-n..n);
```

$$\frac{\binom{b+c}{b}\,\Gamma\,(n+b+1+c)}{\Gamma\,(b+c+1)\,\Gamma\,(n+1)}$$

The Maple procedure

```
>   ratio:=proc(a,k); simpcomb(subs(k=k+1,a)/a); end proc:
```

from the hsum package calculates a_{k+1}/a_k and simplifies this expression according
to Algorithm 2.2; the Maple procedure hyperterm(upper,lower,x,k) generates the hypergeometric term corresponding to the upper parameters upper, the
lower parameters lower, the variable x, and the summation variable k.
   We have for example

```
>   ratio((-1)^k*binomial(n+b,n+k)*binomial(n+c,c+k)*
>   binomial(b+c,b+k),k);
```

$$\frac{(b-k)\,(-n+k)\,(c-k)}{(n+1+k)\,(c+k+1)\,(b+k+1)}$$

```
>   ratio(hyperterm([-n,n+1],[1],(1-x)/2,k),k);
```

$$-\frac{1}{2}\,\frac{(-n+k)\,(n+1+k)\,(-1+x)}{(k+1)^2}$$

```
>   ratio(subs(n=n+1,hyperterm([-n,n+1],[1],(1-x)/2,k))-
>   hyperterm([-n,n+1],[1],(1-x)/2,k),k);
```

$$-\frac{1}{2}\,\frac{(-1+x)\,(n+1+k)\,(-n-1+k)}{k\,(k+1)}$$

Since binomial sums come in quite different disguises, it is an important fact that by
the notion of the generalized hypergeometric function these sums are classified and
hence can be identified. This fact will be stressed in the next chapter. This point of
view has been popularized by Dick Askey and George Andrews.

q-Hypergeometric Identities

An important extension of the hypergeometric function is the *q-hypergeometric
function* (as a general reference for q-hypergeometric functions, see [GR90], and
for an elementary introduction [Gasper97])

$$_r\phi_s \left(\begin{matrix} a_1, a_2, \ldots, a_r \\ b_1, b_2, \ldots, b_s \end{matrix} \,\middle|\, q, x \right) := \sum_{k=0}^{\infty} \frac{(a_1, a_2, \ldots, a_r; q)_k}{(b_1, b_2, \ldots, b_s; q)_k} \frac{x^k}{(q; q)_k} \left((-1)^k q^{\binom{k}{2}} \right)^{1+s-r}$$

where $(a_1, a_2, \ldots, a_r; q)_k$ is a short form for the product $\prod_{j=1}^{r} (a_j; q)_k$, and

$$(a; q)_k := \begin{cases} \prod_{j=0}^{k-1} (1 - aq^j) & \text{if } k > 0 \\ 1 & \text{if } k = 0 \\ \prod_{j=1}^{|k|} (1 - aq^{-j})^{-1} & \text{if } k < 0 \\ \prod_{j=0}^{\infty} (1 - aq^j) & \text{if } k = \infty \end{cases}$$

denotes the *q-Pochhammer symbol*. The q-hypergeometric functions are also called *basic hypergeometric functions* since they come with the base q.

An $_r\phi_s$ series terminates if one of its numerator parameters is of the form q^{-n} with $n \in \mathbb{N}$. In the non-terminating case the q-hypergeometric series converges in its disk of convergence if $|q| < 1$. The additional factor $\left((-1)^k q^{\binom{k}{2}} \right)^{1+s-r}$ (which does not occur in the corresponding definition of the generalized hypergeometric function)[12] is to facilitate a *confluence process*. With this factor one gets the simple formula

$$\lim_{a_r \to \infty} {_r\phi_s} \left(\begin{matrix} a_1, a_2, \ldots, a_r \\ b_1, b_2, \ldots, b_s \end{matrix} \,\middle|\, q, \frac{x}{a_r} \right) = {_{r-1}\phi_s} \left(\begin{matrix} a_1, a_2, \ldots, a_{r-1} \\ b_1, b_2, \ldots, b_s \end{matrix} \,\middle|\, q, x \right).$$

An expression a_k is called a *q-hypergeometric term* if a_{k+1}/a_k is a rational function with respect to q^k, a typical example of which is given by the summand of the q-hypergeometric series. Using the notion of the q-hypergeometric function, series with this property are classified and hence can be identified.

Since for $q \to 1^-$

$$\lim_{q \to 1^-} \frac{(q^a; q)_k}{(q; q)_k} = \frac{(a)_k}{k!}, \tag{2.20}$$

one has

$$\lim_{q \to 1^-} {_r\phi_s} \left(\begin{matrix} q^{a_1}, q^{a_2}, \ldots, q^{a_r} \\ q^{b_1}, q^{b_2}, \ldots, q^{b_s} \end{matrix} \,\middle|\, q, (q-1)^{1+s-r} x \right) = {_r F_s} \left(\begin{matrix} a_1, a_2, \ldots, a_r \\ b_1, b_2, \ldots, b_s \end{matrix} \,\middle|\, x \right)$$

---

[12] Note that basic hypergeometric functions and their properties were already considered in [Bailey35] where the definition of the q-hypergeometric function was given *without* this additional factor, though.

which connects the q-hypergeometric function with the hypergeometric function.
By

$$[k]_q := \frac{1-q^k}{1-q} = 1 + q + \cdots + q^{k-1},$$

$$[k]_q! := \frac{(q;\,q)_k}{(1-q)^k} = [k]_q \cdot [k-1]_q \cdots [1]_q,$$

$$\begin{bmatrix} n \\ k \end{bmatrix}_q := \frac{[n]_q!}{[k]_q! \cdot [n-k]_q!}$$

and

$$\Gamma_q(z) := \frac{(q;\,q)_\infty}{(q^z;\,q)_\infty} (1-q)^{1-z}$$

one defines the q-*brackets* (or q-*numbers*), the q-*factorial*, the q-*binomial coefficient* and the q-*Gamma function*, and there are q-analogues for many hypergeometric identities (see e.g. [GR90]).

We consider an example: Whereas the *binomial theorem* states that

$$\sum_{k=0}^{\infty} \binom{\alpha}{k} x^k = \sum_{k=0}^{\infty} \frac{(-\alpha)_k}{k!} (-x)^k = {}_1F_0 \left(\begin{array}{c} -\alpha \\ - \end{array} \middle| -x \right) = (1+x)^\alpha, \qquad (2.21)$$

(a particular case of which is (2.1)), the q-*binomial theorem* due to Cauchy, Jacobi and Heine is the identity ($|q| < 1$, $|x| < 1$)

$${}_1\phi_0 \left(\begin{array}{c} a \\ - \end{array} \middle| q, x \right) = \sum_{k=0}^{\infty} \frac{(a;\,q)_k}{(q;\,q)_k} x^k = \frac{(ax;\,q)_\infty}{(x;\,q)_\infty}. \qquad (2.22)$$

More details can be found in [Gasper97, GR90], and we will continue these considerations in the later chapters.

Further Reading

For further reading on hypergeometric identities we refer to [AAR99], Chaps. 2–4, and for the q-case to [GR90].

Exercises

Exercise 2.1 *Show that* (2.12) *remains valid in the limit as k tends to the half of a negative integer or zero.* Hint: *Use Exercise 1.1.*

Exercise 2.2 *Prove that the rational term ratio* (2.9) *together with the initial value* $A_0 = 1$ *implies the hypergeometric coefficient formula* (2.8).

Exercise 2.3 (Hypergeometric Differential Equation, see e.g. [Rainville60]) *Show that the generalized hypergeometric function* $F(x) := {}_pF_q \left(\begin{matrix} \alpha_1, \alpha_2, \ldots, \alpha_p \\ \beta_1, \beta_2, \ldots, \beta_q \end{matrix} \,\middle|\, x \right)$ *satisfies the hypergeometric differential equation*

$$\theta(\theta + \beta_1 - 1) \cdots (\theta + \beta_q - 1)F(x) = x(\theta + \alpha_1)(\theta + \alpha_2) \cdots (\theta + \alpha_p)F(x) \quad (2.23)$$

where θ *denotes the differential operator* $\theta f(x) = x f'(x)$. *Hint: Substitute the series into* (2.23), *and equate coefficients.*

Exercise 2.4 (Hypergeometric Derivative Rule, see e.g. [Rainville60]) *Show that the generalized hypergeometric function* $F_n(x) := {}_pF_q \left(\begin{matrix} \alpha_1, \alpha_2, \ldots, \alpha_p \\ \beta_1, \beta_2, \ldots, \beta_q \end{matrix} \,\middle|\, x \right)$ *satisfies the derivative rules*

$$\theta F_n(x) = n \left(F_{n+1}(x) - F_n(x) \right)$$

for any of its numerator parameters $n := \alpha_k$ $(k = 1, \ldots, p)$ *and*

$$\theta F_n(x) = (n - 1) \left(F_{n-1}(x) - F_n(x) \right)$$

for any of its denominator parameters $n := \beta_k$ $(k = 1, \ldots, q)$.

Exercise 2.5 (Hypergeometric Recurrence Equation) *How can Exercises 2.3 and 2.4 be combined to obtain a recurrence equation with respect to any of the parameters of* ${}_pF_q$? *What is the order of this recurrence equation?*

Exercise 2.6 *Use Algorithm 2.2 to determine* a_{k+1}/a_k *for* $a_k := b(n+j, k) - b(n, k)$ *for* $j = 1, \ldots, 3$ *when*

(a) $b(n, k) = \dbinom{n}{k}$,

(b) $b(n, k) = \dbinom{n-k}{k}$,

(c) $b(n, k) = n \dbinom{n}{k}$,

(d) $b(n, k) = (n - k)!$

Exercise 2.7 *In Session 2.11, Maple's result*

$$\sum_{k=0}^{n} (-1)^k \binom{n}{k}^2 = \frac{\sqrt{\pi}\, 2^n}{\Gamma\left(1 + \dfrac{1}{2}n\right) \Gamma\left(\dfrac{1}{2} - \dfrac{1}{2}n\right)}$$

was obtained. Show that this result is equivalent to (2.4).

Exercise 2.8 *Use Algorithm 2.8 to calculate the hypergeometric representations of the sums occurring in (2.1)–(2.5). Which hypergeometric terms are determined by the right-hand sides of these identities?*

Exercise 2.9 *Use an adaptation of Algorithm 2.8 to find hypergeometric representations for $\sum_{k=A}^{B} a_k$ with largest possible summation range, if*

(a) $a_k = k(k-1)(k-2)\binom{n}{k}$,

(b) $a_k = \binom{n-k}{k}$,

(c) $a_k = \frac{1}{k(k-1)(k-2)}\binom{2n}{k}$,

(d) $a_k = \binom{n}{k}\binom{2k}{n}$.

Which shifts are necessary? Which are the actual ranges (A, B) of the hypergeometric representations?

Exercise 2.10 *Show, by a treatment similar to Example 2.5, the identity*

$$2F_1\left(\begin{matrix} -n-1, \, -n/2+1/2 \\ -n/2 - 1/2 \end{matrix} \,\middle|\, x\right) = \sum_{k=0}^{n} \frac{(-n-1)_k \, (-n/2+1/2)_k}{(-n/2 - 1/2)_k} x^k = (1+x)\,(1-x)^n,$$

which is valid for even n.

Exercise 2.11 *The following are the standard series representations of some elementary functions. Use Algorithm 2.8 to give their hypergeometric equivalents.*

(a) $\exp(x) = \sum\limits_{k=0}^{\infty} \frac{x^k}{k!}$,

(b) $\ln(1+x) = \sum\limits_{k=1}^{\infty} \frac{(-1)^{k+1}}{k} x^k$,

(c) $\cos(x) = \sum\limits_{k=0}^{\infty} \frac{(-1)^k}{(2k)!} x^{2k}$,

(d) $(1+x)^\alpha = \sum\limits_{k=0}^{\infty} \binom{\alpha}{k} x^k$,

(e) $\arctan(x) = \sum\limits_{k=0}^{\infty} \frac{(-1)^k}{2k+1} x^{2k+1}$,

(e) $\arcsin(x) = \sum\limits_{k=0}^{\infty} \frac{(-1)^k}{2k+1} \binom{-1/2}{k} x^{2k+1}$.

Exercise 2.12 *The Legendre polynomials have the following two different series representations*

$$P_n(x) = \frac{1}{2^n} \sum_{k=0}^{n} \binom{n}{k}^2 (x-1)^{n-k} (x+1)^k$$

and

$$P_n(x) = \frac{1}{2^n} \sum_{k=0}^{\lfloor n/2 \rfloor} (-1)^k \binom{n}{k} \binom{2n-2k}{n} x^{n-2k}$$

besides the one presented in Example 2.9. Convert these into hypergeometric notation. Here

$$\lfloor x \rfloor := \max \{n \in \mathbb{Z} \mid n \le x\}$$

denotes the floor function.

Which identities between hypergeometric functions correspond to the equality of the three given hypergeometric representations for the Legendre polynomials?

Note that we are not yet able to prove that the three different series representations for the Legendre polynomials represent the same family of functions. This assertion will be proved in Chap. 4, Exercise 4.3.

Exercise 2.13 *Give a hypergeometric representation for the sum and difference of consecutive Legendre polynomials $P_{n+1}(x) \pm P_n(x)$. Try to give one for $P_{n+2}(x) \pm P_n(x)$. What happens?*

Exercise 2.14 (Apéry Numbers, see [Apéry79]) *Convert the Apéry numbers*

$$A_n := \sum_{k=0}^{n} \binom{n}{k}^2 \binom{n+k}{k}^2$$

into hypergeometric notation.

◇ **Exercise 2.15** *Write the Maple function* `hyperterm(upper,lower,x,k)` *(in terms of* `pochhammer`*) that was utilized in Session 3.11.*

Exercise 2.16 (Bieberbach Conjecture, see [deBranges85]) *The following sum was an essential tool in the proof of the* Bieberbach *conjecture by de Branges in 1984* [deBranges85] *(see also Example 7.8)*

$$\sum_{j=k}^{n} (-1)^{k+j} \binom{2j}{j-k} \binom{n+j+1}{n-j} e^{-jt}.$$

Convert into hypergeometric notation under the hypothesis that k denotes a positive integer.

Exercise 2.17 *Convert the identity (see* [GKP94]*, p. 171)*

$$
\sum_{k=-\infty}^{\infty} \frac{(-1)^k}{\binom{2n+2b+2c+2d}{n+b+c+d+k}} \binom{n+b}{n+k}\binom{b+c}{b+k}\binom{c+d}{c+k}\binom{d+n}{d+k} =
$$

$$
\frac{\Gamma(n+b+c+d+1)\Gamma(n+b+c+1)\Gamma(n+b+d+1)\Gamma(n+c+d+1)\Gamma(b+c+d+1)}{n!\,\Gamma(2n+2b+2c+2d+1)\,\Gamma(n+c+1)\,\Gamma(b+d+1)\,\Gamma(b+1)\,\Gamma(c+1)\,\Gamma(d+1)}.
$$

into hypergeometric notation. What are the natural bounds?

Exercise 2.18 *Maple's* expand *procedure expands Γ function terms $\Gamma(a+k)$ for integer k in terms of $\Gamma(a)$. When followed by* normal *to cancel common factors, this gives an alternative way to decide the rationality of expressions involving Γ terms.*

Time the simplification of the expressions $\Gamma(k+1000)/\Gamma(k+999)$, and $\Gamma(k+5000)/\Gamma(k+4999)$ using normal(expand(…)) *and* simpcomb. *Explain!*

Exercise 2.19 *Let a_k denote the kth summand of the generalized hypergeometric function $_pF_q\left(\begin{array}{c}\alpha_1,\ldots,\alpha_p\\ \beta_1,\ldots,\beta_q\end{array}\middle| x\right)$. Show that the following limit procedure generates the mth partial sum*

$$
\sum_{k=0}^{m} a_k = \lim_{\varepsilon \to 0} {}_{p+1}F_{q+1}\left(\begin{array}{c}-m,\alpha_1,\ldots,\alpha_p\\ -m+\varepsilon,\beta_1,\ldots,\beta_q\end{array}\middle| x\right).
$$

Exercise 2.20 *Show that Kummer's confluent hypergeometric function is the following limiting case[13] of Gauss' hypergeometric function*

$$
{}_1F_1\left(\begin{array}{c}a\\ c\end{array}\middle| x\right) = \lim_{b\to\infty} {}_2F_1\left(\begin{array}{c}a,b\\ c\end{array}\middle| \frac{x}{b}\right).
$$

Exercise 2.21 *Prove the following equations for the q-Pochhammer symbol:*

(a) $(a;q)_n = \dfrac{(a;q)_\infty}{(aq^n;q)_\infty}$,

(b) $\dfrac{1-aq^{2n}}{1-a} = \dfrac{(q\sqrt{a};q)_n\,(-q\sqrt{a};q)_n}{(\sqrt{a};q)_n\,(-\sqrt{a};q)_n}$,

(c) $(a;q)_n\,(-a;q)_n = (a^2;q^2)_n$,

(d) $(a;q)_n = (q^{1-n}/a;q)_n\,(-a)^n\,q^{\binom{n}{2}}$.

---

¹³ This is a confluence process, again; hence the name of the function.

Exercise 2.22 *Prove (2.20).*

Exercise 2.23 *Show that for $k, n \in \mathbb{N}$ the relations*

(a) $[k]_q! = \Gamma_q(k+1)$,

(b) $\begin{bmatrix} n \\ k \end{bmatrix}_q = \dfrac{(q;q)_n}{(q;q)_k \cdot (q;q)_{n-k}}$

are valid.

Exercise 2.24 *Work out the connection between the binomial theorem (2.21) and the q-binomial theorem (2.22).*

Exercise 2.25 *Prove the q-binomial theorem (2.22). Hint: Deduce the functional equation*

$$f(a, x) = (1 - ax) f(aq, x)$$

for $f(a, x): = {}_1\phi_0 \left(\begin{matrix} a \\ - \end{matrix} \,\middle|\, q, x \right)$ by series manipulations, and use induction to show that

$$f(a, x) = (ax; q)_n \, f(aq^n, x)$$

which gives

$$f(a, x) = (ax; q)_\infty \, f(0, x)$$

for $n \to \infty$ (see [Gasper97]).

Exercise 2.26 *Show that*

$${}_1\phi_0 \left(\begin{matrix} a \\ - \end{matrix} \,\middle|\, q, x \right) \cdot {}_1\phi_0 \left(\begin{matrix} b \\ - \end{matrix} \,\middle|\, q, ax \right) = {}_1\phi_0 \left(\begin{matrix} ab \\ - \end{matrix} \,\middle|\, q, x \right).$$

References

AAR99. Andrews, G., Askey, R.A., Roy, R.: Special Functions: Encyclopedia of Mathematics and its Applications. Cambridge University Press, Cambridge-New York (1999)

Apéry79. Apéry, R.: Irrationalité de $\zeta(2)$ et $\zeta(3)$. Astérisque **61**, 11–13 (1979)

Bailey35. Bailey, W.N.: Generalized Hypergeometric Series. Cambridge University Press, Cambridge (1935). Reprinted 1964 by Stechert-Hafner Service Agency, New York-London.

deBranges85. De Branges, L.: A proof of the Bieberbach conjecture. Acta Math. **154**, 137–152 (1985)

GR90. Gasper, G., Rahman, M.: Basic Hypergeometric Series (Encyclopedia of Mathematics and its Applications) vol. 35, 2nd edn (2004). Cambridge University Press, London and New York (1990)

Gasper97. Gasper, G.: Elementary derivations of summation and transformation formulas for q-series. In: Ismail, M.E.H., et al. (eds.) Fields Institute Communications, vol. 14, pp. 55–70. American Mathematics Society, Providence (1997)

GKP94. Graham, R.L., Knuth, D.E., Patashnik, O.: Concrete Mathematics: a Foundation for Computer Science, 2nd edn. Addison-Wesley, Reading, Massachussets (1994)

Rainville60. Rainville, E.D.: Special Functions. The MacMillan Co., New York (1960)

Chapter 3
Hypergeometric Database

In this chapter we list some of the major hypergeometric identities. Note that most of these do not require any variables to have integer values. We give examples showing how this database can be used in connection with Algorithm 2.8 to generate binomial identities.

The following identities can be found in the book of Bailey [Bailey35], the hypergeometric "bible". Many more hypergeometric identities are known, but despite this it turns out that most identities that occur "in practice" can be traced back to one of those given here (see e.g. [Roy87]). Since the current chapter will give no more than an idea of how such a database can be used to generate identities, we do not emphasize a "completion" of the given list.

In later chapters, we will give other methods by means of which binomial identities can be discovered without referring to such a database.

Note that, if not otherwise stated, a, b, c, d, e denote arbitrary complex numbers, such that

(a) none of the occurring lower parameters is a nonpositive integer,

and

(b) the hypergeometric sums involved converge.

All series considered have radius of convergence 1, and are evaluated at some boundary point. One can show that for $p = q + 1$ the generalized hypergeometric function $_pF_q$ converges absolutely on the unit circle if

$$\mathrm{Re}\left(\sum_{j=1}^{q} \beta_j - \sum_{j=1}^{p} \alpha_j\right) > 0,$$

(see [Rainville60], Chap. 5). Throughout, the variable n denotes a nonnegative integer.

W. Koepf, *Hypergeometric Summation*, Universitext,
DOI: 10.1007/978-1-4471-6464-7_3, © Springer-Verlag London 2014

Hypergeometric Database

1. **(Gauss)** ([Bailey35], pp. 2–3)

$$_2F_1\left(\begin{array}{c} a, b \\ c \end{array}\middle|\,1\right) = \frac{(c-b)_{-a}}{(c)_{-a}} = \frac{\Gamma(c)\Gamma(c-a-b)}{\Gamma(c-a)\Gamma(c-b)}. \tag{3.1}$$

If $-a = n \in \mathbb{N}_{\geq 0}$, this is the *Chu-Vandermonde* identity

$$_2F_1\left(\begin{array}{c} -n, b \\ c \end{array}\middle|\,1\right) = \frac{(c-b)_n}{(c)_n}.$$

2. **(Kummer)** ([Bailey35], p. 9)

$$_2F_1\left(\begin{array}{c} a, b \\ 1+a-b \end{array}\middle|\,-1\right) = \frac{(1+a)_{-b}}{(1+a/2)_{-b}} = \frac{\Gamma(1+a-b)\Gamma(1+a/2)}{\Gamma(1+a)\Gamma(1+a/2-b)}.$$

3. **(Pfaff-Saalschütz)** ([Bailey35], p. 9)

$$_3F_2\left(\begin{array}{c} a, b, -n \\ c, 1+a+b-c-n \end{array}\middle|\,1\right) = \frac{(c-a)_n\,(c-b)_n}{(c)_n\,(c-a-b)_n}.$$

4. **(Dixon)** ([Bailey35], p. 13)

$$_3F_2\left(\begin{array}{c} a, b, c \\ 1+a-b, 1+a-c \end{array}\middle|\,1\right) = \frac{(1+a)_{-c}\,(1+a/2-b)_{-c}}{(1+a/2)_{-c}\,(1+a-b)_{-c}}$$

$$= \frac{\Gamma(1+a/2)\Gamma(1+a-b)\Gamma(1+a-c)\Gamma(1+a/2-b-c)}{\Gamma(1+a)\Gamma(1+a/2-b)\Gamma(1+a/2-c)\Gamma(1+a-b-c)}.$$

5. **(Watson, Whipple)** ([Bailey35], p. 16)

$$_3F_2\left(\begin{array}{c} a, b, c \\ (a+b+1)/2, 2c \end{array}\middle|\,1\right) = \frac{\sqrt{\pi}\,\Gamma\left(\frac{1+2c}{2}\right)\Gamma\left(\frac{1+a+b}{2}\right)\Gamma\left(\frac{1-a-b+2c}{2}\right)}{\Gamma\left(\frac{1+a}{2}\right)\Gamma\left(\frac{1+b}{2}\right)\Gamma\left(\frac{1-a+2c}{2}\right)\Gamma\left(\frac{1-b+2c}{2}\right)}.$$

6. **(Whipple)** ([Bailey35], p. 16)

$$_3F_2\left(\begin{array}{c} a, 1-a, c \\ e, 1+2c-e \end{array}\middle|\,1\right) = \frac{\pi\,2^{1-2c}\,\Gamma(e)\Gamma(1+2c-e)}{\Gamma\left(\frac{a+e}{2}\right)\Gamma\left(\frac{a+1+2c-e}{2}\right)\Gamma\left(\frac{1-a+e}{2}\right)\Gamma\left(\frac{2+2c-a-e}{2}\right)}.$$

We do not prove any of these identities now. All of them, and many more, will be proved later by several methods.

In order to find a hypergeometric term representation for a given series, when such a list of hypergeometric identities is at hand, one may utilize Algorithm 2.8. This converts the sum under consideration into hypergeometric form so that one can use the above list as a database. If the hypergeometric function is in the list, the job is done. We practice some examples:

Example 3.1 We check first whether identity (2.3) can be found in our list. For $a_k := \binom{n}{k}^2$, we get

$$\frac{a_{k+1}}{a_k} = \frac{(k-n)^2}{(k+1)^2},$$

and therefore

$$\sum_{k=0}^{n} \binom{n}{k}^2 = {}_2F_1\left(\begin{array}{c} -n, -n \\ 1 \end{array} \middle| 1\right).$$

This is obviously a particular case of the Chu-Vandermonde identity ($b = -n, c = 1$), and we have

$$\sum_{k=0}^{n} \binom{n}{k}^2 = \frac{(1+n)_n}{(1)_n} = \frac{(2n)!}{n!^2} = \binom{2n}{n},$$

which proves (2.3).

It is worth remembering that the square binomial sum is a particular case of the Chu-Vandermonde identity.

Example 3.2 To check identity (2.4) for even n, we replace n by $2m$ ($m \in \mathbb{N}$) and get for $a_k := (-1)^k \binom{2m}{k}^2$ the term ratio

$$\frac{a_{k+1}}{a_k} = -\frac{(k-2m)^2}{(k+1)^2},$$

and therefore

$$\sum_{k=0}^{2m} (-1)^k \binom{2m}{k}^2 = {}_2F_1\left(\begin{array}{c} -2m, -2m \\ 1 \end{array} \middle| -1\right).$$

This is *not* a particular case of the Chu-Vandermonde identity, since the ${}_2F_1$ is evaluated at $x = -1$. But, fortunately, Kummer's identity applies with $a = b = -2m$, and gives formally

$$\sum_{k=0}^{2m} (-1)^k \binom{2m}{k}^2 = \frac{(1-2m)_{2m}}{(1-m)_{2m}} = \frac{\Gamma(1-m)}{\Gamma(1-2m)\Gamma(1+m)}.$$

This is a formal result which cannot be a valid representation for our sum because the Pochhammer symbols in both numerator and denominator have a zero factor in common (since m is a positive integer). Equivalently, Γ function terms with negative integer arguments occur. On the other hand, by a continuity argument (Kummer's identity is valid for m near an integer), we get the correct result through a limit computation using the limit deduced in Exercise 1.2. However, we try to keep a rather algebraic viewpoint and avoid limit computations as much as possible. Therefore, to deduce a simpler representation (essentially, by cancelling the common zero factors in both numerator and denominator) for

$$s_m := \frac{(1-2m)_{2m}}{(1-m)_{2m}},$$

we compute (by Algorithm 2.2)

$$\frac{s_{m+1}}{s_m} = -4\frac{m+1/2}{m+1},$$

(you see: it's always the same trick!), and it follows that

$$\sum_{k=0}^{2m}(-1)^k \binom{2m}{k}^2 = s_m = \frac{(1/2)_m}{m!}(-4)^m = \frac{(-1)^m\,(2m)!}{(m!)^2} = (-1)^m \binom{2m}{m},$$

where we used Exercise 1.5 to rewrite the result in terms of factorials and binomial coefficients. Hence, (2.4) is seen to be valid for even n.

Example 3.3 For identity (2.5), we again make a check for even n and replace n by $2m$ $(m \in \mathbb{N})$. Then, we have for $a_k := (-1)^k \binom{2m}{k}^3$, the term ratio

$$\frac{a_{k+1}}{a_k} = \frac{(k-2\,m)^3}{(k+1)^3},$$

so that

$$\sum_{k=0}^{2m}(-1)^k \binom{2m}{k}^3 = {}_3F_2\left(\begin{matrix} -2m, -2m, -2m \\ 1,\,1 \end{matrix} \middle| 1\right).$$

Therefore, to utilize our database, we check whether there is a ${}_3F_2$ entry with unit argument. There are four such entries, namely: Pfaff-Saalschütz's, Dixon's, Watson's and Whipple's identities. First, let us try to match the arguments with those in the Pfaff-Saalschütz identity. Therefore, we set $n = -a = -b = 2m$ and $c = 1$. Thus, we get $1 + a + b - c - n = -2m$, and since this does not equal 1 we don't find a match. On the other hand, with Dixon's identity, the choice $a = b = c = -2m$ is successful.

Formally, we have therefore that

$$\sum_{k=0}^{2m} \binom{2m}{k}^3 = \frac{(1-2m)_{2m}\,(1+m)_{2m}}{(1-m)_{2m}\,(1)_{2m}}.$$

Again, this is only a formal result. As in Example 3.2, we use Algorithm 3.2 to deduce the term ratio

$$\frac{s_{m+1}}{s_m} = -27\frac{(m+1/3)\,(m+2/3)}{(1+m)^2}$$

for

$$s_m := \frac{(1-2m)_{2m}\,(1+m)_{2m}}{(1-m)_{2m}\,(1)_{2m}},$$

to deduce the standard representation

$$s_m = (-27)^m \frac{(1/3)_m\,(2/3)_m}{m!^2}.$$

Using Exercise 1.4, we finally have (2.5), for even n.

This example shows that the cube binomial sum is a particular case of Dixon's identity.

Example 3.4 We continue with Example 3.10, where we discovered that, for $m < n$

$$\sum_{k=0}^{m}(-1)^k \binom{n}{k} = (-1)^m \binom{n}{m}\, {}_2F_1\left(\begin{array}{c}-m,\,1\\ n-m+1\end{array}\middle|\,1\right).$$

We see that this representation is again a particular case of the Chu-Vandermonde identity and leads to

$$\sum_{k=0}^{m}(-1)^k \binom{n}{k} = (-1)^m \binom{n}{m} \frac{(n-m)_m}{(n-m+1)_m} = (-1)^m \frac{n-m}{n} \binom{n}{m}.$$

Note that in Exercise 3.10 this result will be obtained by other means. Furthermore, we will meet this result later in connection with Gosper's algorithm.

Example 3.5 (*Székely Identity, see* [Székely85]). Finally, we try to find a hypergeometric term representation for the series

$$F = \sum_{k=-\infty}^{\infty} a_k := \sum_{k=-\infty}^{\infty} \binom{A+B+C+D+E-k}{E-k}\binom{A+D}{k+D}\binom{B+C}{k+C}$$

$$= \binom{A+C+D+E}{A+C}\binom{B+C+D+E}{C+E},$$

$$\tag{3.2}$$

given first by Székely [Székely85] who proved this identity by combinatorial means (for integer-valued variables).

We get the term ratio

$$\frac{a_{k+1}}{a_k} = \frac{(k-A)\,(k-B)\,(k-E)}{(k-(A+B+C+D+E))\,(k+1+D)\,(k+1+C)}.$$

Since there is no $(k+1)$-term in the denominator, we need an assumption about some of the variables. Assuming C is an integer, we may shift the summation by $-C$ and get

$$F = \binom{A+B+2C+D+E}{E+C}\binom{A+D}{D-C}\,{}_3F_2\left(\begin{array}{c} -A-C,-B-C,-C-E \\ -(A+B+2C+D+E),\,1+D-C \end{array}\bigg| 1\right).$$

This is a particular case of the Pfaff-Saalschütz identity with the choice $n := C + E, a := -A-C, b := -B-C$ and $c := -(A+B+2C+D+E)$ if we further assume E to be an integer.

Hence, the Pfaff-Saalschütz identity gives

$$F = \binom{A+B+2C+D+E}{E+C}\binom{A+D}{D-C}\frac{(-B-C-D-E)_{C+E}\,(-A-C-D-E)_{C+E}}{(-A-B-2C-D-E)_{C+E}\,(-D-E)_{C+E}}.$$

To simplify the result further, we observe that the main integer variable is now $n = C + E$. Therefore, replacing E by $n - C$ and denoting the last term for F by F_n, with the aid of the term ratio

$$\frac{F_{n+1}}{F_n} = \frac{(n+A+D+1)\,(n+B+D+1)}{(n+D+1-C)\,(n+1)},$$

we have the reformulation

$$F = \binom{A+D}{D-C}\frac{(A+D+1)_{C+E}\,(B+D+1)_{C+E}}{(C+E)!\,(D+1-C)_{C+E}}.$$

Converting the Pochhammer symbols and the binomial coefficients to Γ function terms according to (1.5) and (1.11) shows

$$F = \frac{\Gamma(A+C+D+E+1)\Gamma(B+C+D+E+1)}{\Gamma(B+D+1)\Gamma(D+E+1)\Gamma(C+E+1)\Gamma(A+C+1)},$$

which obviously is (3.2).   △

The examples given show that the database can be used efficiently to find hypergeometric term representations of binomial sums. On the other hand, the procedure is not always straightforward but needs limit considerations, or other tricks. Furthermore, it cannot be a safe assumption that we will ever have a complete list of all valid hypergeometric identities at hand. Therefore, we will develop other methods soon.

Session 3.6 Maple knows a lot about hypergeometric functions, and can reproduce some hypergeometric identities.

For this purpose, one may use the procedure `hypergeom (upper,lower,x)` representing the hypergeometric function whose upper and lower parameters are given by the lists `upper` and `lower`, and argument x. We get for example

```
> term:=hypergeom([a,b],[c],1);
```
$$term := {}_2F_1(a, b; c; 1)$$

```
> simplify(term);
```
$$\frac{\Gamma(c)\,\Gamma(c-a-b)}{\Gamma(c-a)\,\Gamma(c-b)}$$

```
> simplify(hypergeom([a,b],[1+a-b],-1));
```
$$\frac{2^{-a}\sqrt{\pi}\,\Gamma(1+a-b)}{\Gamma(1/2+1/2\,a)\,\Gamma(1+1/2\,a-b)}$$

```
> simplify(hypergeom([a,b,-n],[c,1+a+b-c-n],1))
> assuming(n,integer);
```
$${}_3F_2(a, b, -n; c, 1+a+b-c-n; 1)$$

```
> simplify(hypergeom([a,b,c],[1+a-b,1+a-c],1));
```
$$\frac{2^{-a}\sqrt{\pi}\,\Gamma(1+a-b)\,\Gamma(1+a-c)\,\Gamma(1+1/2\,a-b-c)}{\Gamma(1/2+1/2\,a)\,\Gamma(1+1/2\,a-b)\,\Gamma(1+1/2\,a-c)\,\Gamma(1+a-b-c)}$$

```
> simplify(hypergeom([a,b,c],[(a+b+1)/2,2*c],1));
```
$$\frac{\sqrt{\pi}\,\Gamma(c+1/2)\,\Gamma(1/2+1/2\,a+1/2\,b)\;\Gamma(1/2-1/2\,b-1/2\,a+c)}{\Gamma(1/2+1/2\,a)\,\Gamma(1/2+1/2\,b)\,\Gamma(1/2-1/2\,a+c)\,\Gamma(1/2-1/2\,b+c)}$$

```
> simplify(hypergeom([a,1-a,c],[e,1+2*c-e],1));
```
$$2\,\frac{\Gamma(e)\,\Gamma(1+2c-e)\,\pi\,4^{-c}}{\Gamma(1/2-1/2\,a+1/2\,e)\,\Gamma(1-1/2\,a+c-1/2\,e)\,\Gamma(1/2\,a+1/2\,e)\,\Gamma(1/2\,a+1/2+c-1/2\,e)}$$

We see that our hypergeometric database (except for the Pfaff-Saalschütz identity) is accessible in Maple although the output differs modulo an application of the duplication formula.

If Maple cannot represent binomial sums in closed form, it sometimes converts them into hypergeometric notation:

```
> sum(binomial(n,k)^3,k=0..n);
```
$${}_3F_2(-n, -n, -n; 1, 1; -1)$$

Finally, we reproduce the computations for the Székely identity:

```
> summand:=binomial(A+B+C+D+E-k,E-k)*binomial(A+D,k+D)*
> binomial(B+C,k+C);
```
$$summand := \binom{A+B+C+D+E-k}{E-k}\binom{A+D}{k+D}\binom{B+C}{k+C}$$

```
> ratio(summand,k);
```
$$\frac{(E-k)\,(A-k)\,(B-k)}{(k+1+D)\,(k+1+C)\,(A+B+C+D+E-k)}$$

```
>   init:=simplify(eval(summand,k=-C));
```

$$init := \binom{A+B+2C+D+E}{E+C}\binom{A+D}{D-C}$$

```
>   term:=init*
>   hyperterm([-B-C-D-E,-A-C-D-E,1],[-A-B-2*C-D-E,-D-E],1,C+E):
>   term:=subs(E=n-C,term):
>   ratio(term,n);
```

$$-\frac{(A+D+n+1)(B+D+n+1)}{(-D-n-1+C)(n+1)}$$

```
>   expr:=eval(subs(pochhammer=1,n=0,term));
```

$$expr := \binom{A+D}{D-C}$$

```
>   simpcomb(expr*hyperterm([A+D+1,B+D+1],[D+1-C],1,C+E));
```

$$\frac{\Gamma(A+C+D+E+1)\,\Gamma(B+C+D+E+1)}{\Gamma(A+C+1)\,\Gamma(B+D+1)\,\Gamma(1+D+E)\,\Gamma(E+C+1)}$$

The following Maple procedure Sumtohyper(term,k) is an implementation of Algorithm 2.8, for the case that no shift is necessary, and automates the conversion of binomial sums into hypergeometric notation. A complete version (including the shifting) is available in the hsum package.

```
Sumtohyper :=proc(f,k)
local rat,num,den,x,numlist,denlist,init,i,j;
init :=eval(f,k=0);
if init=0 then
  ERROR("shift necessary")
end if;
rat := simpcomb(subs(k=k+1,f)/f);
if not type(rat,ratpoly(anything,k)) then
  ERROR("cannot be converted into hypergeometric form")
end if;
num :=numer(rat);
den :=denom(rat);
numlist :=-[solve(num,k)];
denlist :=-[solve(den,k)];
if not(member(1,denlist,'i')) then
  ERROR("shift necessary or no conversion possible")
end if;
x :=lcoeff(num,k)/lcoeff(den,k);
denlist :=subsop(i=NULL,denlist);
init*Hypergeom(numlist,denlist,x)
end proc:
```

Note that Sumtohyper gives the result using the inert form Hypergeom of hypergeom to avoid automatic evaluation. We get:

```
>   Sumtohyper(binomial(n,k)^2,k);
```

$$Hypergeom([-n,-n],[1],1)$$

```
>    Sumtohyper((-1)^k*binomial(2*m,k)^2,k);
```
$$Hypergeom\left([-2\,m,-2\,m],[1],-1\right)$$
```
>    Sumtohyper((-1)^k*binomial(2*m,k)^3,k);
```
$$Hypergeom\left([-2\,m,-2\,m,-2\,m],[1,1],1\right)$$

Note that the `hsum` package furthermore contains the rather similar function `termtohyper` which uses the same algorithm to rewrite a hypergeometric term in its normal hypergeometric form as the coefficient of a hypergeometric function. For this procedure only the output line `init*Hypergeom(numlist,denlist,x)` has to be changed towards the hypergeometric term `init*hyperterm (numlist, denlist,x,k)`.

Example 3.7 (Hypergeometric Transformations) Here we would like to deduce the Pfaff transformation

$$\frac{1}{(1-x)^a} \cdot {}_2F_1\left(\begin{array}{c} a,b \\ c \end{array}\middle| -\frac{x}{1-x}\right) = {}_2F_1\left(\begin{array}{c} a,c-b \\ c \end{array}\middle| x\right) \tag{3.3}$$

which is valid whenever $|x| < 1/2$. The left-hand side is the series

$$\sum_{k=0}^{\infty} \frac{(a)_k\,(b)_k}{(c)_k\,k!} (-1)^k\,x^k\,(1-x)^{-k-a},$$

and expanding $(1-x)^{-k-a}$ by the binomial theorem, one gets

$$\sum_{k=0}^{\infty}\sum_{j=0}^{\infty} \frac{(a)_k\,(b)_k\,(-1)^k}{(c)_k\,k!}\,\frac{(a+k)_j}{j!}\,x^{j+k}.$$

The coefficient of x^n in this double series is

$$\sum_{k=0}^{n} \frac{(a)_k\,(b)_k\,(-1)^k\,(a+k)_{n-k}}{(c)_k\,k!\,(n-k)!} = \frac{(a)_n}{n!}\sum_{k=0}^{n} \frac{(b)_k\,(-n)_k}{(c)_k\,k!} = \frac{(a)_n\,(c-b)_n}{(c)_n\,n!} \tag{3.4}$$

by an application of the Chu-Vandermonde identity (you might use `termtohyper` to obtain the first equality, see Exercise 3.8). Hence (3.3) is deduced. More hypergeometric transformations of this type are considered in Exercise 3.7. $\triangle$

We would like to point out that there is a Mathematica package written by Christian Krattenthaler [Krattenthaler94a] which contains a database for generalized hypergeometric functions that is much larger than ours. In particular, this package deals with all kinds of hypergeometric transformations like those of Kummer, Pfaff and Euler (see the previous example and Exercise 3.7). With this package these transformations and many more can be carried out automatically.

q-Hypergeometric Database

In the last chapter we introduced q-hypergeometric functions, and we saw a q-analogue of the binomial theorem. Similar q-analogues exist for the entries of our database and for other hypergeometric identities. As an example, we state the q-analogues of the Gauss and of the Pfaff-Saalschütz identities (due to Jacobi and Heine, and to Jackson, respectively):

$$
{}_2\phi_1\left(\left.\begin{matrix} a, b \\ c \end{matrix}\,\right|\, q, \frac{c}{ab}\right) = \frac{(c/a;\, q)_\infty\ (c/b;\, q)_\infty}{(c;\, q)_\infty\ (c/(ab);\, q)_\infty}, \qquad \left|\frac{c}{ab}\right| < 1, \tag{3.5}
$$

$$
{}_3\phi_2\left(\left.\begin{matrix} q^{-n}, a, b \\ c, \frac{ab}{cq^{n-1}} \end{matrix}\,\right|\, q, q\right) = \frac{(c/a;\, q)_n\ (c/b;\, q)_n}{(c;\, q)_n\ (c/(ab);\, q)_n}. \tag{3.6}
$$

Replacing a, b, c by q^a, q^b, and q^c, respectively, and letting $q \to 1$, one obtains the identities of Gauss and Pfaff-Saalschütz.

Note that the method which was developed in this chapter can be adapted to the q-case. In particular, Christian Krattenthaler designed a Mathematica package [Krattenthaler94b] containing a large database for q-hypergeometric functions.

With regard to Maple, the qsum package contains q-analogues of the algorithms discussed here. After loading this package by `read "qsum.mpl";` you have access to a procedure `qsimpcomb(expr)` which performs simplifications to decide the rationality of ratios a_{k+1}/a_k in terms of q^k. The procedure `qratio(expr,k)` computes the term ratio of `expr` w.r.t. k and simplifies it. Furthermore, you can use the procedure `sum2qhyper(expr,q,k)` to convert a q-hypergeometric sum into q-hypergeometric notation. For input purposes the procedures `qpochhammer(a,q,k)`, `qfactorial(k,q)`, `qGAMMA(k,q)`, `qbinomial(n,k,q)`, `qbrackets(k,q)` as well as `qphihyperterm(upper,lower,q,x,k)`, are accessible.

Some examples for the use of the package are given by

```
>    qsimpcomb(q^binomial(k,2)*qbinomial(n,k,q));
```

$$
\frac{q^{1/2k^2}\, qpochhammer\,(q, q, n)}{q^{1/2k}\, qpochhammer\,(q, q, k)\, qpochhammer\,(q, q, n - k)}
$$

```
>    qratio(q^binomial(k,2)*qbinomial(n,k,q),k);
```

$$
-\frac{q^k - q^n}{-1 + qq^k}
$$

```
>    sum2qhyper(q^binomial(k,2)*qbinomial(n,k,q),q,k);
```

$$
\phi\left([q^{-n}], [], q, -q^n\right)
$$

Further Reading

For further reading on the hypergeometric database we refer to[AAR99, Chaps. 3–5], and for the q-case to [GR90]. Prudnikov et al. [PBM90] is an encyclopedia of such identities.

Exercises

Exercise 3.1 *Prove (2.4)–(2.5), for odd n, using the database.*

Exercise 3.2 *Show that the identities*

(a) $\displaystyle\sum_{k=0}^{n} \binom{a}{k}\binom{b}{n-k} = \binom{a+b}{n},$

(b) $\displaystyle\sum_{k=0}^{n} \binom{n}{k}\binom{s}{t-k} = \binom{n+s}{t},$

(c) $\displaystyle\sum_{k=0}^{n} \binom{n}{k}\binom{s}{t+k} = \binom{n+s}{n+t},$

all are special instances of the Chu-Vandermonde identity.

Exercise 3.3 *Find, using the database, hypergeometric terms for the sums*

(a) $\displaystyle\sum_{k=0}^{2n}(-1)^k \binom{2n}{k}\binom{2k}{k}\binom{4n-2k}{2n-k},$

(b) $\displaystyle\sum_{j=k}^{n}(-1)^{k+j} \binom{2j}{j-k}\binom{n+j+1}{n-j},$

(c) $\displaystyle\sum_{k=0}^{n} \binom{n}{k}\binom{2n}{k},$

(d) $\displaystyle\sum_{k=0}^{n}(-1)^k \binom{n}{k}\binom{n+x-k}{x-k}\frac{y}{y+x-k};$

compare Exercise 2.16.

Exercise 3.4 *Prove, using the database, Stanley's identity (see e.g. [Strehl94], (19))*

$$\sum_{k=-\infty}^{\infty} \binom{a}{m-k}\binom{b}{n-k}\binom{a+b+k}{k} = \binom{a+n}{m}\binom{b+m}{n}.$$

Exercise 3.5 *Which of the following binomial sums are special cases of results in our database?*

(a) $\displaystyle\sum_{k=0}^{n} k \binom{n}{k}$,

(b) $\displaystyle\sum_{k=0}^{n} (-1)^k \binom{n}{k}\binom{b+k}{n}$,

(c) $\displaystyle\sum_{k=0}^{n} (-1)^k \binom{n}{k}\binom{2k}{n}$,

(d) $\displaystyle\sum_{k=0}^{n} (-1)^k \binom{n}{k}\binom{3k}{n}$.

In the affirmative cases, give the hypergeometric term results. Hint: If necessary, distinguish between odd and even n.

Exercise 3.6 *Find hypergeometric terms for the following sums using the database.*

(a) $\displaystyle\sum_{k=0}^{n} (-1)^k \binom{n}{k}\binom{2n-k}{m-k}$,

(b) $\displaystyle\sum_{k=0}^{n} (-1)^k \binom{n}{k}\binom{k+s}{t}$.

Exercise 3.7 *Prove the following identities between hypergeometric functions using the database.*

(a) (**Kummer**) $\displaystyle e^x \cdot {}_1F_1\left(\begin{matrix} a \\ b \end{matrix}\,\middle|\, -x\right) = {}_1F_1\left(\begin{matrix} b-a \\ b \end{matrix}\,\middle|\, x\right)$,

(b) (**Euler**) $\displaystyle (1-x)^{a+b-c} \cdot {}_2F_1\left(\begin{matrix} a,b \\ c \end{matrix}\,\middle|\, x\right) = {}_2F_1\left(\begin{matrix} c-a,c-b \\ c \end{matrix}\,\middle|\, x\right)$.

Hint: Use the method of Example 3.7.

Exercise 3.8 *Use* `termtohyper` *to deduce the left equation of (3.4).*

Exercise 3.9 *Use the Pfaff transformation of Example 3.7 to verify the identity*

$$\sum_{k=0}^{n} \binom{n}{k}\binom{-n-1}{k}\left(\frac{1-x}{2}\right)^k = \frac{1}{2^n} \sum_{k=0}^{n} \binom{n}{k}^2 (x+1)^{n-k}(x-1)^k$$

between two representations of the Legendre polynomials.

Exercise 3.10 *Use Exercise 2.19 to simplify* $\displaystyle\sum_{k=0}^{m} (-1)^k \binom{n}{k}$.

Exercise 3.11 *Prove the following identity for the Pochhammer symbols*

$$(x+y)_n = \sum_{k=0}^{n} \binom{n}{k} (x)_k \, (y)_{n-k}.$$

Exercise 3.12 *Show that*

$$s_n := \sum_{k=0}^{n} (-1)^k \binom{n}{k} \binom{2k}{n} = (-2)^n.$$

Hint: *Distinguish between even and odd n.*

Exercise 3.13 *Assume $n = -a \in \mathbb{N}_{\geq 0}$. Bring the identities of Watson and Whipple to a form like identities (2.4)–(2.5), i.e. derive different formulas for even and odd n. (In this form, the results look more natural, and there is no need for the number π to occur.) Hint: Calculate s_{n+2}/s_n for the right-hand sides s_n and read off the results for even and odd n, respectively.*

Exercise 3.14 *A generalized hypergeometric function ${}_pF_q\left(\begin{smallmatrix} \alpha_1,\ldots,\alpha_p \\ \beta_1,\ldots,\beta_q \end{smallmatrix} \Big| x\right)$ is called*

- *k-balanced if $b_1 + b_2 + \cdots + b_q = a_1 + a_2 + \cdots + a_p + k$;*
- *balanced or Saalschützian if it is 1-balanced;*
- *well-poised, if $p = q + 1$, and $1 + a_1 = b_1 + a_2 = b_2 + a_3 = \cdots = b_q + a_p$;*
- *nearly-poised of the first kind, if $p = q+1$, and $b_1+a_2 = b_2+a_3 = \cdots = b_q+a_p$.*
- *nearly-poised of the second kind, if $p = q+1$, and $1+a_1 = b_1+a_2 = b_2+a_3 = \cdots = b_{q-1}+a_{p-1}$.*

Check which of these properties are satisfied by the entries of our database.

Exercise 3.15 *Convert into q-hypergeometric notation:*

(a) $\displaystyle\sum_{k=0}^{n} \begin{bmatrix} n \\ k \end{bmatrix}_q x^k,$

(b) $\displaystyle\sum_{k=0}^{n} \begin{bmatrix} n \\ k \end{bmatrix}_q^2 x^k,$

(c) $\displaystyle\sum_{k=0}^{n} q^{\binom{k+j}{2}} \begin{bmatrix} n \\ k \end{bmatrix}_q,$

(d) $\displaystyle\sum_{k=0}^{n} q^{k^2} \begin{bmatrix} n \\ k \end{bmatrix}_q \begin{bmatrix} n \\ n-k \end{bmatrix}_q,$

(e) $\displaystyle\sum_{k=0}^{2n} q^{k^2} \begin{bmatrix} 2n \\ k \end{bmatrix}_q \begin{bmatrix} n \\ n-k \end{bmatrix}_q,$

(f) $\displaystyle\sum_{k=0}^{2n} q^{k^2} \begin{bmatrix} 2n \\ k \end{bmatrix}_q \begin{bmatrix} 2n \\ 2n-k \end{bmatrix}_q.$

Exercise 3.16 *Show that the identity*

$$\sum_{k=0}^{n} q^{k^2} \begin{bmatrix} n \\ k \end{bmatrix}_q^2 = \frac{(\sqrt{q};\, q)_n \, (-\sqrt{q};\, q)_n \, (-q;\, q)_n}{(q;\, q)_n}$$

is a special case of the q-Gauss identity (3.5) for $a = q^{-n}$,

$$_2\phi_1 \left(\begin{matrix} q^{-n}, b \\ c \end{matrix} \,\middle|\, q, \frac{cq^n}{b} \right) = \frac{(cq^n;\, q)_\infty \, (c/b;\, q)_\infty}{(c;\, q)_\infty \, (cq^n/b;\, q)_\infty} = \frac{(c/b;\, q)_n}{(c;\, q)_n} \tag{3.7}$$

which is the q-analogue of the Chu-Vandermonde identity. Prove the second equality in (3.7).

Exercise 3.17 *Show that by reversing the order of summation, the q-Chu-Vandermonde identity yields the form*

$$_2\phi_1 \left(\begin{matrix} q^{-n}, b \\ c \end{matrix} \,\middle|\, q, q \right) = \frac{(c/b;\, q)_n}{(c;\, q)_n} \, b^n.$$

References

AAR99. Andrews, G., Askey, R.A., Roy, R.: Special Functions. Cambridge University Press, Cambridge. Encyclopedia of Mathematics and its Applications (1999)

Bailey35. Bailey, W.N.: Generalized Hypergeometric Series. Cambridge University Press, Cambridge (reprinted 1964 by Stechert-Hafner Service Agency. London, New York 1935)

GR90. Gasper, G., Rahman, M.: Basic Hypergeometric Series. Encyclopedia of Mathematics and its Applications, vol. 35. Cambridge University Press, London and New York (1990) (2nd edn 2004)

Krattenthaler94a. Krattenthaler, C.: Mathematica Package HYP. Available at http://library.wolfram.com

Krattenthaler94b. Krattenthaler, C.: Mathematica Package HYPQ. Available at http://library.wolfram.com

PBM90. Prudnikov, A.P., Brychkov, Y.A., Marichev, O.I.: Integrals and Series: More Special Functions, vol. 3. Gordon and Breach Publ, New York (1990)

Rainville60. Rainville, E.D.: Special Functions. The MacMillan Co., New York (1960)

Roy87. Roy, R.: Binomial identities and hypergeometric functions. Am. Math. Mon. **94**, 36–46 (1987)

Strehl94. Strehl, V.: Binomial identities—combinatorial and algorithmic aspects. Discrete Math. **136**, 309–346 (1994)

Székely85. Székely, L.A.: Common origin of cubic binomial identities; a generalization of Surányi's proof of the Le Jen Shoo's formula. J. Comb. Theor. Ser. A **40**, 171–174 (1985)

Chapter 4
Holonomic Recurrence Equations

The main *algorithmic* idea for finding hypergeometric term representations of hypergeometric series goes back to Celine Fasenmyer (often called Sister Celine): Her idea is to find a recurrence equation for the sum ([Fasenmyer45, Fasenmyer49], see also [Zeilberger82]). If the resulting recurrence equation can be solved explicitly, you are done. In the 1940s, under the direction of Earl Rainville, Celine Fasenmyer wrote her Ph.D. thesis [Fasenmyer45] on techniques to find such a recurrence equation. Doron Zeilberger [Zeilberger90a, Zeilberger95] extended this idea in the 1990s. In ([Rainville60], Chap. 14), Rainville presented one of Fasenmyer's techniques. Another one (compare [Fasenmyer49]) will be the topic of the present chapter.

Example 4.1 The easiest example for Fasenmyer's algorithm concerns the sum

$$s_n := \sum_{k=0}^{n} \binom{n}{k} = \sum_{k=-\infty}^{\infty} \binom{n}{k}.$$

It is well-known that the binomial coefficients satisfy the *Pascal triangle* recurrence equation

$$\binom{n+1}{k+1} = \binom{n}{k} + \binom{n}{k+1}.$$

Summing this identity for $k = -\infty, \ldots, \infty$ yields

$$\sum_{k=-\infty}^{\infty} \binom{n+1}{k+1} = \sum_{k=-\infty}^{\infty} \binom{n}{k} + \sum_{k=-\infty}^{\infty} \binom{n}{k+1}$$

or equivalently

$$s_{n+1} = 2\, s_n \tag{4.1}$$

W. Koepf, *Hypergeometric Summation*, Universitext,
DOI: 10.1007/978-1-4471-6464-7_4, © Springer-Verlag London 2014

since the two right-hand sums agree because they differ only by a shift of the summation variable. From (4.1) and $s_0 = 1$ one easily deduces $s_n = 2^n$ by induction or by the hypergeometric coefficient formula.

You may not have seen this direct proof for the identity

$$\sum_{k=0}^{n} \binom{n}{k} = 2^n \tag{4.2}$$

before. Observe that this method not only proved (4.2) but the right-hand side was directly computed given the left-hand side.

To illustrate the use of this technique by a less simple example, let us for the moment assume we have found the recurrence equation

$$n\, s_{n+1} - 2\,(n+1)\, s_n = 0$$

for the sum

$$s_n : = \sum_{k=0}^{n} k \binom{n}{k}.$$

If so, then we would have

$$\frac{s_{n+1}}{s_n} = 2\,\frac{n+1}{n}.$$

This equation tells us that a shift by one (putting a $(n+1)$ term in the denominator) makes s_n the coefficient of a generalized hypergeometric series. For $t_n : = s_{n+1}$ we have therefore

$$\frac{t_{n+1}}{t_n} = 2\,\frac{n+2}{n+1},$$

so that using the initial value $t_0 = s_1 = \sum_{k=0}^{1} k \binom{1}{k} = 1$, it follows from the coefficient formula (2.8)–(2.9)

$$t_n = (2)_n\,\frac{2^n}{n!} = (n+1)\,2^n,$$

hence,

$$s_n = t_{n-1} = n\, 2^{n-1} \qquad (n \geq 1). \qquad\qquad \triangle$$

But how do we find a recurrence equation for a hypergeometric sum? Let us give some examples along the lines of Fasenmyer's development.

Example 4.2 We consider the above example sum

$$S_n = \sum_{k=-\infty}^{\infty} F(n, k) \tag{4.3}$$

with

$$F(n, k) = k \binom{n}{k}.$$

Celine Fasenmyer's idea is to deduce, in the first step, a mixed recurrence equation

$$\sum_{i=0}^{I} \sum_{j=0}^{J} a_{ij} \, F(n + j, k + i) = 0 \tag{4.4}$$

for the *summand* $F(n, k)$ with the property that the coefficients $a_{ij} = a_{ij}(n)$ ($i = 0, \ldots, I$, $j = 0, \ldots, J$) are *polynomials* with respect to n and do *not* depend on k. Such a recurrence equation is called k-free; we will see soon why this is an important issue.

Let us choose $I = J = 1$. Then we have the setup

$$a_{00} \, F(n, k) + a_{01} \, F(n + 1, k) + a_{10} \, F(n, k + 1) + a_{11} \, F(n + 1, k + 1) = 0.$$

For simplicity, let's choose $a_{00} = 1$ (which would obviously be a bad choice if a recurrence equation with $a_{00} = 0$ were to exist).

Division by $F(n, k)$ yields

$$1 + a_{01} \frac{F(n + 1, k)}{F(n, k)} + a_{10} \frac{F(n, k + 1)}{F(n, k)} + a_{11} \frac{F(n + 1, k + 1)}{F(n, k)} = 0.$$

By an application of Algorithm 2.2 this results in a purely rational equation whenever $F(n, k)$ satisfies the hypotheses of Algorithm 2.2 with respect to both n and k.

In our case, we substitute the given $F(n, k) = k \binom{n}{k}$ and obtain

$$1 + a_{01} \frac{n + 1}{n + 1 - k} + a_{10} \frac{n - k}{k} + a_{11} \frac{n + 1}{k} = 0$$

which—after multiplication by the common denominator—results in the equation

$$(n + 1 - k) k + a_{01} (n + 1) k + a_{10} (n - k) (n + 1 - k) + a_{11} (n + 1) (n + 1 - k) = 0.$$

To find k-free coefficients a_{ij}, we may consider the resulting expression as a polynomial in k and equate coefficients. This leads to the linear system

$$a_{10} = 1$$
$$n\,a_{10} + (n+1)\,a_{11} = 0$$
$$(n+1)\,a_{01} - (2n+1)a_{10} - (n+1)\,a_{11} = -(n+1)$$

that we might solve with Maple:

```
>   solve({a[1,0]=1,n*a[1,0]+(n+1)*a[1,1]=0,
>   (n+1)*a[0,1]-(2*n+1)*a[1,0]-(n+1)*a[1,1]=-(n+1)},
>   {a[0,1],a[1,0],a[1,1]});
```

$$\left\{ a_{0,1} = 0,\; a_{1,0} = 1,\; a_{1,1} = -\frac{n}{n+1} \right\}$$

therefore leading to the valid recurrence equation

$$(n+1)\,F(n, k+1) + (n+1)\,F(n, k) - n\,F(n+1, k+1) = 0 \tag{4.5}$$

for $F(n, k)$.

Now we come to the second step: To deduce a recurrence equation for the *series*, we sum the k-free recurrence equation (4.5) for $k = -\infty, \ldots, \infty$. Since the two shifted series

$$s_n = \sum_{k=-\infty}^{\infty} F(n, k) = \sum_{k=-\infty}^{\infty} F(n, k+1)$$

have the same value (observe how helpful, again, the bilateral infinite summation bounds are!), we get

$$2\,(n+1)\,s_n - n\,s_{n+1} = 0$$

for s_n, as announced. $\triangle$

Note that the technique—if successful—obviously generates a homogeneous linear recurrence equation with polynomial coefficients for s_n. Such a recurrence equation is called *holonomic*.

Example 4.3 Next, we would like to find a holonomic recurrence equation for the Legendre polynomials. We recall the hypergeometric representation (see Example 2.9)

$$P_n(x) = \sum_{k=-\infty}^{\infty} \binom{n}{k} \binom{-n-1}{k} \left(\frac{1-x}{2} \right)^k .$$

Therefore, we set

$$F(n, k) := \binom{n}{k} \binom{-n-1}{k} \left(\frac{1-x}{2}\right)^k.$$

We note that Fasenmyer's method fails for $I = J = 1$ since the linear system to be solved in this case has no solution (check!). Therefore, we try the next best choice and set $I = 1, J = 2$:

$$0 = a_{00}\, F(n, k) + a_{01}\, F(n+1, k) + a_{02}\, F(n+2, k)$$
$$+ a_{10}\, F(n, k+1) + a_{11}\, F(n+1, k+1) + a_{12}\, F(n+2, k+1).$$

Note that by induction, all of the ratios $\frac{F(n+j,k+i)}{F(n,k)}$ $(i, j \in \mathbb{N}_{\geq 0})$ turn out to be rational and can be treated by Algorithm 2.2. In the present case, division by $F(n, k)$ yields

$$0 = a_{00} + a_{01}\, \frac{n+1+k}{n+1-k} + a_{02}\, \frac{(n+2+k)\,(n+1+k)}{(n+2-k)\,(n+1-k)}$$
$$+ a_{10}\, \frac{(n-k)\,(n+1+k)\,(x-1)}{2\,(k+1)^2} + a_{11}\, \frac{(x-1)\,(n+2+k)\,(n+1+k)}{2\,(k+1)^2}$$
$$+ a_{12}\, \frac{(x-1)\,(k+3+n)\,(n+2+k)\,(n+1+k)}{2\,(k+1)^2\,(n+1-k)}$$

which—after multiplication by the common denominator—results in a large polynomial equation of degree 4 in k. Equating coefficients gives a huge linear system with the astonishingly simple general solution

$$a_{00} = 0,$$
$$a_{01} = \frac{(x-1)\,(2n+3)}{n+1}\, a_{10},$$
$$a_{02} = 0,$$
$$a_{10} = a_{10},$$
$$a_{11} = -\frac{2n+3}{n+1}\, a_{10},$$
$$a_{12} = \frac{n+2}{n+1}\, a_{10}.$$

Therefore, we have found a recurrence equation. Setting $a_{10} = 1$ yields, after multiplication by the common denominator,

$$0 = (n+2)\, F(n+2, k+1) - (2n+3)\, F(n+1, k+1) + (n+1)\, F(n, k+1)$$
$$+ (1-x)\,(2n+3)\, F(n+1, k).$$

Summing with respect to k, finally gives the three-term recurrence equation

$$(n+2)\, P_{n+2}(x) - (2n+3)\, x\, P_{n+1}(x) + (n+1)\, P_n(x) = 0$$

for the Legendre polynomials. $\triangle$

Now we have understood Fasenmyer's recipe and we are prepared to give a detailed description of her method.

Algorithm 4.4 (*Fasenmyer*) The following algorithm searches for a holonomic recurrence equation for series of the form (4.3).

1. (`kfreerecursion`)
   Choose appropriate numbers $I, J \in \mathbb{N}$. Then the following procedure finds a k-free linear recurrence equation with polynomial coefficients (4.4) of order (I, J) if such a recurrence equation is valid.

   (a) Input: $F(n, k)$ satisfying the hypotheses of Algorithm 2.2 with respect to both n and k.

   (b) Use the generic expression (4.4) with as yet undetermined a_{ij} and substitute the given $F(n, k)$.

   (c) Divide by $F(n, k)$ and apply Algorithm 2.2 to rationalize the resulting expression.

   (d) Bring this rational expression into normal form, i.e., put everything on a common denominator, and multiply by it.

   (e) Equate the coefficients with respect to k, i.e., set all coefficients of k-powers equal to zero and solve the resulting linear system for the variables a_{ij} ($i = 0, \ldots, I, j = 0, \ldots, J$).

   (f) If only the trivial solution $a_{ij} \equiv 0$ exists, then no k-free recurrence equation of order (I, J) is valid; exit.

   (g) If a solution exists, substitute it in (4.4) and multiply by the common denominator.

   (h) Output: The k-free recurrence equation for $F(n, k)$ of the last step.

2. (`fasenmyer`)
   Choose an appropriate number J, an upper bound for the order of the resulting recurrence equation. Then the following procedure searches for a holonomic recurrence equation for s_n, given by (4.3).

   (a) Input: The summand $F(n, k)$ satisfying the hypotheses of Algorithm 2.2 with respect to both n and k.

   (b) Apply the procedure `kfreerecursion` to $F(n, k)$, with $I = J$.[1] If this is successful, take the resulting k-free recurrence equation and replace $F(n + j, k + i)$ symbolically by s_{n+j}. This generates a holonomic recurrence equation RE for s_n.

   (c) If the resulting recurrence equation RE is trivial ($0 = 0$) then exit.

   (d) Output: The holonomic recurrence equation RE for s_n of step (b).

Proof (`kfreerecursion`): Obviously, Algorithm 2.2 rationalizes the generic expression in step (c) for the given type of input. Therefore, from now on, rational arithmetic applies and it is clear that the resulting expression is identical to zero

---

[1] This choice is random and might not be the best possible. For a more detailed discussion see Theorem 7.10.

if and only if its numerator is the zero polynomial with respect to k. Therefore, for $F(n, k)$ to satisfy a non-trivial k-free linear recurrence equation with polynomial coefficients, it is necessary and sufficient that the equations system of step (e) has a nontrivial solution. The rest of `kfreerecursion` is straightforward linear algebra.

(`fasenmyer`): Summing the k-free recurrence equation

$$\sum_{i=0}^{I}\sum_{j=0}^{J} a_{ij} \, F(n+j, k+i) = 0$$

for $k = -\infty, \ldots, \infty$, yields

$$0 = \sum_{k=-\infty}^{\infty}\sum_{i=0}^{I}\sum_{j=0}^{J} a_{ij}(n) \, F(n+j, k+i)$$

$$= \sum_{i=0}^{I}\sum_{j=0}^{J} a_{ij}(n) \left(\sum_{k=-\infty}^{\infty} F(n+j, k+i) \right)$$

$$= \sum_{i=0}^{I}\sum_{j=0}^{J} a_{ij}(n) \, s_{n+j} = \sum_{j=0}^{J}\left(\sum_{i=0}^{I} a_{ij}(n) \right) s_{n+j}$$

since $a_{ij}(n)$ does not depend on k. Therefore, the method described obviously leads to a holonomic recurrence equation for s_n if `kfreerecursion` was successful. $\square$

We saw in Example 4.3 that the application of Fasenmyer's method may lead to rather complicated intermediate results, even if the end result is quite simple. This is a typical situation for symbolic (i.e. non-numeric) algorithms and it is faster (and safer) to let Maple do the calculations for us.

Session 4.5 The following Maple procedure `kfreerec(f,k,n,kmax,nmax)` automates the search for the k-free recurrence equation of $F(n, k)$ of order $(I, J) = $ (`kmax`, `nmax`).

```
kfreerec:=proc(f,k,n,kmax,nmax)
local N,ansatz,variables,rec,i,j,l,solution,F,a;
if nargs>5 then F:=args[6] end if;
if nargs>6 then a:=args[7] end if;
N:=(kmax+1)*(nmax+1);
ansatz:=add(add(
   a[i,j]*simpcomb(subs(n=n+j,k=k+i,f)/f),
   j=0..nmax),i=0..kmax);
ansatz:=collect(numer(normal(ansatz)),k);
variables:={seq(seq(a[i,j],j=0..nmax),i=0..kmax)};
solution:={solve({coeffs(ansatz,k)},variables)};
if subs(op(solution),variables)={0} then
```

```
    ERROR(cat("No k-free recurrence equation of order
    (",kmax,",",nmax,") exists"));
  end if;
  rec:=add(add(a[i,j]*F(n+j,k+i),j=0..nmax),i=0..kmax);
  rec:=subs(op(1,solution),rec);
  rec:=numer(normal(rec));
  collect(rec,F,factor)=0
  end proc:
```

Note that the use of optional fifth and sixth arguments, F and a, guarantees that no global variables with these names interfere with our calculations.

Let us try this function on the above problems, and on some similar examples[2]:

```
>   rec:=kfreerec(k*binomial(n,k),k,n,1,1,F,a);
```
$$rec := -(n+1)\,a_{1,1}\,F\,(n,k) - (n+1)\,a_{1,1}\,F\,(n,k+1) + a_{1,1}\,F\,(n+1,k+1)\,n = 0$$
```
>   subs(a[1,1]=1,rec);
```
$$-(n+1)\,F\,(n,k) - (n+1)\,F\,(n,k+1) + F\,(n+1,k+1)\,n = 0$$
```
>   rec:=kfreerec(binomial(n,k),k,n,1,1,F,a);
```
$$rec := -a_{1,1}\,F\,(n,k) - a_{1,1}\,F\,(n,k+1) + a_{1,1}\,F\,(n+1,k+1) = 0$$
```
>   subs(a[1,1]=1,rec);
```

$$-F\,(n,k) - F\,(n,k+1) + F\,(n+1,k+1) = 0$$

The previous computation *generated* the *Pascal triangle* recurrence equation

$$-\binom{n}{k} - \binom{n}{k+1} + \binom{n+1}{k+1} = 0$$

for the binomial coefficients automatically. In a similar fashion one gets for the summand

$$\binom{n}{k}\,x^k\,y^{n-k}$$

```
>   rec:=kfreerec(binomial(n,k)*x^k*y^(n-k),k,n,1,1,F,a);
```
$$rec := -a_{1,1}\,x\,F\,(n,k) - a_{1,1}\,y\,F\,(n,k+1) + a_{1,1}\,F\,(n+1,k+1) = 0$$
```
>   subs(a[1,1]=1,rec);
```

$$-x\,F\,(n,k) - y\,F\,(n,k+1) + F\,(n+1,k+1) = 0$$

Summing this extension of Pascal's triangle recurrence obviously yields

---

[2] Note that for reasons of efficiency expressions, sets, etc. in Maple are sorted by their memory allocation, hence rather randomly. In particular, in different Maple sessions you might get differently sorted results. For this reason, some of the Maple procedures, e.g. `solve`, have random effects. Therefore, executing the examples below in different Maple sessions may result in different surviving variables a_{ij}.

$$(x + y)\, s_n = s_{n+1}$$

for the sum

$$s_n = \sum_{k=0}^{n} \binom{n}{k} x^k\, y^{n-k}$$

and therefore the binomial theorem

$$\sum_{k=0}^{n} \binom{n}{k} x^k\, y^{n-k} = (x + y)^n.$$

Note that using the summand of the left-hand side we therefore generated the closed form on the right-hand side automatically.

Next, we reproduce the result from Example 4.3

```
>   rec:=kfreerec(binomial(n,k)*binomial(-n-1,k)*
>   ((1-x)/2)^k,k,n,1,2,F,a);
```

$$rec := a_{1,0}\,(n+1)\,F\,(n, k+1) - a_{1,0}\,(-1+x)\,(2\,n+3)\,F\,(n+1, k)$$

$$- a_{1,0}\,(2\,n+3)\,F\,(n+1, k+1) + a_{1,0}\,(n+2)\,F\,(n+2, k+1) = 0$$

```
>   legendretermrec:=subs(a[1,0]=1,rec);
```

$$(n+1)\,F(n,\, k+1) - (-1+x)\,(2\,n+3)\,F(n+1,\, k)$$
$$- (2\,n+3)\,F(n+1,\, k+1) + (n+2)\,F(n+2,\, k+1) = 0$$

```
>   rec:=kfreerec((-1)^k*binomial(n,k)/k!*x^k,k,n,1,2,F,a);
```

$$a_{0,\,1}\,(n+1)\,F(n,\, k+1) + a_{0,\,1}\,x\,F(n+1,\, k) - a_{0,\,1}\,(2\,n+3)\,F(n+1,\, k+1)$$
$$+ a_{0,\,1}\,(n+2)\,F(n+2,\, k+1) = 0$$

```
>   laguerretermrec:=subs(a[0,1]=1,rec);
```

$$(n+1)\,F(n,\, k+1) + F(n+1,\, k)\,x - (2\,n+3)\,F(n+1,\, k+1)$$
$$+ (n+2)\,F(n+2,\, k+1) = 0$$

Furthermore, the last example generated a k-free recurrence equation for the summand of the *Laguerre polynomials*

$$L_n(x) := \sum_{k=0}^{n} \frac{(-1)^k}{k!} \binom{n}{k} x^k.$$

The following Maple procedure combines the two steps and calculates a holonomic recurrence equation for the sum using Fasenmyer's method with $I = J = \text{nmax}$.

```
fasenmyer:=proc(f,k,sn,nmax)
local F,a,n,S,i,j,rec;
if type(sn,function) then
  S:=op(0,sn);
  n:=op(1,sn)
else
  n:=sn
end if;
rec:=lhs(kfreerec(f,k,n,nmax,nmax,F,a));
rec:=applyrule(F(n+j::integer,k+i::integer)=S(n+j),rec);
rec:=normal(solve(rec,S(n+nmax)));
rec:=denom(rec)*S(n+nmax)-numer(rec);
collect(rec,S,factor)=0
end proc:
```

By the use of `normal(solve(...))`, factors involving the parameters a_{ij} that are common to the coefficients of all terms S_{n+j} in the resulting recurrence equation are canceled.

Here are some holonomic recurrence equations:

```
>   fasenmyer(k*binomial(n,k),k,s(n),1);
```
$$(-2\,n - 2)\,s\,(n) + ns\,(n + 1) = 0$$

```
>   fasenmyer(binomial(n,k),k,s(n),1);
```
$$s\,(n + 1) - 2\,s\,(n) = 0$$

```
>   legendrerec:=fasenmyer(
>   binomial(n,k)*binomial(-n-1,k)*((1-x)/2)^k,k,s(n),2);
```
$$legendrerec := (n + 1)\,s\,(n) - x\,(2\,n + 3)\,s\,(n + 1) + (n + 2)\,s\,(n + 2) = 0$$

```
>   laguerrerec:=fasenmyer(
>   (-1)^k*binomial(n,k)/k!*x^k,k,s(n),2);
```
$$laguerrerec := (n + 1)\,s\,(n) + (-2\,n - 3 + x)\,s\,(n + 1) + (n + 2)\,s\,(n + 2) = 0$$

Therefore, we have reproduced the three-term recurrence equation for the Legendre polynomials that we saw in Example 4.3, and we have obtained the three-term recurrence equation

$$(n + 2)\,L_{n+2}(x) + (x - 2\,n - 3)\,L_{n+1}(x) + (n + 1)\,L_n(x) = 0$$

for the Laguerre polynomials.

The next example shows that it may happen that no k-free recurrence equation exists.

Example 4.6 (Counterexample, [WZ92]) We consider the sum

$$s_n := \sum_{k=-\infty}^{\infty} \frac{1}{k^2 + n^2 + 1}.$$

We propose that for this input

$$F(n, k) := \frac{1}{k^2 + n^2 + 1}$$

no k-free recurrence equation exists (see [WZ92]). We choose this example even though in the given case the sum under consideration is not finite. It is easy to see, however, that a similar argument applies to the summand of the sum

$$S_n := \sum_{k=-\infty}^{\infty} \frac{1}{k^2 + n^2 + 1} \binom{n}{k} = \sum_{k=0}^{n} \frac{1}{k^2 + n^2 + 1} \binom{n}{k};$$

see Exercise 4.18.

Note that the number of variables a_{ij} always equals $V = (I + 1)(J + 1)$. In the present case, it turns out that the number of equations E to be satisfied by the variables a_{ij} equals $E = 2V - 1$. Therefore, for any choice of I and J there are many more equations than variables and no non-trivial solution can be expected.

To prove this, assume a k-free recurrence equation is valid for $F(n, k)$. This corresponds to an equation of the form

$$\sum_{i=0}^{I} \sum_{j=0}^{J} \frac{a_{ij}(n)}{(k + i)^2 + (n + j)^2 + 1} = 0. \tag{4.6}$$

The left-hand side is a rational function with respect to k. Since any rational function over $\mathbb{C}$ is uniquely determined by a finite number of points, (4.6) is valid for all $k \in \mathbb{C}$. Since not all $a_{ij}(n)$ are identically zero, the left-hand side is a nontrivial sum of meromorphic terms with respect to the variable k. If we take one of the complex poles $k = \tilde{k} \in \mathbb{C}$ of one of the summands, we note that all other summands are finite at $k = \tilde{k}$. Hence for $k = \tilde{k}$ we obtain a contradiction of the form $\infty = 0$, and (4.6) cannot be valid. $\triangle$

It can be shown that for input of a special type, the situation in the above example cannot happen. If $I = J$, for this type of input, the number of linear equations grows linearly with J rather than quadratically and hence, eventually, for large enough J, a k-free recurrence equation must result. In Theorem 7.10, we will prove that for such *proper hypergeometric terms* a k-free linear recurrence equation with polynomial coefficients is satisfied. A hypergeometric term $F(n, k)$ is called proper if it has finite support, and is of the form $F(n, k) = P(n, k) \frac{Q(n,k)}{R(n,k)} w^n z^k$ where $P(n, k)$ is a polynomial (*polynomial part*) and $Q(n, k)$, $R(n, k)$ are Γ-term products with integer-linear arguments (*factorial part*).

If Fasenmyer's algorithm applies, it can prove hypergeometric identities, as Example 4.1 showed, and other identities as well: In Exercise 4.3, it is shown that the three different hypergeometric representations (of the Legendre polynomials) given in Exercise 2.12 define the same functions: just prove that the three different representations satisfy the same holonomic recurrence equation and the same initial values. That's it. This is Zeilberger's paradigm [Zeilberger90a], which will be developed further in later chapters. Another easy adaption of Fasenmyer's algorithm is the computation of differential equations instead of recurrence equations, see Exercise 4.4.

Note that Fasenmyer's algorithm, although very intuitive, has several drawbacks, some of them severe:

1. It does not state for which type of input it will be successful. However, Theorem 7.10 gives a partial but satisfying answer to this question.
2. The procedure kfreerec is a *decision procedure* which will find a k-free recurrence equation of order (I, J) whenever one exists. This is nice. On the other hand, no prior knowledge about a safe choice for I and J is given. Compare Theorem 7.10, again.
3. In the procedure fasenmyer, on the other hand, where the bound J for the order of a proposed recurrence equation is given, no control over a safe choice for I is known. Therefore we arbitrarily set $I = J$. See Theorem 7.10 for a priori bounds for both I and J for proper hypergeometric terms. Moreover, we will discuss this issue in the sequel, and show how it can be resolved intelligently.
4. Nobody can guarantee that fasenmyer will find the holonomic recurrence equation of lowest order valid for s_n. If a holonomic recurrence equation of order J is valid for the sum, whereas for no $I \in \mathbb{N}$ a k-free recurrence equation of order (I, J) for the summand $F(n, k)$ holds—which can happen—this recurrence equation for the sum *cannot be found* by the present method. Quite a few examples of this type are in the exercises. This is therefore a severe problem that can be partially resolved. This will also be discussed in the sequel.
5. The most important problem, however, is the *complexity* problem: The most complicated and time-consuming part of the method illustrated in this chapter is the solution of the linear system. This system has $(I + 1)(J + 1)$ variables. If I is large, this is much greater than J, and seems to be much more than necessary to find a holonomic equation of order J having only J free coefficients. This complexity behavior in the order of the resulting recurrence equation is the most severe drawback of Fasenmyer's method. For our choice $I = J$ we have $(J + 1)^2$ variables to consider.

The last problem (5) of our list will be resolved in Chap. 7. The final tool is the recurrence equation for the sum, which has order J. Zeilberger's algorithm finds it by solving a linear system with (essentially) J variables, in contrast to IJ as the method of this chapter. This might make the difference between being able or not being able to solve the question under consideration.

Nevertheless there are ways to streamline Fasenmyer's approach. Wegschaider [Wegschaider97] developed a more efficient algorithm than Fasenmyer's original method in order to use the algorithm also for multiple summation. He modified Fasenmyer's algorithm in two ways, which we describe now briefly for the one-dimensional case. These concepts can be easily adapted to the multivariate case, e.g. to double or triple sums.

Firstly, Wegschaider utilized an idea by Verbaeten [Verbaeten76] using special polyhedral summation ranges leading to much smaller linear systems and thus to more efficiency. The second improvement uses the fact that recurrences for $F(n, k)$ need not be k-free in order to deduce a holonomic recurrence equation for the sum $s_n = \sum_{k=-\infty}^{\infty} F(n, k)$. For this purpose Wegschaider introduced *certificate recurrence equations* consisting of telescoping parts that vanish by summation and a main part which corresponds roughly to the holonomic recurrence equation for s_n being sought.

We consider the following example illustrating both improvements. Let

$$F(n, k) = (-1)^k \binom{2n}{k}^3. \tag{4.7}$$

Applying Fasenmyer's algorithm to $F(n, k)$ given by (4.7) with $I = 8$ and $J = 3$, we get the following large k-free recurrence equation

$$(384n^5 + 4288n^4 + 18936n^3 + 41318n^2 + 44535n + 18969)(n + 1)^2(2n + 1)^2 F(n, k)$$
$$+ 2(21309 + 48846n + 44563n^2 + 20214n^3 + 4556n^4 + 408n^5)(n + 1)^2(2n + 1)^2 F(n, k + 1)$$
$$- 4(158772 + 369213n + 340279n^2 + 155322n^3 + 35108n^4 + 3144n^5)(n + 1)^2(2n + 1)^2 F(n, k + 2)$$
$$+ 2(926643 + 2157810n + 19906211n^2 + 909162n^3 + 205556n^4 + 18408n^5)(n + 1)^2(2n + 1)^2 F(n, k + 3)$$
$$- 10(255957 + 596199n + 550114n^2 + 251280n^3 + 56816n^4 + 5088n^5)(n + 1)^2(2n + 1)^2 F(n, k + 4)$$
$$+ 2(926643 + 2157810n + 19906211n^2 + 909162n^3 + 205556n^4 + 18408n^5)(n + 1)^2(2n + 1)^2 F(n, k + 5)$$
$$- 4(158772 + 369213n + 340279n^2 + 155322n^3 + 35108n^4 + 3144n^5)(n + 1)^2(2n + 1)^2 F(n, k + 6)$$
$$+ 2(21309 + 48846n + 44563n^2 + 20214n^3 + 4556n^4 + 408n^5)(n + 1)^2(2n + 1)^2 F(n, k + 7)$$
$$+ (384n^5 + 4288n^4 + 18936n^3 + 41318n^2 + 44535n + 18969)(n + 1)^2(2n + 1)^2 F(n, k + 8)$$
$$- (169575 + 5800081n^3 + 3486980n^2 + 5984198n^4 + 1177050n + 3980740n^5$$
$$\quad + 459040n^7 + 1710920n^6 + 4608n^9 + 69888n^8)F(n + 1, k + 2)$$
$$- (10997070 + 350110242n^3 + 214468788n^2 + 356106570n^4 + 74137278n$$
$$\quad + 234442404n^5 + 26732352n^7 + 100057488n^6 + 267840n^9 + 4062240n^8)F(n + 1, k + 3)$$
$$- (108413445 + 3350010438n^3 + 2067361566n^2 + 3388782252n^4 + 721658283n$$
$$\quad + 2222280648n^5 + 252327264n^7 + 945944832n^6 + 2526336n^9 + 38316096n^8)F(n + 1, k + 4)$$
$$- (250652220 + 7652852110n^3 + 4736323490n^2 + 7725210842n^4 + 1659756198n$$
$$\quad + 5058477028n^5 + 573443776n^7 + 2151062672n^6 + 5739840n^9 + 87054240n^8)F(n + 1, k + 5)$$
$$- (108413445 + 3350010438n^3 + 2067361566n^2 + 3388782252n^4 + 721658283n$$
$$\quad + 2222280648n^5 + 252327264n^7 + 945944832n^6 + 2526336n^9 + 38316096n^8)F(n + 1, k + 6)$$
$$- (10997070 + 350110242n^3 + 214468788n^2 + 356106570n^4 + 74137278n + 234442404n^5$$
$$\quad + 26732352n^7 + 100057488n^6 + 267840n^9 + 4062240n^8)F(n + 1, k + 7)$$
$$+ (169575 + 5800081n^3 + 3486980n^2 + 5984198n^4 + 1177050n + 3980740n^5$$
$$\quad + 459040n^7 + 1710920n^6 + 4608n^9 + 69888n^8)F(n + 1, k + 8)$$
$$- (-312606 - 8764049n^3 - 5648475n^2 - 8421402n^4 - 2038176n - 5215700n^5$$
$$\quad - 523040n^7 - 2088744n^6 - 4608n^9 - 74496n^8)F(n + 2, k + 4)$$

$$- (3983364 + 108578372n^3 + 70829130n^2 + 103059462n^4 + 25816932n + 63119108n^5$$
$$+ 6231872n^7 + 25047648n^6 + 54720n^9 + 884640n^8)F(n+2, k+5)$$
$$- (57568644 + 1475731758n^3 + 972932346n^2 + 1393651260n^4 + 361506960n$$
$$+ 852058296n^5 + 84180480n^7 + 338123280n^6 + 739584n^9 + 11956608n^8)F(n+2, k+6)$$
$$- (3983364 + 108578372n^3 + 70829130n^2 + 103059462n^4 + 25816932n + 63119108n^5$$
$$+ 6231872n^7 + 25047648n^6 + 54720n^9 + 884640n^8)F(n+2, k+7)$$
$$- (-312606 - 8764049n^3 - 5648475n^2 - 8421402n^4 - 2038176n - 5215700n^5$$
$$- 523040n^7 - 2088744n^6 - 4608n^9 - 74496n^8)F(n+2, k+8)$$
$$- (720 + 3475n + 6398n^2 + 5624n^3 + 2368n^4 + 384n^5)(n+3)^2(2n+5)^2 F(n+3, k+6)$$
$$- (2(1560n^5 + 9620n^4 + 22942n^3 + 26371n^2 + 14603n + 3120))(n+3)^2(2n+5)^2 F(n+3, k+7)$$
$$- (720 + 3475n + 6398n^2 + 5624n^3 + 2368n^4 + 384n^5)(n+3)^2(2n+5)^2 F(n+3, k+8) = 0$$

for the summand $F(n, k)$. Summing w.r.t. $k \in \mathbb{Z}$ and shifting by one gives a holonomic recurrence equation of order 2

$$9(6n + 5)(3n + 2)(3n + 1)(6n + 1)(36n^3 + 168n^2 + 257n + 128)s_n$$
$$+ (3540 + 160164n^2 + 302838n^3 + 188784n^5 + 41526n$$
$$+ 317592n^4 + 7776n^7 + 59616n^6)s_{n+1}$$
$$+ (36n^3 + 60n^2 + 29n + 3)(n+2)^2(2n+3)^2 s_{n+2} = 0$$

for the famous *Dixon sum* $s_n = \sum_{k=-\infty}^{\infty} F(n, k)$. It turns out that in our example the *structure set* $S := \{(i, j) \in \mathbb{Z}^2 \mid 0 \leq i \leq 8, 0 \leq j \leq 3\}$ is the smallest rectangular region for which Fasenmyer's algorithm succeeds. However, as one can see from the above recurrence for $F(n, k)$, only a certain part of the rectangular grid of the $9 \cdot 4 = 36$ variables in S is needed in the resulting recurrence. If we consider $I = 2$ and $J = 3$, then in step (e) of Fasenmyer's algorithm we get a polynomial in k of degree 24. This leads to a linear system with 25 equations and 12 variables which has no non-trivial solution. The crucial idea of Wegschaider's improvement to Fasenmyer's algorithm uses the fact that the degree of the polynomial does not change if we add certain points to the structure set as shown in Fig. 4.1.

The additional points can be determined directly from the input term $F(n, k)$ by computing certain boundary points and structure lines which is done by the so-called *Verbaeten completion* [Verbaeten76]. This results in a homogeneous linear system of 25 equations and 25 variables. Solving this linear system one obtains a non-trivial solution leading to the *same* k-free holonomic recurrence equation for $F(n, k)$ as for the rectangular structure set S by considering a significantly smaller linear system.

Wegschaider's second improvement is the introduction of holonomic recurrence equations of a more general form, not restricting to k-free recurrences. The recurrence equation

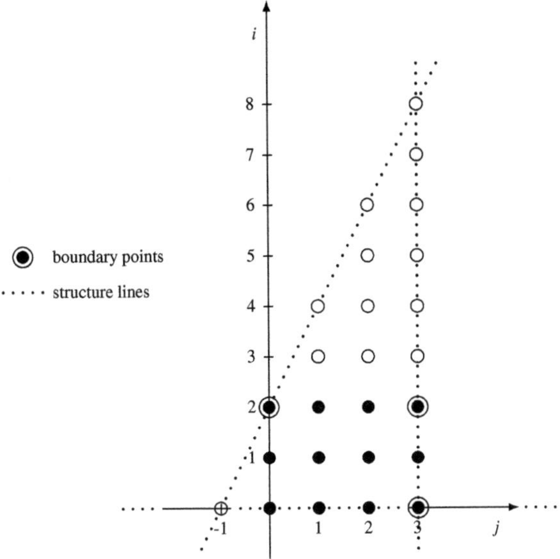

Fig. 4.1 Verbaeten completion of structure sets

$$(-20kn^2 - 30kn - 11k + 96n^2 + 48n + 6 + 56n^3)F(n,k)$$
$$+ (27 + 552n^2 + 336n^3 - 27k + 260n - 48kn^2 - 72kn)F(n,k+1)$$
$$+ (108 + 552n^2 + 240n^3 + 27k + 422n + 48kn^2 + 72kn)F(n,k+2)$$
$$+ (30kn + 11k + 116n + 39 + 96n^2 + 16n^3 + 20kn^2)F(n,k+3)$$
$$+ (-42kn - 16k + 6n - 6 - 28kn^2 + 48n^2 + 40n^3)F(n+1,k+2)$$
$$+ (42kn + 16k + 100n + 42 + 28kn^2 + 48n^2 - 16n^3)F(n+1,k+3) = 0$$

is a recurrence for $F(n,k)$, but the coefficients are *not* independent of the summation variable k. However, it is possible to rewrite this recurrence equation in the following form

$$4(n+1)^2(6n+5)F(n+1,k) + 12(6n+5)(3n+2)(3n+1)F(n,k)$$
$$+ \Delta_k \big(4(n+1)^2(6n+5)F(n+1,k) + 4(n+1)^2(6n+5)F(n+1,k+1)$$
$$+ (-14(k+3)(n+1) + 12n + 16 + 28(k+3)(n+1)^2$$
$$- 16(n+1)^2 - 16(n+1)^3 + 2k)F(n+1,k+2)$$
$$+ (20(k+3)(n+1)^2 - 744(n+1)^2 - 10(k+3)(n+1)$$
$$+ 258n + 236 + 592(n+1)^3 + k)F(n,k)$$
$$+ (-34(k+3)(n+1) + 156n + 152 + 4k - 412(n+1)^2$$
$$+ 256(n+1)^3 + (68(k+3))(n+1)^2)F(n,k+1)$$
$$+ (-10(k+3)(n+1) + 12n + 14 + k - 32(n+1)^2$$
$$+ 16(n+1)^3 + 20(k+3)(n+1)^2)F(n,k+2)\big) = 0,$$

where $\Delta_k a_k = a_{k+1} - a_k$ is the *forward difference operator.* Note that the so-called *main part* (the first line) of this certificate recurrence equation is k-free. After summation w.r.t. k the Δ-part vanishes through telescoping and the resulting recurrence equation for s_n reads

$$3(3\,n+1)(3\,n+2)\,s_n + (n+1)^2\,s_{n+1} = 0.$$

As usual the solution of this first-order recurrence is a hypergeometric term which can be written in terms of products of Pochhammer symbols that can be read off easily from the recurrence. The term ratio is given by

$$\frac{s_{n+1}}{s_n} = -\frac{3(3n+1)(3n+2)}{(n+1)^2} = -\frac{(3n+1)(3n+2)(3n+3)}{(n+1)^3}.$$

Using the initial value $s_0 = \sum_{k=0}^{0}(-1)^k \binom{2n}{k}^3 = 1$, we obtain the result

$$s_n = \sum_{k=0}^{2n}(-1)^k \binom{2n}{k}^3 = (-1)^n 27^n \frac{\left(\frac{1}{3}\right)_n \left(\frac{2}{3}\right)_n \left(\frac{3}{3}\right)_n}{(1)_n^3} = (-1)^n \frac{(3n)!}{n!^3},$$

which is a special case of *Dixon's identity.*

Session 4.7 Let's have a look at the computer generated proof using Sprenger's Maple package `multsum` [Sprenger04] containing an implementation of Wegschaider's algorithm. After loading the package

```
>    read "multsum.mpl";
>    with(multsum):
>    infolevel[multsum]:=3:
```

we define the summand

```
>    term:=(-1)^k*binomial(2*n,k)^3;
```

$$term := (-1)^k \binom{2\,n}{k}^3$$

Fasenmyer's algorithm with rectangular structure set ($I = 8$, $J = 3$) yields the large output that we saw on page 61 and that we suppress:

```
>    rec:=findrec(term,k,n,F,structureset=rect([8,3]),
>    certificate=false,upperbound=0):
```

_krec: structure set: [[-] -] number of equations: 43 number of variables: 36

output suppressed (k-free recurrence of order [[8],3])

Summing w.r.t. k and shifting by 1 gives a holonomic recurrence equation of order 2 for s_n

```
>    shiftrec(sumrec(rec,s));
```

$$\{9\,(6\,n+5)\,(3\,n+2)\,(3\,n+1)\,(6\,n+1)\,\left(36\,n^3+168\,n^2+257\,n+128\right) s\,(n)$$
$$+\,(3540+160164\,n^2+41526\,n+302838\,n^3$$
$$+\,317592\,n^4+188784\,n^5+59616\,n^6+7776\,n^7)s\,(n+1)$$
$$+\,\left(36\,n^3+60\,n^2+29\,n+3\right)(n+2)^2\,(2\,n+3)^2\,s\,(n+2)=0\}$$

The optimal structure set can be determined by

```
>    opt(term,k,n,[[2],3]);
```

$$\{[-1,0],\,[0,0],\,[0,1],\,[0,2],\,[1,0],\,[1,1],\,[1,2],\,[1,3],\,[1,4],$$
$$[2,0],\,[2,1],\,[2,2],\,[2,3],\,[2,4],\,[2,5],\,[2,6],\,[3,0],$$
$$[3,1],\,[3,2],\,[3,3],\,[3,4],\,[3,5],\,[3,6],\,[3,7],\,[3,8]\}$$

Now we apply Wegschaider's improvement with optimal structure set using Verbaeten completion. The following algorithm iterates over increasing optimal structure sets based on rectangular structure sets.

```
>    findrec(term,k,n,F,strategy=3,structureset=optimal,
>    certificate=false,upperkbound=0);
```

_krec: structure set: [[1] 0] number of equations: 4 number of variables: 2
_krec: structure set: [[0] 1] number of equations: 7 number of variables: 4
_krec: structure set: [[1] 1] number of equations: 10 number of variables: 6
_krec: structure set: [[0] 2] number of equations: 13 number of variables: 9
_krec: structure set: [[1] 2] number of equations: 16 number of variables: 12
_krec: structure set: [[0] 3] number of equations: 19 number of variables: 16
_krec: structure set: [[1] 3] number of equations: 22 number of variables: 20
_krec: structure set: [[2] 3] number of equations: 25 number of variables: 25

output suppressed (k-free recurrence of order $[[8],3]$)

Next we consider Wegschaider's algorithm using a *certificate recurrence equation*

```
>    rec:=findrec(term,k,n,F,structureset=optimal,
>    certificate=true,upperkbound=1);
```

_rec: structure set: [[1] 0] number of equations: 4 number of variables: 3
_rec: structure set: [[0] 1] number of equations: 7 number of variables: 6
_rec: structure set: [[1] 1] number of equations: 10 number of variables: 10

$$rec := \{4\,(n+1)^2\,(6\,n+5)\,F\,(n+1,k)+12\,(6\,n+5)\,(3\,n+2)\,(3\,n+1)\,F\,(n,k)$$
$$+\,\Delta_k(4\,(n+1)^2\,(6\,n+5)\,F\,(n+1,k)+4\,(n+1)^2\,(6\,n+5)\,F\,(n+1,k+1)$$
$$+\,(28\,(n+1)^2\,(k+3)-16\,(n+1)^2+2\,k+16-16\,(n+1)^3$$
$$-\,14\,(n+1)\,(k+3)+12\,n)F\,(n+1,k+2)+(258\,n+236+k$$
$$-\,10\,(n+1)\,(k+3)+20\,(n+1)^2\,(k+3)-744\,(n+1)^2$$
$$+\,592\,(n+1)^3)F\,(n,k)+(-412\,(n+1)^2+256\,(n+1)^3+4\,k+152$$
$$+\,68\,(n+1)^2\,(k+3)-34\,(n+1)\,(k+3)+156\,n)F\,(n,k+1)$$
$$+\,(-32\,(n+1)^2+16\,(n+1)^3+k+14+20\,(n+1)^2\,(k+3)$$
$$-\,10\,(n+1)\,(k+3)+12\,n)F\,(n,k+2))=0\}$$

The result is the following non-k-free recurrence equation.

```
>   certorec(rec);
```

$$\{\left(48\,n + 6 - 11\,k - 30\,nk - 20\,n^2k + 96\,n^2 + 56\,n^3\right) F\,(n,k)$$
$$+ \left(27 + 552\,n^2 - 27\,k + 260\,n - 48\,n^2k - 72\,nk + 336\,n^3\right) F\,(n,k+1)$$
$$+ \left(108 + 552\,n^2 + 27\,k + 422\,n + 48\,n^2k + 72\,nk + 240\,n^3\right) F\,(n,k+2)$$
$$+ \left(96\,n^2 + 116\,n + 39 + 16\,n^3 + 11\,k + 20\,n^2k + 30\,nk\right) F\,(n,k+3)$$
$$+ \left(-28\,n^2k + 48\,n^2 - 42\,nk + 6\,n - 16\,k - 6 + 40\,n^3\right) F\,(n+1,k+2)$$
$$+ \left(28\,n^2k + 48\,n^2 + 42\,nk + 100\,n + 16\,k + 42 - 16\,n^3\right) F\,(n+1,k+3) = 0\}$$

Nevertheless, by telescoping this gives a first order holonomic recurrence equation for s_n

```
>   multsumrecursion(term,k,s(n));
```

_rec: structure set: [[1] 0] number of equations: 4 number of variables: 3
_rec: structure set: [[0] 1] number of equations: 7 number of variables: 6
_rec: structure set: [[1] 1] number of equations: 10 number of variables: 10

$$3\,(3\,n+2)\,(3\,n+1)\,s\,(n) + (n+1)^2\,s\,(n+1) = 0$$

which yields a closed form for s_n. The final result can be retrieved completely automatically by a single Maple call:

```
>   multsumrecursion(term,k,s(n),hypersol);
```

_rec: structure set: [[1] 0] number of equations: 4 number of variables: 3
_rec: structure set: [[0] 1] number of equations: 7 number of variables: 6
_rec: structure set: [[1] 1] number of equations: 10 number of variables: 10

$$\frac{s\,(0)\,(-27)^n\,pochhammer\,(2/3,n)\,pochhammer\,(1/3,n)}{(n!)^2}$$

For the determination of the initial value s_0 the natural bounds of the sum must be known. Since these cannot be determined easily in all instances—especially in the multivariate case—the determination of the initial values is therefore left to the user.

Multiple Summation

Fasenmyer's technique can be extended to obtain holonomic recurrence equations for multiple sums

$$s_n = \sum_{k_1=-\infty}^{\infty} \cdots \sum_{k_m=-\infty}^{\infty} F(n, k_1, \ldots, k_m)$$

by using linear algebra to deduce a mixed recurrence equation

$$\sum_{i_1=0}^{I_1} \cdots \sum_{i_m=0}^{I_m} \sum_{j=0}^{J} a_{ij}\, F(n+j, k_1+i_1, \ldots, k_m+i_m) = 0$$

for the summand. Summation then yields a holonomic recurrence equation for s_n. But with increasing number of summation variables m and increasing order of the proposed recurrence equation the number of linear equations to be solved increases dramatically. For $m > 1$ these problems are so severe that only very few results can be obtained with reasonable time and memory resources.

In [WZ92] Wilf and Zeilberger presented a slightly different approach using a multidimensional antidifference. For practical purposes, however, their method still turns out to be rather inefficient.

More efficiently, one can again apply Verbaeten completion to multiple summation. For example, if we consider the hypergeometric term

$$F(n, k_1, k_2) = \binom{n}{k_2}\binom{k_2}{k_1} x^{k_1} y^{k_2-k_1} z^{n-k_2},$$

then the number of equations in the resulting linear system corresponding to the structure sets of a cube ($J = I_1 = I_2 = 3$) and a corresponding larger polyhedral set S are the same; see Fig. 4.2. The black cube contains 64 variables, and the grey structure set S with the same number of equations contains 200 variables. On the other hand, the cube surrounding S has 1,000 variables 800 of which must vanish in the linear equations generated.

Of course, this example is simple enough to be resolved. For $F(n, k_1, k_2)$ one obtains the recurrence

$$F(n+1, k_1+1, k_2+1) - z\,F(n, k_1+1, k_2+1) - y\,F(n, k_1+1, k_2) - x\,F(n, k_1, k_2) = 0$$

which yields for $s_n = \sum_{k_1,k_2=-\infty}^{\infty} F(n, k_1, k_2)$ after summation

$$(x+z+y)\,s_n - s_{n+1} = 0$$

with result

$$s_n = (x+z+y)^n.$$

To show that Wegschaider's second improvement also has a great impact on multiple summation, we consider the following hypergeometric term

$$G(n, k_1, k_2) = \binom{t}{k_1}\binom{u}{k_2}\binom{v}{n-k_1-k_2}$$

with parameters t, u and v. Then

$$(k_1 - t)G(n, k_1, k_2) + (k_2 + 1)G(n+1, k_1+1, k_2+1)$$
$$+(n - k_2 + 1)G(n+1, k_1+1, k_2) + (n - u - v - k_1 - 1)G(n, k_1+1, k_2) = 0$$

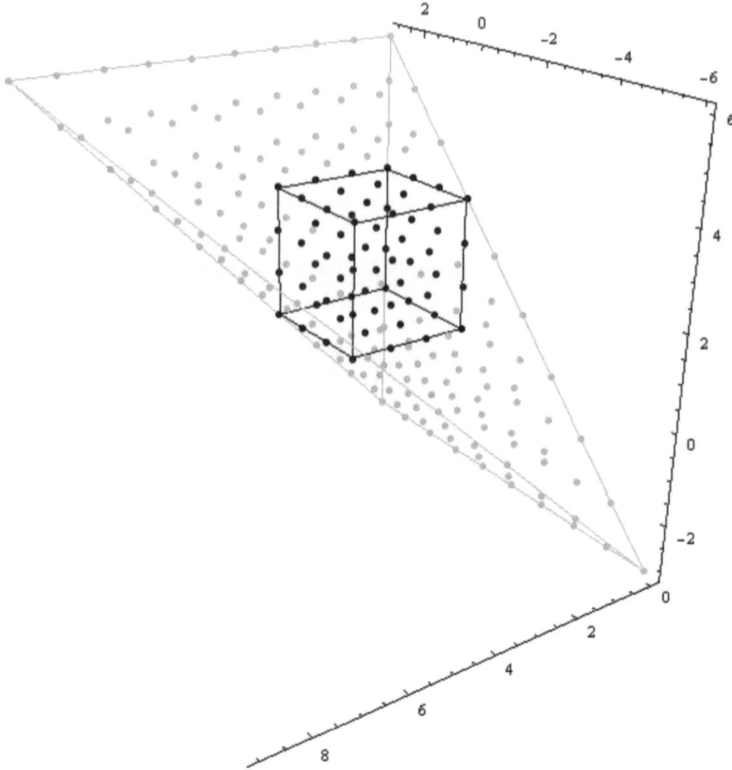

Fig. 4.2 Different structure sets for $F(n, k_1, k_2)$ with the same number of equations

is a recurrence equation for $G(n, k_1, k_2)$, which is obviously not (k_1, k_2)-free. However, one can rewrite this recurrence equation in the following form

$$(n+1)G(n+1, k_1, k_2) + (n-t-u-v)G(n, k_1, k_2)$$
$$+\Delta_{k_1}\big((k_2+1)G(n+1, k_1-1, k_2+1) + (n-k_2+1)G(n+1, k_1-1, k_2)$$
$$+(n-u-v-k_1+1)G(n, k_1-1, k_2)\big)$$
$$+\Delta_{k_2}\big((k_2-1)G(n+1, k_1, k_2-1)\big) = 0.$$

The main part of this certificate recurrence equation (the first line) is (k_1, k_2)-free. After summation w.r.t. k_1 and k_2 the Δ-parts disappear and one obtains the following simple recurrence equation

$$(n+1)s_{n+1} + (n-t-u-v)s_n = 0$$

for the double sum $s_n = \sum_{k_1, k_2=-\infty}^{\infty} G(n, k_1, k_2)$ leading (with $s_0 = 1$) to the closed form

$$s_n = \sum_{k_1=0}^{t} \sum_{k_2=0}^{u} \binom{t}{k_1}\binom{u}{k_2}\binom{v}{n-k_1-k_2} = \frac{(-1)^n}{n!}(-(t+u+v))_n = \binom{t+u+v}{n}.$$

In contrast, the smallest (k_1, k_2)-free recurrence equation for $G(n, k_1, k_2)$ computed by Fasenmyer's algorithm is

$$\begin{aligned}
&(n+3)G(n+3, k_1+1, k_2+1) + (n-v+2)G(n+2, k_1+1, k_2+1) \\
&+ (n-u+2)G(n+2, k_1+1, k_2) + (n-t+2)G(n+2, k_1, k_2+1) \\
&+ (n-u-v+1)G(n+1, k_1+1, k_2) + (n-t-v+1)G(n+1, k_1, k_2+1) \\
&+ (n-t-u+1)G(n+1, k_1, k_2) + (n-t-u-v)G(n, k_1, k_2) = 0,
\end{aligned}$$

which leads to a recurrence equation of order 3 and misses the above simple recurrence of order 1.

Session 4.8 The `multsum` package contains both Verbaeten completion and certificate recurrences in the multivariate case. The examples above are treated by the following Maple commands.

```
>  F:=binomial(n,k2)*binomial(k2,k1)*x^k1*y^(k2-k1)*z^(n-k2);
```
$$F := binomial(n,\, k2)\, binomial(k2,\, k1)\, x^{k1}\, y^{(k2-k1)}\, z^{(n-k2)}$$
```
>  multsumrecursion(F,[k1,k2],s(n));
```
$$(x+z+y)\, s(n) - s(n+1) = 0$$
```
>  multsumrecursion(F,[k1,k2],s(n),hypersol);
```
$$s(0)\, (x+z+y)^n$$
```
>  G:=binomial(t,k1)*binomial(u,k2)*binomial(v,n-k1-k2);
```
$$binomial(t,\, k1)\, binomial(u,\, k2)\, binomial(v,\, n-k1-k2)$$
```
>  multsumrecursion(G,[k1,k2],s(n));
```
$$(-t-u-v+n)\, s(n) + (n+1)\, s(n+1) = 0$$

The whole computation can be done in a single command:

```
>  multsumrecursion(G,[k1,k2],s(n),hypersol);
```
$$binomial(t+u+v,\, n)\, s(0)$$

Next we consider yet another complicated double sum

$$s_n = \sum_{i=-\infty}^{\infty} \sum_{j=-\infty}^{\infty} \binom{j-i}{n-i}\binom{t-u-j+i}{r+u-n+i}\binom{u}{i}\binom{t-u}{j-i}\binom{u-t}{v-j}.$$

We define the summand

```
>  term:=binomial(j-i,n-i)*binomial(t-u-j+i,r+u-n+i)*
>  binomial(u,i)*binomial(t-u,j-i)*binomial(u-t,v-j):
```

and compute the recurrence for the double sum

```
>    RE:=multsumrecursion(term,[i,j],s(n));
```

$$(-v+n)\,(-r+n-2\,u)\,s(n) - (n+1)\,(n-r-v-u+1)\,s(n+1) = 0$$

In one step we get the resulting hypergeometric term

```
>    multsumrecursion(term,[i,j],s(n),hypersol);
```

$$\frac{s(0)\,\mathrm{pochhammer}(-v,\ n)\,\mathrm{pochhammer}(-2\,u-r,\ n)}{n!\,\mathrm{pochhammer}(1-u-v-r,\ n)}$$

hence

$$s_n = \frac{(-v)_n\,(-2\,u-r)_n}{(1-u-v-r)_n\,n!}.$$

Wegschaider's algorithm is able to deduce hypergeometric double or triple sum identities which were out of reach with Fasenmyer's original approach, e.g. identities like [Strehl93, Strehl94]

$$\sum_{k=0}^{n}\sum_{j=0}^{n}\binom{n}{k}\binom{n+1}{k}\binom{k}{j}^3 = \sum_{k=0}^{n}\binom{n}{k}^2\binom{n+k}{k}^2. \qquad (4.8)$$

This identity is settled by showing that both sides satisfy (enough initial values and) the same holonomic recurrence. The common recurrence is obtained by the commands

```
>    term1:=binomial(n,k)*binomial(n+k,k)*binomial(k,j)^3;
```

$$\mathrm{binomial}(n,\ k)\,\mathrm{binomial}(n+k,\ k)\,\mathrm{binomial}(k,\ j)^3$$

```
>    RE1:=multsumrecursion(term1,[k,j],s(n),upperkbound=2);
```

$$s(n)\,(n+1)^3 - (3+2\,n)\,(17\,n^2+39+51\,n)\,s(n+1) + (n+2)^3\,s(n+2) = 0$$

```
>    term2:=binomial(n,k)^2*binomial(n+k,k)^2;
```

$$\mathrm{binomial}(n,\ k)^2\,\mathrm{binomial}(n+k,\ k)^2$$

```
>    RE2:=multsumrecursion(term2,k,s(n),upperkbound=2);
```

$$s(n)\,(n+1)^3 - (3+2\,n)\,(17\,n^2+39+51\,n)\,s(n+1) + (n+2)^3\,s(n+2) = 0$$

Note that the smallest k-free recurrence equation for the summand of the left-hand side of (4.8) has order $J = 7$, and the smallest k-free recurrence equation for the summand of the right-hand side has order $J = 5$, although both sums satisfy the second order recurrence

$$(n+2)^3\,s_{n+2} - (2\,n+3)\left(17\,n^2+51\,n+39\right)s_{n+1} + (n+1)^3\,s_n = 0.$$

The latter identities cannot be proved by iterative summation since the inner sums do not satisfy a hypergeometric recurrence equation. Nevertheless, in many cases, iterative techniques turn out to be successful when dealing with multiple sums. This method is considered in forthcoming chapters.

q-Holonomic Recurrence Equations

In the preceding two chapters we have introduced q-hypergeometric functions, and we have seen q-analogues of the binomial theorem and of the identities of Gauss and Pfaff-Saalschütz.

Similarly, q-analogues for the classical orthogonal polynomials can be given, see e.g. [AW85, KLS10]. The Legendre polynomials $P_n(x)$, e.g., have several q-analogues. The polynomials

$$p_n(x|q) = {}_2\phi_1 \left(\begin{matrix} q^{-n}, q^{n+1} \\ q \end{matrix} \, \Big| \, q, qx \right)$$

are called the *little q-Legendre polynomials*; the polynomials

$$P_n(x; c; q) = {}_3\phi_2 \left(\begin{matrix} q^{-n}, q^{n+1}, x \\ q, cq \end{matrix} \, \Big| \, q, q \right)$$

are called the *big q-Legendre polynomials*; and the polynomials

$$P_n(x; q) = {}_4\phi_3 \left(\begin{matrix} q^{-n}, q^{n+1}, \sqrt{q}\, e^{i\theta}, \sqrt{q}\, e^{-i\theta} \\ q, -q, -q \end{matrix} \, \Big| \, q, q \right), \qquad (x = \cos\theta)$$

and

$$P_n(x|q) = {}_4\phi_3 \left(\begin{matrix} q^{-n}, q^{n+1}, q^{1/4} e^{i\theta}, q^{1/4} e^{-i\theta} \\ q, -\sqrt{q}, -q \end{matrix} \, \Big| \, q, q \right), \qquad (x = \cos\theta)$$

(see [KLS10], Sect. 14.5.1), which are related by

$$P_n(x|q^2) = P_n(x; q), \tag{4.9}$$

are both called the *continuous q-Legendre polynomials*. Similarly the *generalized Laguerre polynomials*

$$L_n^{(\alpha)}(x) := \sum_{k=0}^{n} \frac{(-1)^k}{k!} \binom{n+\alpha}{n-k} x^k = \binom{n+\alpha}{n} {}_1F_1 \left(\begin{matrix} -n \\ 1+\alpha \end{matrix} \, \Big| \, x \right)$$

have the q-analogues

$$L_n^{(\alpha)}(x; q) = \frac{(q^{\alpha+1}; q)_n}{(q; q)_n} \, {}_1\phi_1 \left(\begin{matrix} q^{-n} \\ q^{\alpha+1} \end{matrix} \, \middle| \, q, -xq^{n+\alpha+1} \right)$$

(see e.g. [KLS10], Sect. 14.21), which are called the *q-Laguerre polynomials.* All of these polynomial systems are contained in the so-called *Askey-Wilson scheme* [AW85], [KLS10].

It is rather straightforward to extend Fasenmyer's method to the q-hypergeometric situation. If

$$s_n = \sum_{k=-\infty}^{\infty} F(n, k),$$

and $F(n, k)$ is rational w.r.t. both q^k and q^n, then the same method applies if we expand in powers of $K := q^k$ rather than in powers of k. The resulting recurrence equation is then holonomic in q^n.

The qsum package contains a Maple procedure `qfasenmyer(term,q,k, s(n),kmax,nmax)` to generate this recurrence equation using a k-free recurrence equation of the summand `term` of order (k_{max}, n_{max}) w.r.t. (k, n) if applicable. The request

```
>   qfasenmyer(qbinomial(n,k,q),q,k,s(n),1,2);
```

for example, results in the recurrence equation

$$(1 - q^{(n+1)}) \, s(n) + s(n+2) - 2\,s(n+1) = 0.$$

In the exercises, the reader is asked to find recurrence equations for some of the q-hypergeometric series mentioned.

There is also a multivariate scenario in the q-case which was treated in [Riese03].

Further Reading

For further reading on the univariate case see [Zeilberger82, PWZ96], and on the multivariate case [Wegschaider97, Sprenger04, Riese03].

Exercises

Exercise 4.1 *Prove the identities (2.2)–(2.5) using Fasenmyer's method. Try to prove (2.6). What happens? What does the "Pascal triangle recurrence equation" for the squares of the binomial coefficients look like?*

Exercise 4.2 *Try to prove the entries of our hypergeometric database of Chap. 3 using Fasenmyer's method. Which entries can be proved and which cannot? What is the reason for the failure?*

Exercise 4.3 *Show that the three different hypergeometric representations (see Exercise 2.12)*

$$P_n(x) = \sum_{k=0}^{n} \binom{n}{k} \binom{-n-1}{k} \left(\frac{1-x}{2}\right)^k,$$

$$P_n(x) = \frac{1}{2^n} \sum_{k=0}^{n} \binom{n}{k}^2 (x-1)^{n-k} (x+1)^k,$$

and

$$P_n(x) = \frac{1}{2^n} \sum_{k=0}^{\lfloor n/2 \rfloor} (-1)^k \binom{n}{k} \binom{2n-2k}{n} x^{n-2k}$$

define the same family of functions. Hint: *Generate the same holonomic recurrence equation for all of these representations, and show that they satisfy the same initial values.*

◇ **Exercise 4.4** *Write a Maple procedure* `fasenmyerdiffeq(f,k,s(x), xmax)` *that computes a differential equation for*

$$s(x) = \sum_{k=-\infty}^{\infty} F(x,k)$$

which uses a function `kfreediffeq(f,k,x,kmax,xmax)` *also to be implemented. The function* `fasenmyerdiffeq` *is applicable if $F(x,k)$ is a hypergeometric term w.r.t. k and a hyperexponential term w.r.t. x, compare Chap. 10.*

Apply the procedure to generate a differential equation for the Legendre polynomials, using all three hypergeometric representations given in Exercise 5.3.

Exercise 4.5 *The Legendre polynomials $P_n(x)$ are orthogonal in $[-1, 1]$*

$$\int_{-1}^{1} P_m(x) P_n(x) \, dx = 0 \quad \text{for} \quad m \neq n.$$

Check the orthogonality for $0 \le m < n = 0, 1, 2, 3, 4, 5$.

Exercise 4.6 *Show that the numbers*

$$F_{n+1} := \sum_{k=0}^{\lfloor n/2 \rfloor} \binom{n-k}{k}$$

are the Fibonacci *numbers, i.e.*

$$F_{n+2} - F_{n+1} - F_n = 0, \quad \text{with} \quad F_1 = 1, \quad F_2 = 1.$$

Exercise 4.7 *Show that the sum*

$$s_n := \sum_{k=0}^{n} \binom{n}{k}^3$$

satisfies the holonomic recurrence equation

$$0 = 8\,(3\,n+7)\,(n+1)^2\,s_n + (3\,n+5)\left(15\,n^2 + 55\,n + 48\right) s_{n+1}$$

$$+2\left(9\,n^3 + 57\,n^2 + 116\,n + 74\right) s_{n+2} - (3\,n+4)\,(n+3)^2\,s_{n+3}.$$

Later we will deduce a recurrence equation of lower order for s_n.

Exercise 4.8 *Prove the identity* ($n \in \mathbb{N}_{\geq 0}$)

$$\sum_{k=0}^{n} \binom{n}{k}^3 = \sum_{k=0}^{n} \binom{n}{k}^2 \binom{2k}{n}.$$

Exercise 4.9 *The Laguerre polynomials are orthogonal in* $[0, \infty)$ *with respect to the density* e^{-x}, *i.e.*

$$\int_0^{\infty} L_m(x)\,L_n(x)\,e^{-x}\,dx = 0 \quad \text{for} \quad m \neq n.$$

Check the orthogonality for $0 \leq m < n \leq 5$.

Exercise 4.10 *Deduce a three-term recurrence equation for the generalized Laguerre polynomials*

$$L_n^{(\alpha)}(x) := \sum_{k=0}^{n} \frac{(-1)^k}{k!} \binom{n+\alpha}{n-k} x^k = \binom{n+\alpha}{n}\, {}_1F_1\left(\begin{matrix} -n \\ 1+\alpha \end{matrix}\middle|\, x\right)$$

Show, for $0 \le m < n \le 5$, that $L_m^{(\alpha)}(x)$ and $L_n^{(\alpha)}(x)$ are orthogonal in $[0, \infty)$ with respect to the density $x^\alpha e^{-x}$.

Exercise 4.11 *Find holonomic recurrence equations for the Fasenmyer polynomials ([Fasenmyer47], see also [Rainville60], Chap. 14, Exercise 2) with respect to $n \in \mathbb{N}_{\ge 0}$*

(a) $f_n(x) := {}_2F_2\left(\begin{array}{c} -n,\, n+1 \\ 1,\, 1/2 \end{array}\middle|\, x\right)$,

(b) $g_n(x) := {}_3F_2\left(\begin{array}{c} -n,\, n+1,\, a \\ 1,\, 1/2 \end{array}\middle|\, -x\right)$.

Exercise 4.12 *Find holonomic recurrence equations for the following sequences $(n \in \mathbb{N}_{\ge 0})$:*

(a) $f_n(x) := {}_2F_2\left(\begin{array}{c} -n,\, n+\beta \\ 1,\, 1+\alpha \end{array}\middle|\, x\right)$,

(b) $g_n(x) := (2x)^n\, {}_3F_1\left(\begin{array}{c} -n/2,\, -n/2+1/2,\, 1+\alpha \\ 1+\beta \end{array}\middle|\, -\frac{1}{x^2}\right)$.

Exercise 4.13 *Find a holonomic recurrence equation for the Bateman polynomials (see e.g. [Rainville60], Chap. 18)*

$$Z_n(x) := {}_2F_2\left(\begin{array}{c} -n,\, n+1 \\ 1,\, 1 \end{array}\middle|\, x\right).$$

Exercise 4.14 *Show that the polynomials*

$$f_n(x) := \binom{\lambda}{n}\, {}_1F_1\left(\begin{array}{c} -n \\ 1+\lambda-n \end{array}\middle|\, x\right) = L_n^{(\lambda-n)}(x),$$

which are related to the generalized Laguerre polynomials, do not form a family of orthogonal polynomials, since they do not satisfy a three-term recurrence equation of the type

$$f_n(x) = (A_n + B_n x) f_{n-1}(x) + C_n f_{n-2}(x)$$

for some constants A_n, B_n, C_n (Favard's Theorem, see [Chihara78]).

Exercise 4.15 *Show that for Example 4.6, the number E of equations to be satisfied by the variables a_{ij}, is given by $E = 2V - 1$.*

Exercise 4.16 *Use* `kfreerec` *to deduce the following hypergeometric term representations. Note that the bounds of these sums are not the natural ones.*

(a) $\displaystyle\sum_{k=0}^{m} (-1)^k \binom{n}{k} = (-1)^m \binom{n-1}{m}$,

(b) $\displaystyle\sum_{k=0}^{m}\binom{n+k}{k}=\binom{n+m+1}{n}$,

(c) $\displaystyle\sum_{k=0}^{m}\binom{k}{n}=\binom{m+1}{n+1}$,

(d) $\displaystyle\sum_{k=0}^{m}\frac{1}{2^{k}}\binom{m+k}{k}=2^{m}$.

Exercise 4.17 *Find a holonomic recurrence equation for the Hermite polynomials*

$$H_n(x) := n!\sum_{k=0}^{\lfloor n/2\rfloor}\frac{(-1)^k}{(n-2k)!\,k!}(2x)^{n-2k}.$$

Show, for $0 \le m < n \le 5$, that $H_m(x)$ and $H_n(x)$ are orthogonal in $(-\infty, \infty)$ with respect to the density e^{-x^2}.

Exercise 4.18 *Show that no k-free recurrence equation is valid for the term*

$$F(n,k) := \binom{n}{k}\frac{1}{k^2+n^2+1}\;;$$

compare Example 4.6.

Exercise 4.19 *Try to prove the q-analogues of the Chu-Vandermonde and of the Pfaff-Saalschütz identities (3.7)*

$${}_2\phi_1\left(\begin{matrix}q^{-n},\,b\\c\end{matrix}\;\middle|\;q,\,\frac{cq^n}{b}\right)=\frac{(c/b;\,q)_n}{(c;\,q)_n}$$

and (3.6)

$${}_3\phi_2\left(\begin{matrix}q^{-n},\,a,\,b\\c,\,\frac{ab}{cq^{n-1}}\end{matrix}\;\middle|\;q,\,q\right)=\frac{(c/a;\,q)_n\,(c/b;\,q)_n}{(c;\,q)_n\,(c/(ab);\,q)_n}$$

by the q-Fasenmyer method, using `qfasenmyer`*. What happens?*

Exercise 4.20 *Find three-term recurrence equations for the little q-Legendre polynomials and for the q-Laguerre polynomials defined on p. 71–72.*

References

AW85.   Askey, R., Wilson, J.A.: Some Basic Hypergeometric Polynomials that Generalize Jacobi Polynomials, vol. 319. Memoirs of the American Mathematical Society, Providence (1985)

Chihara78.   Chihara, T.S.: An Introduction to Orthogonal Polynomials. Gordon and Breach Publishers, New York (1978) (reprinted by Dover Publishers, Dover, 2011)

Fasenmyer45.   Fasenmyer, M. C.: Some generalized hypergeometric polynomials. Ph.D. thesis, University of Michigan (1945)

Fasenmyer47.   Fasenmyer, M.C.: Some generalized hypergeometric polynomials. Bull. Amer. Math. Soc. **53**, 806–812 (1947)

Fasenmyer49.   Fasenmyer, M.C.: A note on pure recurrence relations. Amer. Math. Monthly **56**, 14–17 (1949)

KLS10.   Koekoek, R., Lesky, P., Swarttouw, R.F.: Hypergeometric Orthogonal Polynomials and their q-Analogues. Springer Monographs in Mathematics. Springer, Berlin (2010)

PWZ96.   Petkovšek, M., Wilf, H., Zeilberger, D.: $A = B$. A K Peters, Wellesley (1996)

Rainville60.   Rainville, E.D.: Special Functions. The MacMillan Co., New York (1960)

Riese03.   Riese, A.: qMultiSum–a package for proving q-hypergeometric multiple summation identities. J. Symb. Comput. **35**, 349–376 (2003)

Sprenger04.   Sprenger, T.: Algorithmen für mehrfache Summen. Diploma thesis, Universität Kassel (2004)

Strehl93.   Strehl, V.: Binomial sums and identities. Maple Tech. Newslett. **10**, 37–49 (1993)

Strehl94.   Strehl, V.: Binomial identities-combinatorial and algorithmic aspects. Discrete Math. **136**, 309–346 (1994)

Verbaeten76.   Verbaeten, P.: Rekursiebetrekkingen voor lineaire hypergeometrische funkties. Proefschrift voor het doctoraat in de toegepaste wetenschapen. Katholieke Universiteit te Leuven, Heverlee (1976)

Wegschaider97.   Wegschaider, K.: Computer generated proofs of binomial multi-sum identities. Diploma thesis, RISC Linz (1997)

WZ92.   Wilf, H.S., Zeilberger, D.: An algorithmic proof theory for hypergeometric (ordinary and "q") multisum/integral identities. Invent. Math. **108**, 575–633 (1992)

Zeilberger82.   Zeilberger, D.: Sister Celine's technique and its generalizations. J. Math. Anal. Appl. **85**, 114–145 (1982)

Zeilberger90a.   Zeilberger, D.: A holonomic systems approach to special functions identities. J. Comput. Appl. Math. **32**, 321–368 (1990)

Zeilberger95.   Zeilberger, D.: Three recitations on holonomic systems and hypergeometric series. In: Foata, D. (ed.) Proceedings of the 24th Séminaire Lotharingen. Publ. I. R. M. A. Strasbourg, pp 5–37 (reprinted J. Symb. Comput. 20, 1995, 699–724)

Chapter 5
Gosper's Algorithm

For a moment, let's have a break from searching for hypergeometric term solutions and recurrence equations of infinite series. Instead, we will deal with sums with variable limits of summation, an interesting topic in itself. Later, this will prove to be a useful tool in discovering an algorithmic method for infinite sums.

Let us motivate the topic of the present chapter by considering the issue of integration. Thanks to the fundamental theorem of calculus, the knowledge of an antiderivative, i.e., a function $F(x)$ with the property

$$F'(x) = f(x)$$

makes the evaluation of any definite integral of f easy, according to the simple rule[1]

$$\int_a^b f(x)\, dx = F(b) - F(a).$$

Therefore, a large *database* of antiderivatives makes definite integration a solvable task. On the other hand, no database is complete, and all of us know that many tricks may be needed if an integrand cannot be found in the database.

To avoid those problems, an algorithmic theory of integration in elementary terms has been developed by Risch [Risch69, Risch70], and others (see [Bronstein96], and [GCL92, Chaps. 11, 12]). This theory, however, is rather difficult.

It turns out that a discrete analogue of Risch's algorithm for indefinite summation, an algorithm due to Gosper ([Gosper78], see also [GKP94], and [Koepf95a]), is much simpler.

Gosper's algorithm deals with the question of how to find a (forward) *antidifference* s_k for given a_k, i.e. a sequence s_k for which

---

[1] Of course this is only so if f is absolutely continuous. Otherwise the Fundamental Theorem of Calculus might not be applicable.

W. Koepf, *Hypergeometric Summation*, Universitext,
DOI: 10.1007/978-1-4471-6464-7_5, © Springer-Verlag London 2014

$$a_k = \Delta s_k = s_{k+1} - s_k, \tag{5.1}$$

in the particular case that s_k is a hypergeometric term, i.e.

$$\frac{s_{k+1}}{s_k} \in \mathbb{Q}(k). \tag{5.2}$$

Finding an antidifference s_k according to (5.1) is called *indefinite summation.*

Note that—similar to the case of integration—once an antidifference s_k of a_k is known, definite summation is easy since by telescoping we get

$$\sum_{k=m}^{n} a_k = (s_{n+1} - s_n) + (s_n - s_{n-1}) + \cdots + (s_{m+1} - s_m) = s_{n+1} - s_m,$$

by an evaluation at the (shifted) limits of summation.

Note that if a hypergeometric term antidifference s_k exists, we call the input function a_k *Gosper-summable.* This function must then itself be a hypergeometric term, since by (5.1) and (5.2),

$$\frac{a_{k+1}}{a_k} = \frac{s_{k+2} - s_{k+1}}{s_{k+1} - s_k} = \frac{s_{k+1}}{s_k} \frac{\frac{s_{k+2}}{s_{k+1}} - 1}{\frac{s_{k+1}}{s_k} - 1} = \frac{u_k}{v_k} \in \mathbb{Q}(k), \tag{5.3}$$

with polynomials $u_k, v_k \in \mathbb{Q}[k]$ which can be found by Algorithm 2.2.

What follows is a short overview of Gosper's algorithm.

In the first step, Gosper uses a representation lemma for rational functions to express a_{k+1}/a_k in terms of some specific polynomials.

The idea behind this step comes from the following observation: If we calculate a_k from $s_k = (2k)!/k!$, for example, we get

$$a_k = s_{k+1} - s_k = \frac{(2k+2)!}{(k+1)!} - \frac{(2k)!}{k!} = (4k+1) \cdot \frac{(2k)!}{k!},$$

i.e. a product of a polynomial $p_k = (4k+1)$ and a factorial term $b_k = \frac{(2k)!}{k!}$ for which $\frac{b_{k+1}}{b_k}$ is rational, hence there are $q_k, r_k \in \mathbb{Q}[k]$ with $\frac{b_{k+1}}{b_k} = \frac{q_{k+1}}{r_{k+1}}$.

Gosper then shows that such a representation with the property

$$\gcd(q_k, r_{k+j}) = 1 \quad \text{for all } j \in \mathbb{N}_{\geq 0}, \tag{5.4}$$

can generally be found and gives an algorithm to generate it (Lemma 5.1 and Algorithm 5.2). The *greatest common divisor* of the two polynomials q_k and r_k is denoted by $\gcd(q_k, r_k)$.[2]

---

[2] Relation $\gcd(f, g) = 1$ means that there is no nontrivial, hence nonconstant, common divisor.

Therefore for a_k we have the relation[3]

$$\frac{a_{k+1}}{a_k} = \frac{p_{k+1}}{p_k} \frac{q_{k+1}}{r_{k+1}}, \tag{5.5}$$

with p_k corresponding to its polynomial part and (q_k, r_k) to its factorial part.

Gosper finally defines the function f_k by the equation

$$f_k := \frac{s_{k+1}}{a_{k+1}} \frac{p_{k+1}}{r_{k+1}} \quad \text{or} \quad s_k = \frac{r_k}{p_k} f_{k-1} a_k, \tag{5.6}$$

from which one sees immediately that

$$f_k = \frac{s_{k+1}}{a_{k+1}} \frac{p_{k+1}}{r_{k+1}} = \frac{s_{k+1}}{s_{k+2} - s_{k+1}} \frac{p_{k+1}}{r_{k+1}} = \frac{1}{\frac{s_{k+2}}{s_{k+1}} - 1} \frac{p_{k+1}}{r_{k+1}}$$

is rational. Using property (5.4), Gosper proves the essential fact that f_k is a polynomial (Lemma 5.4).

It follows by definition that the polynomial f_k satisfies

$$a_k = s_{k+1} - s_k = \frac{r_{k+1}}{p_{k+1}} f_k \, a_{k+1} - \frac{r_k}{p_k} f_{k-1} a_k.$$

Multiplying by p_k/a_k, and using (5.5), gives the inhomogeneous linear recurrence equation

$$p_k = \frac{a_{k+1}}{a_k} \frac{p_k}{p_{k+1}} r_{k+1} f_k - r_k f_{k-1} = q_{k+1} f_k - r_k f_{k-1} \tag{5.7}$$

for f_k. Using (5.7), Gosper gives an upper bound for the degree of f_k in terms of p_k, q_k, and r_k (Lemma 5.5). This yields a method for calculating f_k by introducing the appropriate generic polynomial, equating coefficients, and solving the corresponding linear system, so that we finally find s_k, given by (5.6).

If one of the steps to find the polynomial f_k fails, then the algorithm *has proved* that no hypergeometric term antidifference s_k of a_k exists, which may give valuable information. Thus in this situation, Gosper's algorithm did not fail, but instead decided the nonexistence of a hypergeometric term antidifference.

Having seen a general overview of Gosper's algorithm, we will now go on to prove each of its steps.

Lemma 5.1 and Algorithm *The functions p_k, q_k and r_k in (5.5) can be chosen such that*

$$\gcd(q_k, r_{k+j}) = 1 \quad \text{for all } j \in \mathbb{N}_{\geq 0}. \tag{5.8}$$

---

[3] We use the same functions p_k, q_k, and r_k as Gosper did, even though he worked with the backward antidifference. It may mean that we have some more shifts here than necessary.

Proof We start with the initial choice $p_k := 1$, $q_k := u_{k-1}$, and $r_k := v_{k-1}$. Now either (5.8) is satisfied, and so we are done, or (5.8) is not valid. In the latter case there is some $j \in \mathbb{N}_{\geq 0}$ such that

$$\gcd(q_k, r_{k+j}) = g_k \quad \text{with} \ \deg(g_k) > 0. \tag{5.9}$$

Let J denote the set of all $j \in \mathbb{N}_{\geq 0}$ such that (5.9) is valid. The construction of the finite set J is considered in Algorithm 5.2.

For any $j \in J$, we may eliminate the common factor g_k choosing new functions p_k', q_k' and r_k' where

$$p_k' = p_k \, g_k \, g_{k-1} \cdots g_{k-j+1}, \quad q_k' = \frac{q_k}{g_k}, \quad \text{and} \ r_k' = \frac{r_k}{g_{k-j}},$$

since then

$$\frac{p_{k+1}'}{p_k'} \frac{q_{k+1}'}{r_{k+1}'} = \frac{p_{k+1} \, g_{k+1} \, g_k \cdots g_{k-j+2}}{p_k \, g_k \, g_{k-1} \cdots g_{k-j+1}} \frac{q_{k+1}}{g_{k+1}} \frac{g_{k-j+1}}{r_{k+1}} = \frac{p_{k+1}}{p_k} \frac{q_{k+1}}{r_{k+1}}.$$

By (5.9) it follows that

$$\gcd(q_k', r_{k+j}') = \gcd\left(\frac{q_k}{g_k}, \frac{r_{k+j}}{g_k}\right) = 1$$

so that for the triple (p_k', q_k', r_k'), (5.9) no longer is valid. Therefore, applying the rewriting technique to all $j \in J$ leads to (p_k, q_k, r_k) satisfying (5.8). Note that if J contains more than one number, then the gcd condition has to be checked again before continuing with each of the rewriting steps since it may happen that one of the previous rewritings had already done the job. $\square$

For any two polynomials q_k and r_k, the number[4]

$$\mathrm{disp}(q_k, r_k) := \max\left\{ j \in \mathbb{N}_{\geq 0} \ | \ \gcd(q_k, r_{k+j}) = g_k \ \text{with} \ \deg(g_k) > 0 \right\}$$

is called the *dispersion* of q_k and r_k (note that this notion is *not* symmetric!). Furthermore, the set

$$J := \left\{ j \in \mathbb{N}_{\geq 0} \ | \ \gcd(q_k, r_{k+j}) = g_k \ \text{with} \ \deg(g_k) > 0 \right\}$$

is called the *dispersion set* of q_k and r_k. Note that the dispersion of two polynomials gives information about their *shift structure* (answering the question: Does an integer shift generate a common divisor?).

---

[4] For the sake of completeness let the maximum of the empty set equal $-\infty$.

The next algorithm shows how to find the dispersion set J. The reader who is familiar with polynomial *resultants* (see e.g. [DST88], Appendix, [GCL92, Chap. 7], [Koepf06, Sect. 7.5]) might observe that J is the set of nonnegative integer roots of the resultant of q_k and r_{k+j} with respect to k.

With Maple, this set can be found roughly using

```
dispersionset:=proc(q,r,k)
  isolve(resultant(q,subs(k=k+j,r),k),j)
end proc:
```

(which calculates all integer roots, not just the nonnegative ones, though). This is a short and elegant description. Note, however, that this way of calculating the dispersion set is not very efficient, particularly because the variable j that is used makes the problem a two-variable problem even though the dispersion set only contains information about polynomials of one variable. Further, the resultant of two polynomials q_k, and r_{k+j} of degrees m and n with respect to k, respectively, is a polynomial of degree mn with respect to j.

Since the dispersion set can be calculated more efficiently using *rational factorization*, this approach will be dealt with here. It is an important fact that rational factorization can be done algorithmically (see e.g. [Zippel93, GCL92, Chap. 8]). Moreover, computer algebra systems like Maple contain (for our purposes) rather efficient implementations for the rational factorization of polynomials, which need about the same time as is needed to find all integer roots of a polynomial. Since the resultant computation generates a polynomial (in j) of much higher degree, the calculation of its integer roots is typically more time consuming than the factorization of the original polynomials q_k, and r_{k+j}.

Therefore the following algorithm, first implemented by Koornwinder in his Maple package `zeilb` [Koornwinder93] and described in an article by Man and Wright [MW94], makes the entire resultant computation superfluous, and gives a fairly efficient method to obtain the dispersion set.

Algorithm 5.2 (`dispersionset`) The following algorithm calculates the dispersion set of two polynomials q_k and r_k:

1. Input: Two polynomials q_k and $r_k \in \mathbb{Q}[k]$.
2. Factorize q_k and r_k over $\mathbb{Q}$. (This is exactly what Maple's `factor` command does.)
3. Set $J := \{\}$. For each pair of factors s_k of q_k, and t_k of r_k calculate $D :=$ `primedispersion(s,t,k)` by the following steps:

   (a) If the degrees m of s_k, and n of t_k are different, return $D := \{\}$; exit.
   (b) Calculate the coefficients $a :=$ `coeff(s,k,n)`, $b :=$ `coeff(s,k, n-1)`, $c :=$ `coeff(t,k,n)`, and $d :=$ `coeff(t,k,n-1)`.

(c) If $j := \frac{bc-ad}{acn}$ is not a nonnegative integer, then return $D := \{\}$; exit.

(d) Check whether $cs_k - at_{k+j} \equiv 0$. If this is the case, then set $D := \{j\}$, else set $D := \{\}$. Return D.

$J := J \cup D$.

4. Output: J.

Proof First, we show that the dispersion of two polynomials q_k and r_k which are relatively prime (over the rationals) is given by the subroutine `primedispersion (q,r,k)`.

Assume that q_k, and r_k have dispersion $j \geq 0$. We propose that in this case q_k and r_k must have the same degree and their ratio must be constant, since the relation

$$\gcd(q_k, r_{k+j}) = g_k \neq 1,$$

implies that q_k has the rational factor g_k, and r_k has the rational factor g_{k-j}. Therefore, since by assumption q_k and r_k are relatively prime, the degree of g_k must be equal to the (common) degree n of q_k and r_k, proving our assertion.

Hence, the two polynomials

$$q_k = ak^n + bk^{n-1} + \cdots ,$$

and

$$r_k = ck^n + dk^{n-1} + \cdots$$

have dispersion $j \in \mathbb{N}_{\geq 0}$ if and only if

$$\frac{c}{a} q_k \equiv r_{k+j} = c(k+j)^n + d(k+j)^{n-1} + \cdots = ck^n + (cnj + d) k^{n-1} + \cdots \quad (5.10)$$

using the binomial expansion. The resulting identity can be valid only if the coefficients of k^{n-1} on both sides match, implying

$$\frac{bc}{a} = cnj + d, \qquad \text{or} \qquad j = \frac{bc - ad}{acn}. \qquad (5.11)$$

Therefore j, given by (5.11), must be a nonnegative integer. In the affirmative case, r_{k+j} is completely known, and (5.10) can be checked.

To calculate the complete dispersion set of q_k, and r_k, the algorithm uses the subroutine `primedispersion` for any pair of rational factors of q_k, and r_k, and all dispersion entries found are collected in the set J. $\qquad \square$

Session 5.3 We implement the dispersion calculation described in Algorithm 5.2 in Maple. The first procedure

```
primedispersion:=proc(q,r,k)
local f,g,n,a,b,c,d,j;
```

```
f:=collect(q,k);
g:=collect(r,k);
n:=degree(f,k);
if n=0 or n<>degree(g,k) then return {} end if;
a:=coeff(f,k,n);
b:=coeff(f,k,n-1);
c:=coeff(g,k,n);
d:=coeff(g,k,n-1);
j:=normal((b*c-a*d)/(a*c*n));
if not type(j,nonnegint) then return {} end if;
if collect(c*f-a*subs(k=k+j,g),k)=0 then
  return {j}
else
  return {};
end if;
end proc: # primedispersion
```

calculates the dispersion of two polynomials of the same degree, which do not have proper rational factors. Therefore the procedure

```
dispersionset:=proc(q,r,k)
local f,g,i,j,result,tmp,op1,op2;
f:=map2(op,1,op(2,factors(q)));
g:=map2(op,1,op(2,factors(r)));
result:={};
for op1 in f do
  for op2 in g do
    tmp:=primedispersion(op1,op2,k);
    if tmp<>{} then result:=result union {op(1,tmp)}
    end if;
  end do;
end do;
return result;
end proc: # dispersionset
```

completes the implementation of Algorithm 5.2.

Next we do an example computation to show the different timings between the resultant-based and the resultant-free algorithms. We give two polynomials in factored form so that we know the dispersion set in advance, and expand them. Read off the dispersion set by a brief look at the input before looking for Maple's answer!

```
>   q:=expand(k*(3*k^2+a)*(k-2536)*(k^3+a));
```

$$3\,k^7 - 2533\,a\,k^4 - 7608\,k^6 - 7608\,k^3\,a + a\,k^5 + a^2\,k^2 - 2536\,a^2\,k$$

```
>   r:=expand(subs(k=k-345,q));
```

$$258519719195803125\,k - 40460071839750\,a + 484174775025\,a\,k$$
$$+ 4678182\,k^3\,a - 2211703920\,k^2\,a - 4258\,a\,k^4 - 3226\,a^2\,k$$
$$+ 993945\,a^2 + a\,k^5 + a^2\,k^2 - 17894813625\,k^4$$
$$- 19246510771\,59375\,k^2 + 7735770995625\,k^3 - 14853\,k^6$$
$$+ 23247135\,k^5 + 3\,k^7 - 145740106798\,29421875$$

```
> TIME:=time(): dispersionset(q,r,k); time()-TIME;
```
$$\{345,\ 2881\}$$
$$0.015$$
```
> TIME:=time(): res:=resultant(q,subs(k=k+j,r),k):
> time()-TIME;
```
$$0.281$$
```
> TIME:=time(): isolve(res): time()-TIME;
```
$$0.187$$
```
> degree(res,j);
```
$$49$$

Here we see that the resultant-free calculation finds the two shifts quite efficiently, whereas the resultant computation is much more time consuming, particularly since several variables are involved. The computation returns a polynomial of degree 49 in j. Have a look at it, and observe its huge coefficients! It is not surprising that the calculation of its integer roots is also rather time consuming: This computation alone needs much longer than that of the entire resultant-free dispersion computation!

The next step in Gosper's proof is to check that f_k, defined by (5.6), is a polynomial.

Lemma 5.4 *The function f_k, defined by (5.6–5.8), is a polynomial.*

Proof Since we know that f_k is rational, we may assume

$$f_k = \frac{c_k}{d_k}$$

with polynomials c_k, d_k such that d_k has positive degree and

$$\gcd\,(c_k, d_k) = 1 = \gcd\,(c_{k-1}, d_{k-1}). \tag{5.12}$$

Then after multiplication by $d_k\,d_{k-1}$ the recurrence Eq. (5.7) gives the main equation

$$d_k\,d_{k-1}\,p_k = d_k\,d_{k-1}(q_{k+1}\,f_k - r_k\,f_{k-1}) = c_k\,d_{k-1}\,q_{k+1} - c_{k-1}\,d_k\,r_k. \tag{5.13}$$

Now let $j \geq 0$ be the dispersion of d_k with itself, i.e. the largest integer such that

$$\gcd\,(d_k, d_{k+j}) = g_k \neq 1. \tag{5.14}$$

Since j is maximal, and d_{k+j} is a multiple of g_k, we have

$$\gcd\,(d_{k-1}, d_{k+j}) = 1 = \gcd\,(d_{k-1}, g_k). \tag{5.15}$$

Therefore, shifting k by $-(j+1)$ in (5.14), we get

$$\gcd\,(d_{k-(j+1)}, d_{k-1}) = g_{k-(j+1)} \neq 1, \tag{5.16}$$

and, shifting k by $-j$ in the left side of (5.15), we get

$$\gcd\,(d_{k-(j+1)}, d_k) = 1 = \gcd\,(g_{k-(j+1)}, d_k) \tag{5.17}$$

since $d_{k-(j+1)}$ is a multiple of $g_{k-(j+1)}$.

Now, we take a careful look at the main Eq. (5.13). We divide this equation by g_k and get

$$\frac{d_k\,d_{k-1}\,p_k}{g_k} = \frac{c_k\,d_{k-1}\,q_{k+1}}{g_k} - \frac{c_{k-1}\,d_k\,r_k}{g_k}. \tag{5.18}$$

Since, by (5.14), d_k is divisible by g_k, the left-hand side of (5.18) is a polynomial. The same is true for the rightmost term, so that it follows that $c_k\,d_{k-1}\,q_{k+1}$ is divisible by g_k. By (5.15), d_{k-1} and g_k are relatively prime. By (5.14), d_k is divisible by g_k, so that, by (5.12), c_k and g_k are relatively prime, too. Therefore q_{k+1} must be divisible by g_k, and hence q_k is divisible by g_{k-1}.

Next, we divide (5.13) by $g_{k-(j+1)}$. We get

$$\frac{d_k\,d_{k-1}\,p_k}{g_{k-(j+1)}} = \frac{c_k\,d_{k-1}\,q_{k+1}}{g_{k-(j+1)}} - \frac{c_{k-1}\,d_k\,r_k}{g_{k-(j+1)}}. \tag{5.19}$$

Since, by (5.16), d_{k-1} is divisible by $g_{k-(j+1)}$, the left-hand side of (5.19) is a polynomial. The same is true for the middle term. It follows that $c_{k-1}\,d_k\,r_k$ is divisible by $g_{k-(j+1)}$. By (5.17), d_k and $g_{k-(j+1)}$ are relatively prime. By (5.16), d_{k-1} is divisible by $g_{k-(j+1)}$, so that, by (5.12) c_{k-1} and $g_{k-(j+1)}$ are relatively prime, too. Therefore r_k must be divisible by $g_{k-(j+1)}$, and hence r_{k+j} is divisible by g_{k-1}.

So we have finally found that both q_k and r_{k+j} are divisible by g_{k-1}. This contradicts the main condition (5.8), however, and therefore f_k must be a polynomial.

$\square$

The last step of Gosper's algorithm gives an *a priori bound* for the degree of f_k. Remember, as soon as such a bound is known in advance, a generic polynomial can be substituted into the main Eq. (5.7), and f_k can be calculated by equating coefficients. If in the following lemma the degree bound is negative or if linear algebra does not provide a suitable f_k, then we can deduce that such an f_k does not exist.

Lemma 5.5 and Algorithm *An upper bound for the degree of f_k is determined by the following algorithm.*

1. *If* $\deg(q_{k+1} + r_k) \leq \deg(q_{k+1} - r_k)$, *then*

$$\deg f_k = \deg p_k - \deg(q_{k+1} - r_k).$$

2. *If* $n := \deg(q_{k+1} + r_k) > \deg(q_{k+1} - r_k)$, *then let a denote the coefficient of* k^n *in the polynomial* $q_{k+1} + r_k$, *and b denote the coefficient of* k^{n-1} *in* $q_{k+1} - r_k$.

    (a) *If* $-2b/a$ *is not a nonnegative integer, then*

$$\deg f_k = \deg p_k - n + 1.$$

    (b) *If* $-2b/a$ *is a nonnegative integer, then*

$$\deg f_k \leq \max\{-2b/a, \deg p_k - n + 1\}.$$

Proof We rewrite the main Eq. (5.7) as

$$p_k = (q_{k+1} - r_k)\frac{f_k + f_{k-1}}{2} + (q_{k+1} + r_k)\frac{f_k - f_{k-1}}{2}. \tag{5.20}$$

Observe that for any polynomial $f_k \neq 0$, one has the identity

$$\deg(f_k - f_{k-1}) = \deg(f_k + f_{k-1}) - 1, \tag{5.21}$$

since the highest order term in $f_k - f_{k-1}$ cancels, whereas the highest order term of $f_k + f_{k-1}$ is not equal to 0.[5]

Therefore, if $\deg(q_{k+1} + r_k) \leq \deg(q_{k+1} - r_k)$, then the second summand in (5.20) is definitely of lower degree than the first one. This gives 1.

If $\deg(q_{k+1} + r_k) > \deg(q_{k+1} - r_k)$, then the situation is a little bit more complicated. Let m be the degree of f_k, and $f_k = ck^m + \cdots$. Then, by (5.20),

$$p_k = \left(b + a\frac{m}{2}\right) c\, k^{n+m-1} + \text{ terms of lower order,}$$

proving 2.   □

This finishes the proof of Gosper's algorithm.

Next, we give some example classes to which Gosper's algorithm can be applied.

Example 5.6 (Polynomials) Any polynomial a_k is Gosper-summable, since any polynomial a_k has a polynomial antidifference s_k (proof by induction!). Assuming we do not cancel common factors in a_{k+1}/a_k (compare with Exercise 5.16), then the application of Gosper's algorithm leads to the initial choice $p_k = 1$, $q_k = a_k$ and $r_k = a_{k-1}$. Therefore, the dispersion set contains the number 1, so that the rewriting

---

[5] For convenience, we set $\deg(0) := -1$. Then (5.21) remains valid for non-zero constant f_k.

according to Lemma 5.1 leads to the final choice $p_k = a_k$, and $q_k = r_k = 1$, satisfying the conditions of Lemma 5.1. Hence, the main equation is equivalent to the difference equation

$$a_k = f_k - f_{k-1} \qquad (5.22)$$

for f_k, and, according to (5.6), we also have $f_k = s_{k+1}$. Since $q_{k+1} - r_k \equiv 0$, case (2b) of Lemma 5.5 applies, and the degree bound turns out to be equal to $\deg p_k + 1 = \deg a_k + 1$. Note that by (5.21) this can be directly seen from (5.22). So, in the final step, the coefficients of s_k are determined by solving a linear system with $\deg a_k + 2$ variables.

Note, however, that for polynomials there exist faster algorithms than Gosper's to deduce their antidifference; see Exercise 5.1.

Example 5.7 (Rational Functions) Not every rational function a_k is Gosper-summable. Indeed, Gosper's algorithm *proves* that $a_k = 1/k$ does not have a hypergeometric term antidifference: We get the representation $p_k = 1$, $q_k = k - 1$ and $r_k = k$ satisfying the conditions of Lemma 5.1, and the degree bound corresponding to Lemma 5.5 is equal to zero. Therefore, we have the setup $f_k = c$ for some constant c, and the main equation reads

$$1 = ck - ck = 0,$$

so that obviously no solution exists.

Therefore the *harmonic numbers*

$$H_k = \sum_{j=1}^{k} \frac{1}{j}$$

do *not* constitute a hypergeometric term. Note that H_{k-1} is a discrete antidifference of a_k since obviously $H_k - H_{k-1} = \frac{1}{k}$.

If, on the other hand, some rational a_k has a rational antidifference s_k, then it is Gosper-summable, and the application of Gosper's algorithm is a safe way to determine s_k. As an example, we consider

$$a_k = \frac{1}{k} - \frac{1}{k+1} = \frac{1}{k(k+1)}.$$

Obviously, s_k can be determined by telescoping here (do it!). Instead, we use Gosper's algorithm: In the present case the initial choice of $p_k = 1$, $q_k = k - 1$ and $r_k = k + 1$ is final. Again case (2b) of Lemma 5.5 applies and gives the degree bound 1 for f_k. With the generic polynomial $f_k = a + bk$, the main equation reads

$$1 = k(a + bk) - (k + 1)(a + b(k - 1)) = b - a.$$

This generates a solution space of dimension 1, and we may set $b = 0$ (to make the order of f_k the lowest possible). Equating coefficients, we get $a = -1$.

Therefore $f_k = -1$. By (5.6), we finally have

$$s_k = \frac{r_k}{p_k}\, f_{k-1} a_k = -(k+1)\, \frac{1}{k(k+1)} = -\frac{1}{k}.$$

Example 5.8 (Binomial Theorem) We try now to see whether the expression $a_k := \binom{n}{k}$ is Gosper-summable. In the affirmative case, this would imply a hypergeometric term formula for

$$\sum_{k=0}^{m} \binom{n}{k}$$

for arbitrary m. Such a formula would extend the binomial identity (2.1) which is the special case when the upper bound $m = n$ is the natural one.

We get

$$\frac{a_{k+1}}{a_k} = \frac{n-k}{k+1},$$

and therefore the initial choice $p_k = 1$, $q_k = n - k + 1$, and $r_k = k$, which is final.

In Lemma 5.5, case (1) applies, and the degree bound yields the value -1. Hence no polynomial f_k satisfying the main equation can exist, and therefore a_k is not Gosper-summable.

Now we change the problem a little, and consider the alternating sum with $a_k := (-1)^k \binom{n}{k}$. Note that in Example 3.4, we discovered the formula

$$\sum_{k=0}^{m} a_k = \sum_{k=0}^{m} (-1)^k \binom{n}{k} = (-1)^m \frac{n-m}{n} \binom{n}{m} \tag{5.23}$$

by other means, which shows that a_k is indeed Gosper-summable. But how does Gosper's algorithm deduce this result?

Using

$$\frac{a_{k+1}}{a_k} = \frac{k-n}{k+1},$$

we have $p_k = 1$, $q_k = k - n - 1$, and $r_k = k$. In the current situation, Lemma 5.5 leads to the case (2a) and the degree bound 0 for f_k. Substituting $f_k = c$ into the main equation leads to the identity

$$1 = (k-n)\, c - k\, c = -nc,$$

with the solution $c = -1/n$, so that $f_k = -1/n$. Finally, by (5.6), we have

$$s_k = \frac{r_k}{p_k}\, f_{k-1}\, a_k = -\frac{k}{n}\, a_k = -\frac{k}{n}(-1)^k \binom{n}{k}$$

in agreement with (5.23) (check!). $\triangle$

We summarize the results of this chapter in:

Algorithm 5.9 (*Gosper*) Given a_k, the following algorithm decides whether there is a hypergeometric term antidifference s_k, and returns it in the affirmative case:

1. Input: $a_k \neq 0$, a hypergeometric term.
2. Calculate the term ratio a_{k+1}/a_k. Use Algorithm 2.2 to find $u_k, v_k \in \mathbb{Q}[k]$ for which
$$\frac{a_{k+1}}{a_k} = \frac{u_k}{v_k}.$$

3. Calculate p_k, q_k and $r_k \in \mathbb{Q}[k]$ with the property (5.8) by Algorithm 5.1.
4. Use Algorithm 5.5 to determine the degree bound M for the polynomial $f_k \in \mathbb{Q}[k]$. If $M < 0$, then return "no hypergeometric term solution exists"; exit.
5. Substitute the generic polynomial
$$f_k = b_0 + b_1 k + b_2 k^2 + \cdots + b_M k^M$$

in the functional equation

$$p_k = q_{k+1}\, f_k - r_k\, f_{k-1}$$

for f_k, equate coefficients, and solve the resulting linear system for the unknowns b_l $(l = 0, \ldots, M)$.
6. If there is no solution, then return "no hypergeometric term solution exists"; exit.
7. Output: $s_k = \frac{r_k}{p_k}\, f_{k-1}\, a_k$. $\square$

Gosper's algorithm shows in particular that whenever a_k possesses a hypergeometric term antidifference s_k then, by (5.6), s_k has to be a rational multiple of a_k:

$$s_k = R_k\, a_k \quad \text{with} \quad R_k = \frac{r_k}{p_k}\, f_{k-1}.$$

We call R_k the *rational certificate* of the hypergeometric term a_k: Given R_k, it is easy to check by pure rational arithmetic (which is fast) whether $s_k = R_k\, a_k$ is an antidifference of a_k by simply checking that

$$s_{k+1} - s_k = R_{k+1}\, a_{k+1} - R_k\, a_k = a_k$$

or equivalently

$$\frac{R_k + 1}{R_{k+1}} = \frac{a_{k+1}}{a_k} \tag{5.24}$$

without knowing where the information about R_k comes from. Therefore for both the application of Gosper's algorithm and for a fast proof of its result, given R_k, (besides polynomial arithmetic) it is only necessary to decide the rationality of a_{k+1}/a_k which is done by Algorithm 2.2.

Finally, we consider the following question: Assume you are given a rational function R_k which is known to be the rational certificate of some hypergeometric term a_k. Is a_k then uniquely determined by R_k? It is clear that for a multiple $\tilde{a}_k = c a_k$ of a_k we get the same R_k. The following corollary states that the converse is also true.

Corollary 5.10 *Let the rational certificate R_k of a hypergeometric term a_k be given. Then a_k is uniquely determined by R_k up to a constant factor.*

Proof This follows immediately from (5.24): Given R_k, (5.24) is a first order recurrence equation for a_k, determining this hypergeometric term uniquely up to a multiplicative constant. $\qquad\square$

Session 5.11 Gosper's algorithm is implemented in Maple and is used as a subroutine of the `sum` command.[6] By the assignment `infolevel[SumTools]:=5`, intermediate information is given. We get for the polynomial $a_k = k^3$

```
>    infolevel[SumTools]:=5:

>    s:=SumTools[Hypergeometric][Gosper](k^3,k);

Gosper: Step 1 of Gosper's algorithm

Gosper: Step 2 of Gosper's algorithm
PolynomialNormalForm: construct the polynomial normal form
Gosper: Step 3 of Gosper's algorithm
Gosper: find non-zero polynomial solution
Gosper: upper bound 4
Gosper: size of the system: 4 equations, 5 unknowns
```

$$\frac{1}{4} k^2 (1 - 2k + k^2)$$

The result can be verified by computing $s_{k+1} - s_k$:

```
>    expand(eval(s,k=k+1)-s);
```

$$k^3$$

We reproduce the results of Examples 5.7 and 5.8.

```
>    SumTools[Hypergeometric][Gosper](1/k,k);

Gosper: Step 1 of Gosper's algorithm
Gosper: Step 2 of Gosper's algorithm
PolynomialNormalForm: construct the polynomial normal form
```

---

[6] The current author did the implementation of Version V.4 which is available in the `sumtools` package. However, this version is now superseded by the `SumTools` package.

```
Gosper: Step 3 of Gosper's algorithm
Gosper: find non-zero polynomial solution
Gosper: upper bound 0
Gosper: size of the system: 1 equations, 1 unknowns

Error, (in SumTools:-Hypergeometric:-Gosper) no solution found

>   SumTools[Hypergeometric][Gosper](binomial(n,k),k);

Gosper: Step 1 of Gosper's algorithm
Gosper: Step 2 of Gosper's algorithm
PolynomialNormalForm: construct the polynomial normal form

Gosper: Step 3 of Gosper's algorithm
Gosper: find non-zero polynomial solution

Error, (in SumTools:-Hypergeometric:-Gosper) no solution found

>   SumTools[Hypergeometric][Gosper]((-1)^k*

>   binomial(n,k),k);
```

$$0.281$$

```
Gosper: Step 1 of Gosper's algorithm

Gosper: Step 2 of Gosper's algorithm
PolynomialNormalForm: construct the polynomial normal form
Gosper: Step 3 of Gosper's algorithm
Gosper: find non-zero polynomial solution
Gosper: upper bound 0
Gosper: size of the system: 1 equations, 1 unknowns
```

$$-\frac{k\,(-1)^k\,\mathrm{binomial}(n,\ k)}{n}$$

We give a final example:

```
>   SumTools[Hypergeometric][Gosper](k*k!,k);

Gosper: Step 1 of Gosper's algorithm

Gosper: Step 2 of Gosper's algorithm
PolynomialNormalForm: construct the polynomial normal form
Gosper: Step 3 of Gosper's algorithm
Gosper: find non-zero polynomial solution
Gosper: upper bound 0
Gosper: size of the system: 1 equations, 1 unknowns
```

$$k!$$

The hsum package also contains a procedure gosper which is decomposed by the
algorithms dispersionset and degreebound, compare Exercises 5.3–5.6.

We would like to remark that many more summation algorithms have been published.
Karr [Karr81] gave a difference field analogue to Risch integration, and hence considered summation in a rather general framework, see also [Schneider01, Schneider04, Schneider08].

Paule [Paule95] presented an algorithm for indefinite hypergeometric summation
(slightly different from Gosper's) based on *greatest factorial factorization* which

extends to more general cases. Paule's approach can be regarded as a deduction of Gosper's algorithm. Böing ([Böing98, Lemma 3.2]) pointed out how the dispersion arises quite naturally when searching for a hypergeometric type antidifference; see, e.g., [Koepf06]. Furthermore, [Koepf06] contains a simplified variant of Gosper's algorithm which abandons the rewriting procedure of Lemma 5.1.

Abramov, Bronstein, and Petkovšek [ABP95] presented a method of obtaining f_k as a solution of the recurrence Eq. (5.7) by solving a linear system whose size is not proportional to the degree of f_k, and iterative computations. Hence if the degree bound for f_k is large, this method is advantageous.

Linearization of Gosper's Algorithm

Although the set of hypergeometric terms is closed under multiplication and taking reciprocals, it is not closed under addition; e.g. $2^k + 1$ is not a hypergeometric term although 2^k and 1 are.

As a result, Gosper's algorithm is nonlinear, i.e., if a_k and b_k are Gosper-summable, then $a_k + b_k$ is not necessarily Gosper-summable. Moreover, $a_k + b_k$ might be Gosper-summable, although the individual summands a_k and b_k are not; e.g., for $a_k = \frac{1}{k+1}$ and $b_k = -\frac{1}{k}$. These are important disadvantages.

Given an arbitrary linear combination of hypergeometric terms one would like to decide whether or not there is an antidifference of the same type. Petkovšek, Wilf and Zeilberger [PWZ96] found a way to *linearize* Gosper's algorithm in this way.

The essential tool is to consider *similarity* under hypergeometric terms. Two hypergeometric terms a_k and b_k are called *similar* if their ratio $a_k/b_k \in \mathbb{Q}(k)$ is a rational function of k. This notion divides the family of hypergeometric terms into equivalence classes one of which consists of the rational functions.

Now, if a linear combination of hypergeometric terms is given, it can be written as a linear combination of pairwise dissimilar ones. Petkovšek, Wilf and Zeilberger [PWZ96] show that an application of Gosper's algorithm to these dissimilar summands decides whether or not the given linear combination of hypergeometric terms has an antidifference of the same type; this applies if and only if Gosper's algorithm is successful for each of the dissimilar summands.

q-Gosper Algorithm

Just as Gosper's algorithm finds a hypergeometric term antidifference whenever one exists, there is a corresponding q-analogue (see [Koornwinder93], compare [PR97]) which finds a q-hypergeometric term antidifference whenever one exists.

An implementation [BK99] based on Koornwinder's work [Koornwinder93] and the present book is available through the `qsum` package containing the Maple procedure `qgosper(term,q,k)` for this purpose.

The request

```
>   qgosper((-1)^k*q^(k*(k-1)/2)*qbinomial(n,k,q),q,k);
```

e.g., results in the *q*-hypergeometric antidifference

$$\frac{(1-q^k)\,(-1)^k\,q^{(1/2\,k\,(k-1))}\,\text{qbinomial}(n,\,k,\,q)}{q^n-1}.$$

In [Riese96] Gosper's algorithm was generalized to the bibasic case, i.e., to (p,q)-hypergeometric terms that are hypergeometric with respect to two bases p and q, and in [Böing98] to *q*-hypergeometric terms that are hypergeometric with respect to a finite number of bases $q = (q_1, \ldots, q_m)$, see also [BK99]. The qsum package contains an implementation of this algorithm. The following computation gives [GR90], Appendix II, Formula (34):

```
>   qgosper((1-a*p^k*q^k)/(1-a)/c^k*qpochhammer(a,p,k)/
>   qpochhammer(a*p/c,p,k)*qpochhammer(c,q,k)/
>   qpochhammer(q,q,k),[p,q],k=0..n);
```

$$\text{qpochhammer}(c,\,q,\,n+1)\,\text{qpochhammer}(a,\,p,\,n+1)\,(-c+p^{(n+1)}\,a)$$

$$(-1+q^{(n+1)})\,c^{(-n-1)}\Big/\big(\text{qpochhammer}(q,\,q,\,n+1)$$

$$\text{qpochhammer}(\tfrac{a\,p}{c},\,p,\,n+1)\,(-1+a)\,(-1+c)\big)$$

Further Reading

For further reading on Gosper's algorithm see [GKP94, PWZ96, Koepf06], and for the *q*-case [Koornwinder93, PR97] and [BK99].

Exercises

Exercise 5.1 *Let Δ denote the forward difference operator*

$$\Delta s_k = s_{k+1} - s_k$$

and for $m \in \mathbb{Z}$, let

$$k^{\underline{m}} := \frac{k!}{(k-m)!}$$

denote the falling factorial. Note that $k^{\underline{m}} = k(k-1)\cdots(k-m+1)$ for $m \in \mathbb{N}$.
  Show that

$$\Delta k^{\underline{m}} = m \cdot k^{\underline{m-1}},$$

where the difference operator is assumed to operate with respect to the variable k.
Show that therefore

$$\sum_{k=a}^{b-1} k^{\underline{m}} = \frac{k^{\underline{m+1}}}{m+1}\Bigg|_{k=a}^{k=b} = \frac{b^{\underline{m+1}} - a^{\underline{m+1}}}{m+1} \quad for\ m \neq -1.$$

Observe the astonishing analogy to the formulas

$$\frac{d}{dx} x^m = m\,x^{m-1} \quad and \quad \int_a^b x^m\,dx = \frac{b^{m+1} - a^{m+1}}{m+1} \quad (m \neq -1).$$

The above information can be used to find the antidifference for any polynomial without solving a linear system, just by rewriting. How?

Exercise 5.2 *(Summation by Parts) Show the following rule of summation by parts (which is similar to the integration by parts formula):*

$$\sum_{k=a}^{b} u_k\,\Delta v_k = u_k v_k \Big|_{k=a}^{k=b+1} - \sum_{k=a}^{b} v_{k+1}\,\Delta u_k.$$

Use it to find an antidifference of the harmonic numbers

$$H_k = \sum_{j=1}^{k} \frac{1}{j}$$

that we introduced in Example 5.7, in terms of themselves.

$\diamond$ **Exercise 5.3**   *Write a Maple procedure* `update(p,q,r,k)` *that uses* `dispersionset` *to update the functions* (p_k, q_k, r_k) *according to Lemma 5.1.*

$\diamond$ **Exercise 5.4**   *Write a Maple procedure* `degreebound(p,q,r,k)` *determining the degree bound for* f_k *according to Lemma 5.5. Check the function with the examples of this chapter and use it for the later exercises!*

$\diamond$ **Exercise 5.5**   *Write a Maple procedure* `findf(p,q,r,k)` *that solves the linear system to find* f_k, *and returns this polynomial.*

$\diamond$ **Exercise 5.6**   *Write a Maple procedure* `gosper(a,k)` *that applies Gosper's algorithm to* a_k.

Exercise 5.7 *Calculate the antidifference solution for* $a_k = \frac{1}{k(k+1)}$, *see Example 5.7.*

Exercise 5.8 *Find antidifferences for the following sequences.*

(a) $a_k := \frac{1}{k(k+10)}$,

(b) $a_k := \frac{1}{k(k+1)(k+2)\ldots(k+10)}$,

(c) $a_k := \frac{6k+3}{4k^4+8k^3+8k^2+4k+3}$,

(d) $a_k := \frac{2^k(k^3-3k^2-3k-1)}{k^3(k+1)^3}$,

(e) $a_k := (k+n)(k+n)!$,

(f) $a_k := \frac{n\,\Gamma(k+n+2)}{(k+n+1)\Gamma(k+2)\Gamma(n+1)}$,

(g) $a_k := \binom{k}{n}$,

(h) $a_k := \frac{(-1)^k}{\binom{n}{k}}$,

(i) $a_k := \frac{\binom{m}{k}}{\binom{n}{k}}$,

(j) $a_k := k\binom{m-k-1}{m-n-1}$,

(k) $a_k := (k)_n$,

(l) $a_k := (k-n)_n$.

Exercise 5.9 *Check whether the summands of identities (2.3–2.6) are Gosper-summable.*

Exercise 5.10 *Show that if $a_k = t_{k+m} - t_k$ for some hypergeometric term t_k, and $m \in \mathbb{N}$, then a_k is Gosper-summable. Determine the antidifference s_k in terms of t_k.*

Exercise 5.11 *In SIAM Review 36, 1994, Problem 94-2 [OK94], the following question was posed:*
  Determine the infinite sum

$$S = \sum_{n=1}^{\infty} \frac{(-1)^{n+1}(4n+1)(2n-1)!!}{2^n(2n-1)(n+1)!},$$

  where $(2n-1)!! = 1 \cdot 3 \ldots (2n-1)$.
Solve the problem using Gosper's algorithm. For the limit computation, use Stirling's formula

$$\lim_{n\to\infty} \frac{n!}{\sqrt{n}\,(n/e)^n} = \sqrt{2\pi}.$$

Give another proof using the hypergeometric database.

Exercise 5.12 *Are the sequences*

(a) $a_k := \frac{1}{2^{n+1}} \binom{n+1}{k} - \frac{1}{2^n} \binom{n}{k}$,

(b) $a_k := \frac{(n+1)!^2}{(2n+2)!} \binom{n+1}{k}^2 - \frac{n!^2}{(2n)!} \binom{n}{k}^2$

Gosper-summable? Try to give an interpretation with regard to identities (2.1) and (2.3)! Hint: Sum from $k = -\infty$ to $k = \infty$.

Exercise 5.13 *Characterize those hypergeometric terms a_k for which the rational certificate R_k is given by (α constant)*

(a) $R_k = \frac{\alpha}{\alpha-1}$,
(b) $R_k = k$,
(c) $R_k = k^2$,
(d) $R_k = 1/k$,
(e) $R_k = (k-1)/k$,
(f) $R_k = (k+1)/k$.

Exercise 5.14 *Assume, $s_k, \tilde{s}_k$ are the antidifferences of two hypergeometric terms $a_k, \tilde{a}_k$ with rational certificates $R_k, \tilde{R}_k$. Then $\tilde{s}_k = 1/s_k$ implies $\tilde{R}_k = 1 - R_k$.*

Exercise 5.15 *Calculate*

(a) $\displaystyle \sum_{k=0}^{n} \frac{\binom{n}{k}\binom{n+1}{k}}{\binom{2n}{2k}}$,

(b) $\displaystyle \sum_{k=0}^{n} \frac{\binom{2k}{k}\binom{2n-2k}{n-k}}{(n-k+1)(k+1)}$.

Exercise 5.16 *Assume, in the application of Gosper's algorithm to a polynomial a_k (Example 5.6), that a_{k+1}/a_k is reduced. Show that then the dispersion calculation according to Lemma 5.1 "repairs" this, and generates $p_k = a_k$, and $q_k = r_k = 1$ as in the non-reduced case.*

Exercise 5.17 *Show that if f_k in Gosper's algorithm is not uniquely determined, it contains exactly one parameter; this happens if and only if a_k is rational. Hence in general the solution space has dimension 0 or 1.*

Exercise 5.18 *In Example 5.8 it was shown that $a_k := \binom{n}{k}$ does not possess a hypergeometric term antidifference for arbitrary n. Show that, on the other hand, a_k has a hypergeometric term antidifference for fixed negative $n \in \mathbb{Z}$. Determine the degree bound of f_k. How does it depend on the particular value of n?*

Exercise 5.19 *It is well-known that*

$$\sum_{k=1}^{\infty} \frac{1}{k^2} = \frac{\pi^2}{6}.$$

From this result it follows that $1/k^2$ does not have a hypergeometric term antidifference. Why? Hint: The number π^2 is irrational.

Exercise 5.20 *The expression*

$$a_k := \frac{\prod\limits_{j=1}^{k-1} \left(bj^2 + cj + d\right)}{\prod\limits_{j=1}^{k} \left(bj^2 + cj + e\right)}$$

is Gosper-summable [Gosper78]. Try your Gosper implementation of Exercise 5.6 on this input. Does it work? If it does not, why not? If necessary, modify your implementation to work for this example.[7]

Exercise 5.21 *In SIAM Review 38, 1996, Problem 96-16 [IR96], the following question was posed:*
    Define

$$S_n(p) = \sum_{j=0}^{n} \left\{ \binom{pn+p+1}{pj+p-1} - \binom{pn+p+1}{pj+p-2} \right\}$$

    for integers $n \geq 0$ and $p \geq 1$.
    Evaluate $S_n(p)$ for $p = 1, 2, 3, 4, 5, 6$.
    Solve this problem for $p = 1$ and $p = 2$ using Gosper's algorithm. For a solution for $p = 3, 4, 5, 6$ see Exercise 7.34.

Exercise 5.22 *For the following expressions s_k, construct $a_k := s_{k+1} - s_k$ $(a_k := s_{k+2} - s_k)$, and apply the q-analogue of Gosper's algorithm to find the antidifference of a_k. Verify the results.*

(a) $s_k := (a; q)_k$,

(b) $s_k := \begin{bmatrix} n \\ k \end{bmatrix}_q$,

(c) $s_k := [k]_q$.

Exercise 5.23 *Find a q-hypergeometric antidifference (or just a hypergeometric antidifference) for $a_k := q^{jk}$ where j is assumed to be any integer. Then replace j iteratively by $1, 2, 3, \ldots$, and apply the q-Gosper algorithm. Describe what happens and explain!*

---

[7] It might be inefficient to use the modified implementation in the general case, though.

Exercise 5.24 *Find a q-hypergeometric antidifference for*

(a) $a_k := q^{jk}\,(n;\,q)_k \quad (j = 1, \ldots, 5)$,

(b) $a_k := \dfrac{q^{k+j}-q^k}{(q^{k+j}-1)(q^k-1)} \quad (j = 1, 2, 3)$,

(c) $a_k := (-1)^k\,q^{\binom{k}{2}} \begin{bmatrix} n \\ k \end{bmatrix}_q$,

(d) $a_k := q^{jk}\,[k]_q \quad (j = 1, \ldots, 5)$,

(e) $a_k := q^{\binom{k}{2}}\,[k]_q$,

(f) $a_k := q^{-\frac{k(k+1)}{2}}\,[k]_q$.

References

ABP95. Abramov, S.A., Bronstein, M., Petkovšek, M.: On polynomial solutions of linear operator equations. In: Proceedings of ISSAC 95, pp. 290–296. ACM Press, New York (1995)

Böing98. Böing, H.: Theorie und Anwendungen zur q-hypergeometrischen Summation. Diploma thesis, Freie Universität Berlin (1998)

BK99. Böing, H., Koepf, W.: Algorithms for q-hypergeometric summation in computer algebra. J. Symbolic Comput. **28**, 777–799 (1999)

Bronstein96. Bronstein, M.: Introduction to Symbolic Integration. Algorithms and Computation in Mathematics. Springer, Berlin (1996)

DST88. Davenport, J.H., Siret, Y., Tournier, E.: Computer-Algebra: Systems and Algorithms for Algebraic Computation. Academic Press, London (1988)

GR90. Gasper, G., Rahman, M.: Basic Hypergeometric Series. Encyclopedia of Mathematics and its Applications, vol. 35, 2nd edn. Cambridge University Press, London and New York (1990). (Second Edition 2004)

GCL92. Geddes, K.O., Czapor, S.R., Labahn, G.: Algorithms for Computer Algebra. Kluwer Academic Publishers, Boston/Dordrecht/London (1992)

Gosper78. Gosper Jr, R.W.: Decision procedure for indefinite hypergeometric summation. Proc. Natl. Acad. Sci. USA **75**, 40–42 (1978)

GKP94. Graham, R.L., Knuth, D.E., Patashnik, O.: Concrete Mathematics, 2nd edn. In: A Foundation for Computer Science. Addison-Wesley, Reading, Massachussets (1994)

IR96. Ierley, G.R., Ruehr, O.G.: Problem 96–16. SIAM Rev. **38**, 668 (1996)

Karr81. Karr, M.: Summation in finite terms. J. ACM **28**, 305–350 (1981)

Koepf95a. Koepf, W.: Algorithms for m-fold hypergeometric summation. J. Symbolic Comput. **20**, 399–417 (1995)

Koepf06. Koepf, W.: Computeralgebra. Springer, Berlin (2006)

Koornwinder93. Koornwinder, T.H.: On Zeilberger's algorithm and its q-analogue. J. Comput. Appl. Math. **48**, 91–111 (1993)

MW94. Man, Y.-K., Wright, F.J.: Fast polynomial dispersion computation and its application to indefinite summation. In: Proceedings of ISSAC 94, pp. 175–180. ACM Press, New York (1994)

OK94. Overhauser, A.W., Kim, Y.I.: Problem 94–2. SIAM Rev. **36**, 107 (1994)

Paule95. Paule, P.: Greatest factorial factorization and symbolic summation. J. Symbolic Comput. **20**, 235–268 (1995)

PR97. Paule, P., Riese, A.: A Mathematica q-analogue of Zeilberger's algorithm based on an algebraically motivated approach to q-hypergeometric telescoping. In: Ismail, M.E.H., et al. (eds.) Fields Institute Communications, vol. 14, pp. 179–210. American Mathematical Society, Providence (1997)

PWZ96. Petkovšek, M., Wilf, H., Zeilberger, D.: $A = B$. A K Peters, Wellesley (1996)

Riese96. Riese, A.: A generalization of Gosper's algorithm to bibasic hypergeometric summation. Electron. J. Comb. 3 (1996)

Risch69. Risch, R.: The problem of integration in finite terms. Trans. Amer. Math. Soc. **139**, 167–189 (1969)

Risch70. Risch, R.: The solution of the problem of integration in finite terms. Bull. Amer. Math. Soc. **76**, 605–608 (1970)

Schneider01. Schneider, C.: Symbolic summation in difference fields. PhD thesis, RISC Linz, Johannes Kepler Universität Linz (2001)

Schneider04. Schneider, C.: Symbolic summation with single-nested sum extensions. In: Proceedings of ISSAC 04, pp. 282–289. ACM Press, New York (2004)

Schneider08. Schneider, C.: A refined difference field theory for symbolic summation. J. Symbolic Comput. **43**, 611–644 (2008)

Zippel93. Zippel, R.: Effective Polynomial Computation. Kluwer Academic Publishers, Dordrecht (1993)

Chapter 6
The Wilf-Zeilberger Method

In this chapter, we study the connection between Gosper's algorithm and definite sums. Firstly, we give a direct application of Gosper's algorithm to definite summation. Throughout this chapter k denotes a summation variable, n denotes a nonnegative integer variable, and $F(n, k)$ denotes a hypergeometric term with respect to both n and k, i.e.

$$\frac{F(n+1, k)}{F(n, k)} \quad \text{and} \quad \frac{F(n, k+1)}{F(n, k)}$$

are rational functions with respect to both n and k. Further, we assume that $F(n, k)$ has *finite support*. The latter means that for any fixed $n \in \mathbb{N}_{\geq 0}$ we have $F(n, k) \neq 0$ only for finitely many $k \in \mathbb{Z}$.

For those summands, we have

Theorem 6.1 *Let $F(n, k)$ be a hypergeometric term with respect to both n and k, Gosper-summable with respect to k with an antidifference $s_k = G(n, k)$ that is finite for all $k \in \mathbb{Z}$. Furthermore, let $F(n, k)$ be well-defined for all $n \in \mathbb{N}_{\geq 0}$, and have finite support. Then*

$$\sum_{k=-\infty}^{\infty} F(n, k) = 0 \tag{6.1}$$

for all but (possibly) finitely many $n \in \mathbb{N}_{\geq 0}$. In detail: If $G(n, k) = R(n, k)\, F(n, k)$ is an antidifference of $F(n, k)$ with respect to k, then (6.1) is valid for all $n \in \mathbb{N}_{\geq 0}$ for which the denominator of $R(n, k)$ is not identical to zero.

Proof By assumption, $F(n, k)$ is Gosper-summable with respect to k, so there is a (hypergeometric term) antidifference $G(n, k)$:

$$F(n, k) = G(n, k+1) - G(n, k).$$

Summing over all k leads to

W. Koepf, *Hypergeometric Summation*, Universitext,
DOI: 10.1007/978-1-4471-6464-7_6, © Springer-Verlag London 2014

$$\sum_{k=-\infty}^{\infty} F(n,k) = \sum_{k=-\infty}^{\infty} \Big(G(n,k+1) - G(n,k)\Big) = 0$$

since the right-hand side is telescoping. Note that because $F(n,k)$ has finite support, the sum under investigation is finite. The only problem which might occur, is that $G(n,k)$ has a singularity at a certain $n \in \mathbb{N}_{\geq 0}$. Since $G(n,k) = R(n,k)\,F(n,k)$ is a rational multiple of $F(n,k)$, the only singularities of $G(n,k)$ are the poles of $R(n,k)$. $\square$

Note that if the denominator of $R(n,k)$ for some integer $k \in \mathbb{Z}$ (depending on n) has a zero, this must be compensated by a zero of $F(n,k)$ since otherwise the antidifference $G(n,k) = R(n,k)\,F(n,k)$ is not finite for this particular k. Therefore, the zeros $n \in \mathbb{N}_{\geq 0}$ of the denominator of $R(n,k)$ that are mentioned in the theorem are those which are independent of k.

Example 6.2 In Example 5.8, we showed that $F(n,k) = (-1)^k \binom{n}{k}$ is Gosper-summable with

$$G(n,k) = -\frac{k}{n}\,F(n,k).\tag{6.2}$$

It follows immediately from Theorem 6.1 that for $n \in \mathbb{N}$ (but not (!) for $n = 0$ since this is a pole of (6.2))

$$\sum_{k=0}^{n}(-1)^k\binom{n}{k} = 0,$$

(compare with (2.2)).

On the other hand, for $n = 0$ we have

$$\sum_{k=0}^{0}(-1)^k\binom{n}{k} = 1.$$

As another example we consider

$$F(n,k) = \frac{\binom{n}{k}\binom{n+1}{k}}{\binom{2n}{2k}}.$$

First we must define $F(n,k)$ for $k > n$ and for $k < 0$ since in this case the given representation is of the form $0/0$ (or ∞/∞ if we replace the binomial coefficients by Γ terms).

In these cases it is better to define $F(n, k)$ so that the formula

$$\frac{F(n, k + 1)}{F(n, k)} = \frac{(n + 1 - k)\,(2\,k + 1)}{(2\,n - 2\,k - 1)\,(k + 1)}$$

for the term ratio remains valid. From the value $F(n, n) = n + 1$, we get then

$$F(n, n + 1) = \left.\frac{F(n, k + 1)}{F(n, k)}\right|_{k=n} \cdot F(n, n) = -\frac{2\,n + 1}{n + 1}(n + 1) = -(2n + 1)$$

and

$$F(n, n + 2) = \left.\frac{F(n, k + 1)}{F(n, k)}\right|_{k=n+1} \cdot F(n, n + 1) = 0$$

so that $F(n, k) = 0$ for $k \geq n + 2$. Similarly, one shows that $F(n, k) = 0$ for $k < 0$. Therefore $F(n, k)$ has finite support.

This fact is taken care of by the computation

```
>    termtohyper(
>    binomial(n,k)*binomial(n+1,k)/binomial(2*n,2*k),k);
```

$$\frac{\operatorname{pochhammer}(-n - 1,\, k)\,(2\,k)!}{4^k\,(k!)^2\,\operatorname{pochhammer}(-n + \frac{1}{2},\, k)}$$

which uses Algorithm 2.8 and shows that we have a sum from $k = 0$ to $k = n + 1$.

It turns out that $F(n, k)$ is Gosper-summable with

$$G(n, k) = \frac{(2\,n - 2\,k + 1)\,k}{n + 1} F(n, k)$$

(check!). Therefore by Theorem 6.1 we have

$$\sum_{k=0}^{n+1} \frac{\binom{n}{k}\binom{n + 1}{k}}{\binom{2n}{2k}} = 0$$

for all $n \in \mathbb{N}_{\geq 0}$; compare Exercise 5.15. $\triangle$

As an immediate consequence of Theorem 6.1, we have

Corollary 6.3 *Let $F(n, k)$ be a hypergeometric term with respect to both n and k, well-defined for all $n \in \mathbb{N}_{\geq 0}$, and having finite support. If $\sum_{k=-\infty}^{\infty} F(n, k)$ can be represented by hypergeometric terms which are not identically zero, then $F(n, k)$ is not finitely Gosper-summable with respect to k.*

Example 6.4 From the identities (2.1)–(2.6), we immediately observe that none of

$$\binom{n}{k}, \quad \binom{n}{k}^2, \quad (-1)^k\binom{n}{k}^2, \quad (-1)^k\binom{n}{k}^3, \quad (-1)^k\binom{n+b}{n+k}\binom{n+c}{c+k}\binom{b+c}{b+k}$$

are Gosper-summable functions according to Corollary 1; compare this with Example 5.8.   △

Note that the corollary shows that results on definite series like (2.1)–(2.6) can *never* be obtained by a *direct* application of Gosper's algorithm!

On the other hand, some very nice examples of an application of Gosper's algorithm are given by the *Wilf-Zeilberger method* for definite summation ([WZ90a, WZ90b]; see also [PWZ96, Koepf95a]).

The Wilf-Zeilberger method (*WZ method* for short) is a clever application of Gosper's algorithm to *prove identities* of the form

$$s_n := \sum_{k=-\infty}^{\infty} F(n,k) = 1 \tag{6.3}$$

for which $F(n,k)$ is a hypergeometric term with respect to both n and k with finite support.

To prove a statement of the form (6.3) by the WZ method, one applies Gosper's algorithm to the expression

$$a_k := F(n+1,k) - F(n,k)$$

(rather than to $F(n,k)$ itself!) with respect to the variable k. If successful, this generates $G(n,k)$ with

$$a_k = F(n+1,k) - F(n,k) = G(n,k+1) - G(n,k), \tag{6.4}$$

and summing over all k leads to

$$s_{n+1} - s_n = \sum_{k=-\infty}^{\infty} \Big(F(n+1,k) - F(n,k)\Big) = \sum_{k=-\infty}^{\infty} \Big(G(n,k+1) - G(n,k)\Big) = 0$$

since the right-hand side is again telescoping. Therefore, s_n is constant, $s_n = s_0$, and if we are able to prove $s_0 = 1$, we are done. Note that $s_0 = 1$ can generally be proved since $F(n,k)$ has finite support, so that no questions concerning convergence arise.

We summarize the above in

Algorithm 6.5 (*Wilf-Zeilberger Method to Prove Hypergeometric Identities*) This method proves the identity (6.3) for $n > N$, where $N \in \mathbb{N}_{\geq 0}$ is determined by step 4.

1. Input: $F(n, k)$, a hypergeometric term with respect to both n and k with finite support (with respect to the summation variable k).
2. Define $a_k := F(n + 1, k) - F(n, k)$.
3. Apply Gosper's algorithm to a_k. If successful, a hypergeometric term antidifference $G(n, k) = \tilde{R}(n, k) a_k$ is generated. If not, the method fails; exit.
4. Let N be the largest nonnegative integer root of the denominator $\tilde{Q} \in \mathbb{Q}[n, k]$ of $\tilde{R}(n, k)$ (found by a rational factorization), i.e. $\tilde{Q}(N, k) \equiv 0$.
5. If $s_{N+1} := \sum_{k=-\infty}^{\infty} F(N + 1, k) = 1$, the procedure has proved (6.3) for $n > N$.
6. This initial value, s_{N+1}, can be evaluated by an application of Algorithm 2.8 which provides the information about suitable natural bounds of the sum.

The method described is very simple, and is a direct application of Gosper's algorithm. It has the disadvantage, however, that we have no prior knowledge about for which input it will work. As a test suite for the capabilities of the given method we use the set of hypergeometric identities found in the book of Bailey [Bailey35] which is reproduced in Table 6.1.

Since the WZ method works only if n is an integer, we can only try to prove the statements of Bailey's list in Table 6.1 if one of the upper parameters of the hypergeometric series involved is assumed to be a negative integer.

Note that the rationality of a_{k+1}/a_k for the WZ method is generally decided by Algorithm 2.2 since

$$\frac{a_{k+1}}{a_k} = \frac{F(n + 1, k + 1) - F(n, k + 1)}{F(n + 1, k) - F(n, k)} = \frac{F(n, k + 1)}{F(n, k)} \cdot \frac{\dfrac{F(n + 1, k + 1)}{F(n, k + 1)} - 1}{\dfrac{F(n + 1, k)}{F(n, k)} - 1}.$$

When Gosper's algorithm generates the function $G(n, k)$, it also finds the rational function

$$R(n, k) := \frac{G(n, k)}{F(n, k)}.$$

$R(n, k)$ is rational since $G(n, k)$ is a rational multiple of $a_k = F(n + 1, k) - F(n, k)$, $G(n, k) = r(n, k) \cdot (F(n + 1, k) - F(n, k))$, say, so that

$$R(n, k) = \frac{G(n, k)}{F(n, k)} = r(n, k)\frac{F(n + 1, k) - F(n, k)}{F(n, k)} = r(n, k)\left(\frac{F(n + 1, k)}{F(n, k)} - 1\right)$$

is also rational. $R(n, k)$ is called the *WZ certificate* of $F(n, k)$. Once the WZ certificate of a hypergeometric expression $F(n, k)$ is known, it is a matter of purely rational arithmetic to decide on the validity of the statement to be proved (6.3). This is quick and easy and can, in principle, be done by hand since one only has to show (6.4) which after division by $F(n, k)$ is equivalent (modulo a possible application of Algorithm 2.2) to the purely rational identity

Table 6.1 Bailey's hypergeometric database

Page	Theorem	Identity
2 3	Chu-V. Gauss	$\displaystyle {}_2F_1\left(\begin{matrix} a,b \\ c \end{matrix}\,\middle|\,1\right) = \frac{(c-b)_{-a}}{(c)_{-a}} = \frac{\Gamma(c)\Gamma(c-a-b)}{\Gamma(c-a)\Gamma(c-b)}$
9	Pfaff-S.	$\displaystyle {}_3F_2\left(\begin{matrix} a,b,-n \\ c,1+a+b-c-n \end{matrix}\,\middle|\,1\right) = \frac{(c-a)_n\,(c-b)_n}{(c)_n\,(c-a-b)_n}$
9	Kummer	$\displaystyle {}_2F_1\left(\begin{matrix} a,b \\ 1+a-b \end{matrix}\,\middle|\,-1\right) = \frac{(1+a)_{-b}}{(1+a/2)_{-b}} = \frac{\Gamma(1+a-b)\Gamma(1+a/2)}{\Gamma(1+a)\Gamma(1+a/2-b)}$
11	Gauss	$\displaystyle {}_2F_1\left(\begin{matrix} a,b \\ (a+b+1)/2 \end{matrix}\,\middle|\,\tfrac{1}{2}\right) = \frac{\Gamma(1/2)\Gamma((a+b+1)/2)}{\Gamma((a+1)/2)\Gamma((b+1)/2)}$
11	Bailey	$\displaystyle {}_2F_1\left(\begin{matrix} a,1-a \\ c \end{matrix}\,\middle|\,\tfrac{1}{2}\right) = \frac{\Gamma(c/2)\Gamma((c+1)/2)}{\Gamma((a+c)/2)\Gamma((1-a+c)/2)}$
13	Dixon	$\displaystyle {}_3F_2\left(\begin{matrix} a,b,c \\ 1+a-b,1+a-c \end{matrix}\,\middle|\,1\right) = \frac{(1+a)_{-c}\,(1+a/2-b)_{-c}}{(1+a/2)_{-c}\,(1+a-b)_{-c}}$ $\displaystyle \qquad = \frac{\Gamma(1+a/2)\Gamma(1+a-b)\Gamma(1+a-c)\Gamma(1+a/2-b-c)}{\Gamma(1+a)\Gamma(1+a/2-b)\Gamma(1+a/2-c)\Gamma(1+a-b-c)}$
16	Watson Whipple	$\displaystyle {}_3F_2\left(\begin{matrix} a,b,c \\ (a+b+1)/2,2c \end{matrix}\,\middle|\,1\right) = \frac{\Gamma(\tfrac{1}{2})\Gamma(\frac{1+2c}{2})\Gamma(\frac{1+a+b}{2})\Gamma(\frac{1-a-b+2c}{2})}{\Gamma(\frac{1+a}{2})\Gamma(\frac{1+b}{2})\Gamma(\frac{1-a+2c}{2})\Gamma(\frac{1-b+2c}{2})}$
16	Whipple	$\displaystyle {}_3F_2\left(\begin{matrix} a,1-a,c \\ e,1+2c-e \end{matrix}\,\middle|\,1\right) = \frac{\pi\,2^{1-2c}\,\Gamma(e)\Gamma(1+2c-e)}{\Gamma(\frac{a+e}{2})\Gamma(\frac{a+1+2c-e}{2})\Gamma(\frac{1-a+e}{2})\Gamma(\frac{2+2c-a-e}{2})}$
26	Dougall	$\displaystyle {}_7F_6\left(\begin{matrix} a,1+a/2,b,c,d,1+2a-b-c-d+n,-n \\ a/2,1+a-b,1+a-c,1+a-d,b+c+d-a-n,1+a+n \end{matrix}\,\middle|\,1\right)$ $\displaystyle \qquad = \frac{(1+a)_n\,(1+a-b-c)_n\,(1+a-b-d)_n\,(1+a-c-d)_n}{(1+a-b)_n(1+a-c)_n\,(1+a-d)_n\,(1+a-b-c-d)_n}$
25	Dougall	$\displaystyle {}_5F_4\left(\begin{matrix} a,1+a/2,c,d,e \\ a/2,1+a-c,1+a-d,1+a-e \end{matrix}\,\middle|\,1\right) = \frac{(1+a)_{-e}(1+a-c-d)_{-e}}{(1+a-c)_{-e}(1+a-d)_{-e}}$
27		$\displaystyle \qquad = \frac{\Gamma(1+a-c)\Gamma(1+a-d)\Gamma(1+a-e)\Gamma(1+a-c-d-e)}{\Gamma(1+a)\Gamma(1+a-d-e)\Gamma(1+a-c-e)\Gamma(1+a-c-d)}$
28	Whipple	$\displaystyle {}_4F_3\left(\begin{matrix} a,1+a/2,d,e \\ a/2,1+a-d,1+a-e \end{matrix}\,\middle|\,-1\right) = \frac{(1+a)_{-e}}{(1+a-d)_{-e}} = \frac{\Gamma(1+a-d)\Gamma(1+a-e)}{\Gamma(1+a)\Gamma(1+a-d-e)}$
30	Bailey	$\displaystyle {}_3F_2\left(\begin{matrix} a,1+a/2,-n \\ a/2,w \end{matrix}\,\middle|\,1\right) = \frac{(w-a-1-n)(w-a)_{n-1}}{(w)_n}$
30	Bailey	$\displaystyle {}_3F_2\left(\begin{matrix} a,b,-n \\ 1+a-b,1+2b-n \end{matrix}\,\middle|\,1\right) = \frac{(a-2b)_n(1+a/2-b)_n(-b)_n}{(1+a-b)_n(a/2-b)_n(-2b)_n}$
30	Bailey	$\displaystyle {}_4F_3\left(\begin{matrix} a,1+a/2,b,-n \\ a/2,1+a-b,1+2b-n \end{matrix}\,\middle|\,1\right) = \frac{(a-2b)_n(-b)_n}{(1+a-b)_n(-2b)_n}$
30	Bailey	$\displaystyle {}_4F_3\left(\begin{matrix} a,1+a/2,b,-n \\ a/2,1+a-b,2+2b-n \end{matrix}\,\middle|\,1\right) = \frac{(a-2b-1)_n(1/2+a/2-b)_n(-b-1)_n}{(1+a-b)_n(a/2-b-1/2)_n(-2b-1)_n}$

$$\frac{F(n+1,k)}{F(n,k)} - 1 + R(n,k) - R(n,k+1)\frac{F(n,k+1)}{F(n,k)} = 0. \qquad (6.5)$$

Example 6.6 (*Binomial Identity*) As an example, to prove the binomial identity (2.1) in the form

$$s_n := \sum_{k=0}^{n} F(n,k) = \sum_{k=0}^{n} \frac{1}{2^n}\binom{n}{k} = 1 \qquad (6.6)$$

by the WZ method, we set

$$a_k := \frac{1}{2^{n+1}}\binom{n+1}{k} - \frac{1}{2^n}\binom{n}{k};$$

compare Example 2.5 and Exercise 5.12. Algorithm 2.2 yields

$$\frac{a_{k+1}}{a_k} = \frac{F(n+1,k+1) - F(n,k+1)}{F(n+1,k) - F(n,k)} = \frac{(n-k+1)(n-2k-1)}{(n+1-2k)(k+1)}.$$

Therefore Gosper's algorithm can be applied (do it!), and results in

$$G(n,k) = \frac{k}{n+1-2k}\left(\frac{1}{2^{n+1}}\binom{n+1}{k} - \frac{1}{2^n}\binom{n}{k}\right).$$

This proves (6.6) since $s_0 = \sum_{k=0}^{0} 1 = 1$. Note that the denominator zero $k = \frac{n+1}{2}$ of the rational factor $\tilde{R}(n,k) = k/(n+1-2k)$, which is an integer for odd values of n, is compensated for by a zero of a_k at this point so that $G(n,k)$ is finite there, thus the argument remains valid.

The WZ certificate is

$$R(n,k) = \frac{G(n,k)}{F(n,k)} = -\frac{k}{2(n+1-k)}$$

(check!), and the proof of identity (6.6) is therefore reduced to simplifying the rational expression

$$\frac{F(n+1,k)}{F(n,k)} - 1 + R(n,k) - R(n,k+1)\frac{F(n,k+1)}{F(n,k)}$$
$$= \frac{n+1}{2(n+1-k)} - 1 - \frac{k}{2(n+1-k)} + \frac{k+1}{2(n-k)}\cdot\frac{n-k}{k+1}$$

to zero, which is trivial.   △

Table 6.2 is a complete list of those identities in Bailey's list (Table 6.1) that can be treated by the given method, together with their rational certificates. Using the latter the reader may easily prove these identities. Observe that therefore Table 6.2 consists of simple *proofs* for all of these statements. You might check any of these by

Table 6.2 The WZ Method

Identity	$nR(n,k)$
Chu-Vandermonde	$-a - \dfrac{(c+k-1)\,k}{(c-b+n)\,(n+1-k)}$
Pfaff-Saalschütz	$n - \dfrac{k\,(c+k-1)\,(c-b+n-a-k)}{(c-b+n)\,(n+1-k)\,(c-a+n)}$
Kummer	$-b - \dfrac{k}{2(n+1-k)}$
Dixon	$-c - \dfrac{(a-b+k)\,k}{(2+a-2b+2n)\,(n+1-k)}$
Watson Whipple	$-c - \dfrac{(2n-k+1)\,k\,(a+b-1+2k)}{2\,(1+b+2n)\,(1+a+2n)\,(n+1-k)}$
Whipple	$-c - \dfrac{(e+k-1)\,k\,(2n+e-k)}{(n+1-k)\,(a+2n+e)\,(-a+1+2n+e)}$
Dougall	n $-\dfrac{(2n+2a+2-c-d-b)\,(1+a-b-c-d+n-k)\,k\,(a-d+k)\,(a-c+k)\,(a-b+k)}{(1+2a-b-c-d+n)\,(1+a-c-d+n)\,(1+a-b-d+n)\,(1+a-b-c+n)\,(a+2k)\,(n+1-k)}$
Dougall	$-e - \dfrac{(a-c+k)\,(a-d+k)\,k}{(1+a-c-d+n)\,(a+2k)\,(n+1-k)}$
Whipple	$-e - \dfrac{k\,(a-d+k)}{(a+2k)\,(n+1-k)}$
Bailey	$n\dfrac{\left(na+a^2-wa+a-2+2nk-2n+4k+2ka-2kw\right)\,(w+k-1)\,k}{(a+2k)\,(n+2+a-w)\,(w-a-1+n)\,(k-n-1)}$
Bailey	$n - \dfrac{\left(ba-2b^2+n+2bn+1-k\right)\,(-2b+n-k)\,k\,(a-b+k)}{(-b+n)\,(2+a-2b+2n)\,b\,(a-2b+n)\,(n+1-k)}$
Bailey	$n\dfrac{(n-ba-k+1-2kb)\,(-2b+n-k)\,k\,(a-b+k)}{(-b+n)\,b\,(a+2k)\,(a-2b+n)\,(n+1-k)}$
Bailey	$n - (2b-ab+2nb+2nab-2n-2n^2+a^2b-2ab^2+4nbk+2abk+2k$ $+4nk-4bk-4b^2k-2k^2)\,k\,(a-b+k)\,(1+2b-n+k)/$ $((n+1-k)\,b\,(a+2k)\,(b+1-n)\,(a-2b-1+n)\,(a-2b+1+2n))$

simply verifying (6.5). Note how simple the rational certificate of Dougall's identity is compared to the much more complicated rational certificate of the last of Bailey's identities and also note the contrast with the complexity of the corresponding inputs.

Note that neither the statements of Gauss and Bailey with argument $x = 1/2$ ([Bailey35], p. 11) can be deduced with respect to any of the parameters involved, nor can Watson's Theorem ([Bailey35], p. 16) be proved by the WZ method with respect to Watson's original integer parameter a, nor can the method be applied to Whipple's Theorem ([Bailey35], p. 16) concerning parameters a or b since, in all these cases, the term ratio a_{k+1}/a_k is not rational.

In particular, the results of Table 6.2 constitute the first proof of our hypergeometric database given in Chap. 2, except for Watson's identity (with respect to the parameters a or b).

Session 6.7 We can use the Gosper implementation `gosper(a,k)` of the `hsum` package which for Gosper-summable a_k returns a hypergeometric term antidifference. We recommend, however, that you write such a Maple procedure on your own; see Exercises 5.3–5.6.

The procedure

```
WZcertificate:=proc(F,k,n)
local a,gos;
a:=subs(n=n+1,F)-F;
try
   gos:=gosper(a,k);
catch:
   error 'WZ method fails';
end try;
return simpcomb(gos/F);
end proc:
```

calculates the rational certificate of the WZ method when applied to $F(n,k)$, if this is applicable. Here, we use Maple's `try`, `catch` and `error` commands to give our own error message in case `gosper` does not return a hypergeometric term antidifference.

The calculation

```
>   WZcertificate(binomial(n,k)/2^n,k,n);
```

$$\frac{1}{2}\frac{k}{-n-1+k}$$

confirms the result of Example 6.5, and

```
>   WZcertificate(binomial(n,k)*x^k/(1+x)^n,k,n);
```

$$\frac{k}{(-n-1+k)(1+x)}$$

yields the binomial theorem.[1] On the other hand, the calculation

```
>   WZcertificate(binomial(n,k),k,n);
    Error, (in WZcertificate) WZ method fails
```

shows that the WZ method is not applicable for $F(n,k) = \binom{n}{k}$. This is not surprising since we know that the sum

$$\sum_{k=-\infty}^{\infty}\binom{n}{k}$$

---

[1] What happens if $x = -1$?

is *not* constant with respect to n.

We prove some of the identities of Chap. 1:

```
>    WZcertificate(binomial(n,k)^2/binomial(2*n,n),k,n);
```

$$\frac{1}{2} \frac{(-3n - 3 + 2k) k^2}{(2n + 1)(-n - 1 + k)^2}$$

```
>    WZcertificate(
>    (-1)^k*binomial(n+b,n+k)*binomial(n+c,c+k)*binomial(b+c,b+k)/
>    (GAMMA(b+c+n+1)/(n!*GAMMA(b+1)*GAMMA(c+1))),k,n);
```

$$\frac{1}{2} \frac{(c + k)(b + k)}{(b + c + n + 1)(-n - 1 + k)}$$

The final statement proves Dixon's identity in binomial form (1.6) where no conversion to hypergeometric form was necessary!

Note that the success of the procedure WZcertificate *proves* that $s_n = \sum_{k=-\infty}^{\infty} F(n, k)$ is constant! You can find this constant by merely considering the initial value $s_0 = \sum_{k=-\infty}^{\infty} F(0, k)$. If $F(n, k)$ has finite support this sum is finite and can thus be evaluated in finite terms since its bounds can be found by Algorithm 2.8.

Here is an example for which the WZ method does not work (compare [PS95]). Although

$$\sum_{k=-\infty}^{\infty} \frac{(-1)^k}{(-3)^n} \binom{n}{k} \binom{3k}{n} = 1$$

is a true statement for $n \in \mathbb{N}_{\geq 0}$ (see Session 7.4 and Exercise 7.14) supported by the calculation

```
>    seq(
>    add((-1)^k/(-3)^n*binomial(n,k)*binomial(3*k,n),k=0..n),
>    n=0..25);
```

$$1, 1, 1, 1, 1, 1, 1, 1, 1, 1, 1, 1, 1, 1, 1, 1, 1, 1, 1, 1, 1, 1, 1, 1, 1, 1$$

the WZ method fails:

```
>    WZcertificate(
>    (-1)^k/(-3)^n*binomial(n,k)*binomial(3*k,n),k,n);
```

```
Error, (in WZcertificate) WZ method fails
```

Note that the main identity (6.4) has a remarkable symmetry. In the WZ approach, (6.4) is then summed with respect to k to obtain a formula for $\sum_{k\in\mathbb{Z}} F(n, k)$. There seems to be no reason why the roles of n and k should not be interchangeable to obtain a formula for $\sum_{n\in\mathbb{Z}} G(n, k)$. Under very mild restrictions this is indeed the case, and the resulting formula is called the *companion identity*. Hence by proving one identity, a new identity is generated. Wilf and Zeilberger ([WZ90a, WZ90b]) also introduced the notion of the *dual identity*, and Gessel [Gessel95] used a related approach to generate new identities from known ones in an almost algorithmic way. More details about these topics can be found in [PWZ96, Chapter II 7].

q-WZ Method

In the previous chapters we have already seen q-hypergeometric identities. As in the hypergeometric case, such identities can be proven by an application of the q-analogue of Gosper's algorithm.

The procedure

```
qWZcertificate:=proc(F,q,k,n)
local a,gos;
a:=subs(n=n+1,F)-F;
try
  gos:=qgosper(a,q,k);
catch:
  error `q-WZ method fails`;
end try;
return qsimpcomb(gos/F);
end proc:
```

which uses the qgosper implementation of the qsum package calculates the certificate, rational in q^n, of the q-WZ method when applied to $F(n, k)$, if this is applicable. For example, the request

```
>   qWZcertificate(
>   qphihyperterm([q^(-n),b],[c],q,c/b*q^n,k)/
>   (qpochhammer(c/b,q,n)/qpochhammer(c,q,n)),q,k,n);
```

results in

$$- \frac{q^n \left(-1 + q^k\right) \left(q - q^k c\right) b}{\left(q^n q - q^k\right) \left(-c q^n + b\right)} \tag{6.7}$$

which gives the proof certificate for the q-Chu-Vandermonde identity (3.7). This proves the q-Chu-Vandermonde identity and at the same time gives a manual proof at your disposal which uses only rational arithmetic; see Exercise 6.9.

Further Reading

For further reading on the WZ method see [Gessel95], [PWZ96], [Wilf05] and [Koepf06].

Exercises

Exercise 6.1 *Prove the identity*

$$\sum_{k=-\infty}^{\infty} \frac{(-1)^k}{\binom{2n+2b+2c+2d}{n+b+c+d+k}} \binom{n+b}{n+k}\binom{b+c}{b+k}\binom{c+d}{c+k}\binom{d+n}{d+k} =$$

$$\frac{\Gamma(n+b+c+d+1)\,\Gamma(n+b+c+1)\,\Gamma(n+b+d+1)\,\Gamma(n+c+d+1)\,\Gamma(b+c+d+1)}{n!\,\Gamma(2n+2b+2c+2d+1)\,\Gamma(n+c+1)\,\Gamma(b+d+1)\,\Gamma(b+1)\,\Gamma(c+1)\,\Gamma(d+1)}$$

(see Exercise 2.17) by the WZ method.

$\diamond$ **Exercise 6.2** *Write a Maple procedure* `checksum(F,R,k,n)` *that checks whether $\sum_{k=-\infty}^{\infty} F(n,k) = 1$, using the rational certificate $R(n,k)$; see (6.5).*

Use this procedure to prove the identities of Bailey's list using the rational certificates of Table 6.2 (and no Gosper implementation!).

Exercise 6.3 *Prove Székely's and Stanley's identity (Example 3.5 and Exercise 3.4) by the WZ method.*

Exercise 6.4 **(Clausen Identity)** *Prove the following hypergeometric identity due to Clausen*

$$_4F_3\left(\begin{matrix} a,b,1/2-a-b-n,-n \\ 1/2+a+b,1-a-n,1-b-n \end{matrix}\middle| 1\right) = \frac{(2a)_n\,(a+b)_n\,(2b)_n}{(2a+2b)_n\,(a)_n\,(b)_n}.$$

Exercise 6.5 *In Corollary 5.10, we saw that the rational certificate of Gosper's algorithm is essentially unique. Show that the same is not true for the rational WZ certificate. Hint: A counterexample can be found in this chapter.*

Exercise 6.6 *Prove the identities of Exercise 3.2 by the WZ method.*

Exercise 6.7 *Try to prove the following identities by the WZ method:*

(a) $\displaystyle\sum_{k=-n}^{n} (-1)^k \binom{2n}{n+k}\binom{2b}{b+k}\binom{2c}{c+k} = \frac{(n+b+c)!\,(2n)!\,(2b)!\,(2c)!}{(n+b)!\,(b+c)!\,(n+c)!\,n!\,b!\,c!},$

(b) $\displaystyle\sum_{k=0}^{m} \binom{n}{m+k}\binom{m+k}{2k} 4^k = \frac{4^m \binom{n}{m}\binom{n-1/2}{m}}{\binom{2m}{m}},$

(c) $\displaystyle\sum_{k=0}^{n} \binom{m-r+s}{k}\binom{n+r-s}{n-k}\binom{r+k}{n+m} = \binom{r}{m}\binom{s}{n},$

(d) $\displaystyle\sum_{k=-n}^{n} (-1)^k \binom{n+a}{n+k}\binom{a+n}{a+k} = \binom{n+a}{n},$

(e) $\displaystyle\sum_{k=1}^{n} k\binom{n}{k}\binom{s}{k} = s\binom{n+s-1}{n-1},$

(f) $\displaystyle\sum_{k=0}^{n} \frac{(-1)^k}{k+x}\binom{n}{k} = \frac{1}{x\binom{n+x}{n}}.$

Specify carefully any restrictions on the parameters occurring!

Exercise 6.8 *Assume that $F(n,k)$ is Gosper-summable with respect to k. Then the Maple procedure* WZcertificate *always finds a rational certificate for the WZ method. Why?*

Exercise 6.9 *Describe in detail how the proof certificate (6.7) can be used to prove the q-Chu-Vandermonde identity by rational arithmetic.*

Exercise 6.10 *Use the q-WZ method to prove the q-binomial theorem (2.22) for $a = q^{-n}$*

$$\sum_{k=0}^{\infty} \frac{(q^{-n}; q)_k}{(q; q)_k} x^k = \frac{(q^{-n}x; q)_{\infty}}{(x; q)_{\infty}} = (q/x; q)_n \, q^{-\binom{n+1}{2}} (-x)^n,$$

the q-Pfaff-Saalschütz identity (3.6)

$$_3\phi_2 \left(\begin{matrix} q^{-n}, a, b \\ c, \frac{ab}{cq^{n-1}} \end{matrix} \middle| \, q, q \right) = \frac{(c/a; q)_n \, (c/b; q)_n}{(c; q)_n \, (c/(ab); q)_n},$$

the q-Kummer identity

$$_2\phi_1 \left(\begin{matrix} a, b \\ aq/b \end{matrix} \middle| \, q, -\frac{q}{b} \right) = \frac{(-q; q)_{\infty} \, (aq, aq^2/b^2; q^2)_{\infty}}{(-q/b, aq/b; q)_{\infty}},$$

([GR90], Appendix (II.9)) for $b = q^{-n}$, the q-Dixon identity

$$_4\phi_3 \left(\begin{matrix} a, -q\sqrt{a}, b, q^{-n} \\ -\sqrt{a}, aq/b, aq^{1+n} \end{matrix} \middle| \, q, \frac{q^{1+n}\sqrt{a}}{b} \right) = \frac{(aq, q\sqrt{a}/b; q)_n}{(q\sqrt{a}, aq/b; q)_n},$$

([GR90], Appendix (II.14)), and Jackson's identity

$$_8\phi_7 \left(\begin{matrix} a, q\sqrt{a}, -q\sqrt{a}, b, c, d, e, q^{-n} \\ \sqrt{a}, -\sqrt{a}, aq/b, aq/c, aq/d, aq/e, aq^{1+n} \end{matrix} \middle| \, q, q \right)$$
$$= \frac{(aq, aq/(bc), aq/(bd), aq/(cd); q)_n}{(aq/b, aq/c, aq/d, aq/(bcd); q)_n},$$

where $a^2q = bcdeq^{-n}$ ([GR90], Appendix (II.22)), which is a q-analogue of Dougall's identity.

References

Bailey35. Bailey, W. N.: Generalized Hypergeometric Series. Cambridge University Press, Cambridge (1935) (reprinted 1964 by Stechert-Hafner Service Agency, New York-London).

GR90. Gasper, G., Rahman, M.: Basic Hypergeometric Series. Encyclopedia of Mathematics and its Applications, 35, Cambridge University Press, London and New York (1990) (Second Edition 2004).

Gessel95. Gessel, I.: Finding identities with the WZ method. J. Symbolic Comput. **20**, 537–566 (1995)

Koepf95a. Koepf, W.: Algorithms for m-fold hypergeometric summation. J. Symbolic Comput. **20**, 399–417 (1995)

Koepf06. Koepf, W.: Computeralgebra. Springer, Berlin (2006)

PS95. Paule, P., Schorn, M.: A Mathematica version of Zeilberger's algorithm for proving binomial coefficient identities. J. Symbolic Comput. **20**, 673–698 (1995)

PWZ96. Petkovšek, M., Wilf, H., Zeilberger, D.: $A = B$. A K Peters, Wellesley (1996)

Wilf05. Wilf, H.S.: Generatingfunctionology, 3rd edn. A K Peters, Wellesley (2005)

WZ90a. Wilf, H.S., Zeilberger, D.: Rational functions certify combinatorial identities. J. Amer. Math. Soc. **3**, 147–158 (1990)

WZ90b. Wilf, H.S., Zeilberger, D.: Towards computerized proofs of identities. Bull. Amer. Math. Soc. **23**, 77–83 (1990)

Chapter 7
Zeilberger's Algorithm

In this chapter, we introduce Zeilberger's extension of Gosper's algorithm ([Zeilberger90b, Zeilberger91a, Zeilberger91b]), using which one can not only prove hypergeometric identities but also sum definite series in many cases, if they represent hypergeometric terms.

Like Fasenmyer, Zeilberger deals with the question of how to determine a holonomic recurrence equation

$$\sum_{j=0}^{J} P_j(n)\, s_{n+j} = 0 \tag{7.1}$$

with polynomials $P_j \in \mathbb{Q}[n]$, for sums

$$s_n := \sum_{k=-\infty}^{\infty} F(n,k) \tag{7.2}$$

where $F(n,k)$ is a hypergeometric term with respect to both n and k. We assume further that $F(n,k)$ has finite support.

If $F(n,k)$ is Gosper-summable with respect to k, then we get a hypergeometric term form representation for any sum

$$\sum_{k=a}^{b} F(n,k),$$

in particular if a and b are the natural bounds of $F(n,k)$. On the other hand, as we saw in Theorem 6.1, this information is worthless since we get $s_n \equiv 0$ in all these cases. Therefore it is *never* possible to find a nonzero hypergeometric term representation for s_n, by a direct application of Gosper's algorithm.

Zeilberger's idea is to apply Gosper's algorithm in the following non-obvious way: For suitable $J = 1, 2, \ldots$ set

W. Koepf, *Hypergeometric Summation*, Universitext,
DOI: 10.1007/978-1-4471-6464-7_7, © Springer-Verlag London 2014

$$a_k := F(n, k) + \sum_{j=1}^{J} \sigma_j(n)\, F(n + j, k) \tag{7.3}$$

with as yet undetermined variables σ_j depending on n, but not depending on k. Then

$$\frac{a_{k+1}}{a_k} = \frac{F(n, k + 1) + \sum_{j=1}^{J} \sigma_j(n)\, F(n + j, k + 1)}{F(n, k) + \sum_{j=1}^{J} \sigma_j(n)\, F(n + j, k)}$$

$$= \frac{F(n, k + 1)}{F(n, k)} \cdot \frac{1 + \sum_{j=1}^{J} \sigma_j(n)\, \dfrac{F(n + j, k + 1)}{F(n, k + 1)}}{1 + \sum_{j=1}^{J} \sigma_j(n)\, \dfrac{F(n + j, k)}{F(n, k)}} \in \mathbb{Q}(n, k) \tag{7.4}$$

turns out to be rational with respect to k, so Gosper's algorithm is applicable.

The first step of Gosper's algorithm generates three polynomials p_k, q_k and r_k. Note that since the variables σ_j $(j = 1, \ldots, J)$ are unknowns, the dispersion set always contains the value 1, and after the rewriting, the unknowns σ_j occur linearly in p_k. Next, we calculate the degree bound M for f_k. This can be done by a mechanical application of Lemma 5.5, computing over $\mathbb{Q}(n)$, and thus *ignoring* possible values of n for which the bound might be lower, since we search for a bound which is valid for *all n*.

Zeilberger's crucial observation is the following: If we take the appropriate generic polynomial

$$f_k = b_0 + b_1 k + b_2 k^2 + \cdots + b_M k^M,$$

and equate coefficients in the functional Eq. (5.7) for f_k, we get a *linear system* in the coefficients b_l $(l = 0, \ldots, M)$ of f_k *and the unknowns* σ_j $(j = 1, \ldots, J)$. Rather than solving this system for b_l $(l = 0, \ldots, M)$ alone, we may search for a solution for b_l $(l = 0, \ldots, M)$ and σ_j $(j = 1, \ldots, J)$ at the same time!

If this procedure is successful, it provides us with an antidifference $G(n, k)$ of a_k (depending on n), and a set of rational functions $\sigma_j(n) \in \mathbb{Q}(n)$ $(j = 1, \ldots, J)$ such that

$$G(n, k + 1) - G(n, k) = a_k = F(n, k) + \sum_{j=1}^{J} \sigma_j(n)\, F(n + j, k). \tag{7.5}$$

Therefore, by summation

$$\sum_{k=-\infty}^{\infty} a_k = \sum_{k=-\infty}^{\infty} \left(F(n,k) + \sum_{j=1}^{J} \sigma_j(n)\, F(n+j,k) \right)$$

$$= s_n + \sum_{j=1}^{J} \sigma_j(n)\, s_{n+j} = \sum_{k=-\infty}^{\infty} \Big(G(n,k+1) - G(n,k) \Big) = 0,$$

since the right-hand side is a telescoping sum. After multiplication by the common denominator this establishes the desired recurrence Eq. (7.1).

Note that if this procedure is not successful, this does *not* prove that such a holonomic recurrence equation is not valid. Zeilberger's luck is that this happens rather rarely.

Example 7.1 Let's try again to find a holonomic recurrence equation of first order for

$$s_n := \sum_{k=0}^{n} \binom{n}{k}.$$

In Chap. 4, Exercise 4.1, we found such a recurrence equation by Fasenmyer's method. Now, we use Zeilberger's algorithm.

We set

$$F(n,k) := \binom{n}{k}$$

and

$$a_k := F(n,k) + \sigma_1\, F(n+1,k) = \binom{n}{k} + \sigma_1 \binom{n+1}{k}.$$

By Algorithm 2.2, we get

$$\frac{a_{k+1}}{a_k} = \frac{(n+1-k)\,(n-k+\sigma_1 n + \sigma_1)}{(n+1-k+\sigma_1 n + \sigma_1)\,(k+1)}. \tag{7.6}$$

The dispersion calculation shows that we have $p_k = n+1-k+\sigma_1 n + \sigma_1$, $q_k = n+2-k$ and $r_k = k$ (observe the shift of the factors containing σ_1 in the numerator and denominator of (7.6)). The degree bound for f_k turns out to be zero. Substituting the generic polynomial $f_k = b_0$ in (5.7) yields the identity

$$n+1-k+\sigma_1 n + \sigma_1 = (n+1-k)b_0 - k\, b_0.$$

Equating coefficients of like powers of k gives the linear equations

$$\begin{aligned} -1 + 2\,b_0 &= 0 \\ n+1 + \sigma_1 n + \sigma_1 - (n+1)\,b_0 &= 0 \end{aligned}$$

that we can solve for the unknowns $\{b_0, \sigma_1\}$ to give the solution

$$\{b_0 = 1/2, \sigma_1 = -1/2\}.$$

Therefore $f_k = 1/2$, but the more important information for us is that $\sigma_1 = -1/2$ which produces the recurrence equation

$$s_n - \frac{1}{2}s_{n+1} = 0.$$

Using $s_0 = 1$, we conclude that $s_n = 2^n$. $\qquad\qquad\triangle$

Session 7.2 We apply our previous Maple procedures (compare Exercises 5.3–5.6), available from `hsum.mpl`, to the above example.

```
>   F:=binomial(n,k);
```

$$binomial(n, k)$$

```
>   A:=F+sigma[1]*subs(n=n+1,F);
```

$$binomial(n, k) + \sigma_1 binomial(n + 1, k)$$

```
>   rat:=ratio(A,k);
```

$$-\frac{(-n - 1 + k)(-n + k - \sigma_1 n - \sigma_1)}{(-n - 1 + k - \sigma_1 n - \sigma_1)(k + 1)}$$

```
>   p:=1:  q:=subs(k=k-1,numer(rat)):  r:=subs(k=k-1,denom(rat)):
```

```
>   upd:=update(p,q,r,k);
```

$$[-n - 1 + k - \sigma_1 n - \sigma_1, n + 2 - k, k]$$

```
>   p:=op(1,upd):  q:=op(2,upd):  r:=op(3,upd):
```

```
>   degreebound(p,q,r,k);
```

$$0$$

```
>   f:=b[0];
```

$$b_0$$

```
>   rec:=collect(subs(k=k+1,q)*f-r*subs(k=k-1,f)-p,k);
```

$$(-2 b_0 - 1) k + (n + 1) b_0 + \sigma_1 n + n + 1 + \sigma_1$$

```
>   sol:={solve({coeffs(rec,k)},{sigma[1],b[0]})};
```

$$\{\{b_0 = \frac{-1}{2}, \sigma_1 = \frac{-1}{2}\}\}$$

```
> sigma[1]:=subs(op(1,sol),sigma[1]);
```

$$\frac{-1}{2}$$

```
> S(n)+sigma[1]*S(n+1)=0;
```

$$S(n) - \frac{1}{2}\,S(n+1) = 0$$

These steps can be combined by the following procedure searching for a holonomic recurrence equation of order one:

```
zeilberger:=proc(F,k,sn)
local n,A,S,sigma,rat,p,q,r,upd,deg,f,b,j,var,rec,sol,num,den;
option remember;
if type(sn,function) then S:=op(0,sn); n:=op(1,sn)else n:=sn end if;
A:=F+sigma[1]*subs(n=n+1,F);
rat:=ratio(A,k);
if not type(rat,ratpoly(anything,k)) then
   error 'Algorithm not applicable'
end if;
p:=1: q:=subs(k=k-1,numer(rat)): r:=subs(k=k-1,denom(rat)):
upd:=update(p,q,r,k);
p:=op(1,upd): q:=op(2,upd): r:=op(3,upd):
deg:=degreebound(p,q,r,k);
if deg<0 then
   error 'Algorithm finds no recurrence equation of first order'
end if;
f:=add(b[j]*k^j,j=0..deg);
var:={sigma[1],seq(b[j],j=0..deg)};
rec:=collect(subs(k=k+1,q)*f-r*subs(k=k-1,f)-p,k);
sol:={solve({coeffs(rec,k)},var)};
if sol={} then
   error 'Algorithm finds no recurrence equation of first order'
end if;
sigma[1]:=subs(op(1,sol),sigma[1]);
sigma[1]:=normal(sigma[1]);
for j from 0 to deg do
   sigma[1]:=subs(b[j]=0,sigma[1]);
end do;
num:=factor(numer(sigma[1]));
den:=factor(denom(sigma[1]));
return den*S(n)+num*S(n+1)=0;
end proc:
```

Therefore, in just one step we get

```
> zeilberger(binomial(n,k),k,s(n));
```

$$2\,s(n) - s(n+1) = 0$$

and similarly

```
> zeilberger(binomial(n,k)^2,k,s(n));
```

$$(4n + 2)\, s(n) + (-n - 1)\, s(n + 1) = 0$$

```
>   zeilberger((-1)^k*
>   binomial(n+b,n+k)*binomial(n+c,c+k)*binomial(b+c, b+k),
>   k,s(n));
```

$$(b + c + n + 1)\, s(n) + (-n - 1)\, s(n + 1) = 0$$

This gives a very short proof for Dixon's theorem. (Finish this proof!). Moreover, given the left-hand side of Dixon's theorem, we have deduced its right-hand side. What about Dougall's Theorem? Can we derive the right-hand side of Dougall's identity directly from the left-hand sum? Here is the application:

```
>   zeilberger(hyperterm([a,1+a/2,b,c,d,1+2*a-b-c-d+n,-n],
>   [a/2,1+a-b,1+a-c,1+a-d,1+a-(1+2*a-b-c-d+n),1+a+n],1,k),
>   k,s(n));
```

$$(1 + a + n)\,(a - d - c + 1 + n)\,(a + 1 + n - b - d)\,(a - c + n + 1 - b)\, s(n) -$$
$$(n + 1 + a - d)\,(a - c + 1 + n)\,(a + 1 + n - b)\,(1 + a - b - c - d + n)$$
$$s(n + 1) = 0$$

which yields an immediate result!

To get the hypergeometric term solutions automatically, we use

```
closedform:=proc(F,k,sn)
local zeilberg,S,n,rat,num,den,lc,numlist,denlist,j,i,init,cert;
if type(sn,function) then S:=op(0,sn); n:=op(1,sn) else n:=sn end if;
init:=eval(eval(F,k=0),n=0);
if init=0 then
  error `Shift necessary`
end if;
zeilberg:=zeilberger(F,k,S(n));
rat:=normal(solve(zeilberg,S(n+1))/S(n));
num:=numer(rat);
den:=denom(rat);
lc:=lcoeff(num,n)/lcoeff(den,n);
numlist:=normal([solve(num,n)]);
numlist:=[seq(-j,j=numlist)];
denlist:=normal([solve(den,n)]);
denlist:=[seq(-j,j=denlist)];
if member(1,denlist,'i') then
  denlist:= subsop(i=NULL,denlist)
else
  numlist:= [op(numlist),1]
end if;
return init*hyperterm(numlist,denlist,lc,n);
end proc:
```

Here, we assume that the sum goes from $k = 0$ to $k = n,$[1] and $F(0, 0)$ is well-defined. In particular $F(0, 0) \neq 0$; otherwise an appropriate shift must be applied in advance.

---

[1] The sum can also have bounds $[-n, \ldots, n]$ as the Dixon example has. We need mainly $s_0 = F(0, 0) \neq 0$.

For the above examples we get

```
>   closedform(binomial(n,k),k,n);
```
$$2^n$$

```
>   closedform(binomial(n,k)^2,k,n);
```
$$\frac{(2\,n)!}{(n!)^2}$$

```
>   closedform((-1)^k*binomial(n+b,n+k)*
>   binomial(n+c,c+k)*binomial(b+c,b+k),k,n);
```
$$\frac{\mathrm{binomial}(b+c,\,b)\,\mathrm{pochhammer}(b+c+1,\,n)}{n!}$$

```
>   closedform(hyperterm([a,1+a/2,b,c,d,1+2*a-b-c-d+n,-n],
>   [a/2,1+a-b,1+a-c,1+a-d,1+a-(1+2*a-b-c-d+n),1+a+n],1,k),k,n);
```

$$\mathrm{pochhammer}(1+a,\,n)\,\mathrm{pochhammer}(a-d-c+1,\,n)$$
$$\mathrm{pochhammer}(a+1-b-d,\,n)\,\mathrm{pochhammer}(a-c+1-b,\,n)/($$
$$\mathrm{pochhammer}(1+a-d,\,n)\,\mathrm{pochhammer}(1+a-c,\,n)$$
$$\mathrm{pochhammer}(1+a-b,\,n)\,\mathrm{pochhammer}(1+a-b-c-d,\,n))$$

We do not yet have a proof for Watson's identity with respect to integer $-a = m \in \mathbb{N}_{\geq 0}$, and arbitrary $b, c \in \mathbb{Q}$. Let's find one using Zeilberger's algorithm! We will distinguish between even and odd m. First let $m = 2n, n \in \mathbb{N}_{\geq 0}$. Here we get

```
>   closedform(hyperterm([-2*n,b,c],[(-2*n+b+1)/2,2*c],1,k),k,n);
```

$$\frac{(2\,n)!\,4^{(-n)}\,\mathrm{pochhammer}\!\left(c+\frac{1}{2}-\frac{1}{2}b,\,n\right)}{\mathrm{pochhammer}\!\left(\frac{1}{2}+c,\,n\right)\,\mathrm{pochhammer}\!\left(\frac{1}{2}-\frac{1}{2}b,\,n\right)n!}$$

Similarly, for odd m, we get using $m = 2n + 1$

```
>   closedform(
>   hyperterm([-2*n-1,b,c],[(-2*n-1+b+1)/2,2*c],1,k),k,n);
```

$$\frac{n!\,\mathrm{pochhammer}\!\left(1+c-\frac{1}{2}b,\,n\right)}{\mathrm{pochhammer}(c+1,\,n)\,\mathrm{pochhammer}\!\left(1-\frac{1}{2}b,\,n\right)}$$

These results can be combined easily to prove Watson's identity!

We see that the application of Zeilberger's algorithm has the advantage over the WZ method that the right-hand side of the hypergeometric identity of interest does not have to be known in advance, but instead is *generated* by the algorithm. Therefore, Zeilberger's algorithm can be used to calculate the values of definite sums rather than only to prove identities.

In particular all hypergeometric identities that we had proved using the WZ method can be generated by Zeilberger's algorithm.

Moreover, one might be lucky and the returned recurrence equation can be solved explicitly even though it is not of first order. For example, with this method, we obtain identities (2.4), (2.5) directly without the a priori distinction between even and odd values of n.

Session 7.3 The Maple procedure

```
MAXORDER:=5:

sumrecursion:=proc(F,k,sn)
local n,S,b,sigma,rat,p,q,r,upd,deg,f,i,j,jj,l,var,req,sol,
num,den,J,a; option remember;
if type(sn,function) then S:=op(0,sn); n:=op(1,sn) else n:=sn end if;
for J from 1 to MAXORDER do
  a:=F+add(sigma[j]*subs(n=n+j,F),j=1..J);
  rat:=ratio(a,k);
  if not type(rat,ratpoly(anything,k)) then
    error 'Algorithm not applicable'
  end if;
  p:=1: q:=subs(k=k-1,numer(rat)): r:=subs(k=k-1,denom(rat)):
  upd:=update(p,q,r,k);
  p:=op(1,upd): q:=op(2,upd): r:=op(3,upd):
  deg:=degreebound(p,q,r,k);
  if deg>=0 then
    f:=add(b[j]*k^j,j=0..deg);
   var:={seq(sigma[jj],jj=1..J),seq(b[jj],jj=0..deg)};
    req:=collect(subs(k=k+1,q)*f-r*subs(k=k-1,f)-p,k);
    sol:={solve({coeffs(req,k)},var)};
    if not(sol={} or
    {seq(op(2,op(1,op(1,sol))),l=1..nops(op(1,sol)))}={0}) then
      req:=S(n)+add(sigma[j]*S(n+j),j=1..J);
      req:=subs(op(1,sol),req);
      req:=numer(normal(req));
      req:=collect(req,[seq(S(n+J-j),j=0..J)]);
      return map(factor,req)=0;
    end if;
  end if;
end do;
error
cat('Algorithm finds no recurrence equation of order
<= ',MAXORDER);
end proc:
```

uses Zeilberger's method iteratively increasing the order J. It results in the holonomic recurrence equation of lowest order that can be obtained by this method; see Exercise 7.14. Although for a specific type of input we know a priori that this algorithm terminates (see Theorem 7.10), we search only up to order MAXORDER, which is a global variable that can be adjusted. The default value for MAXORDER is set to 5 since higher order recurrence equations are often too complicated to be useful.

The calculations

```
>   sumrecursion((-1)^k*binomial(n,k)^2,k,s(n));
```

$$(n+2)\,s(n+2) + 4\,s(n)\,(n+1) = 0$$

```
> sumrecursion((-1)^k*binomial(n,k)^3,k,s(n));
```

$$(n + 2)^2 \, s(n + 2) + 3\, s(n)\,(3\,n + 4)\,(3\,n + 2) = 0$$

for example, imply identities (2.4), (2.5) in a straightforward manner, using two initial values rather than one!

With Zeilberger's algorithm at hand, one has the possibility to prove identities of another type: If we want to show that two sums represent the same function, we prove that they satisfy the same holonomic recurrence equation and the same initial values. That's it. This is Zeilberger's paradigm. Here are some examples.

Session 7.4 To prove the identity

$$s_n = \sum_{k=0}^{n} \binom{n}{k}^3 = \sum_{k=0}^{n} \binom{n}{k}^2 \binom{2k}{n} \tag{7.7}$$

(compare Exercise 4.8), we calculate the common holonomic recurrence equation

```
> sumrecursion(binomial(n,k)^3,k,s(n));
```

$$-(n + 2)^2 \, s(n + 2) + (7\,n^2 + 21\,n + 16)\, s(n + 1) + 8\, s(n)\,(n + 1)^2 = 0$$

```
> sumrecursion(binomial(n,k)^2*binomial(2*k,n),k,s(n));
```

$$-(n + 2)^2 \, s(n + 2) + (7\,n^2 + 21\,n + 16)\, s(n + 1) + 8\, s(n)\,(n + 1)^2 = 0$$

and check that two initial values $s_0 = 1$ and $s_1 = 2$ agree. That's all you need to do! Note that (7.7) is not at all trivial, since for $n = 1$ it reads $1 + 1 = 0 + 2$.

Another example of the same type is given by the three different representations of the Legendre polynomials that we met earlier:

```
> sumrecursion(
> binomial(n,k)*binomial(-n-1,k)*((1-x)/2)^k,k,s(n));
```

$$(n + 2)\, s(n + 2) - (2\,n + 3)\, x\, s(n + 1) + s(n)\,(n + 1) = 0$$

```
> sumrecursion(
> 1/2^n*binomial(n,k)^2*(x-1)^(n-k)*(x+1)^k,k,s(n));
```

$$(n + 2)\, s(n + 2) - (2\,n + 3)\, x\, s(n + 1) + s(n)\,(n + 1) = 0$$

```
> sumrecursion(1/2^n*(-1)^k*binomial(n,k)*
> binomial(2*n-2*k,n)*x^(n-2*k),k,s(n));
```

$$(n + 2)\, s(n + 2) - (2\,n + 3)\, x s(n + 1) + s(n)\,(n + 1) = 0$$

Finally, we present a holonomic recurrence equation for the so-called Apéry numbers

$$A_n := \sum_{k=0}^{n} \binom{n}{k}^2 \binom{n+k}{k}^2, \tag{7.8}$$

namely

$$(n+2)^3 A_{n+2} - (2n+3)\left(17n^2 + 51n + 39\right) A_{n+1} + (n+1)^3 A_n = 0. \tag{7.9}$$

This played a crucial role in Apéry's proof [Apéry79] of the irrationality of the number

$$\zeta(3) = \sum_{k=1}^{\infty} \frac{1}{k^3}.$$

Even the proof of this recurrence equation is not at all trivial (without Zeilberger's algorithm: try it yourself!). Concerning this question, you might have a look in van der Poorten's [vanderPoorten78] entertaining presentation of Apéry's proof and its history.[2] On the other hand, it seems to be much simpler to prove such a recurrence equation than to *derive* it from scratch. Let's have a look at Zeilberger's approach. We get

```
> sumrecursion(binomial(n,k)^2*binomial(n+k,k)^2,k,s(n));
```

$$(n+2)^3 s(n+2) - (2n+3)(17n^2 + 51n + 39) s(n+1) + (n+1)^3 s(n) = 0$$

Hence Zeilberger's algorithm finds the recurrence equation completely automatically!

Note that—unlike Fasenmyer's approach—Zeilberger's algorithm almost always generates the lowest order recurrence equation in reasonable time.

Here is an example from [PS95] which shows that this is not always the case, though; see Exercise 7.14 as well. For the sum

$$s_n := \sum_{k=-\infty}^{\infty} (-1)^k \binom{n}{k} \binom{3k}{n},$$

we have

```
> sumrecursion((-1)^k*binomial(n,k)*binomial(3*k,n),k,s(n));
```

$$2(2n+3) s(n+2) + 3(5n+7) s(n+1) + 9 s(n)(n+1) = 0$$

The sum has the value $(-3)^n$ as we shall soon show. Therefore s_n satisfies the first order recurrence equation $s_{n+1} + 3 s_n = 0$. Hence, we see that Zeilberger's algorithm does not always generate the recurrence equation of lowest order.

---

[2] "To convince ourselves of the validity of Apéry's proof we need only complete the following exercise: show that (7.9) is valid for (7.8). Neither Cohen nor I had been able to prove this in the intervening two months…".

On the other hand, the identity $s_n = (-3)^n$ is easily checked using the recurrence equation that is returned by Zeilberger's algorithm: Just substitute $s_n = (-3)^n$ into the recurrence equation obtained, check whether the recurrence equation is satisfied, and check enough initial values. Then you are done because typically a holonomic recurrence equation of order N together with N initial values determines s_n uniquely.[3]

In the given case, we get

$$2(3 + 2n)(-3)^{n+2} + 3(5n + 7)(-3)^{n+1} + 9(n + 1)(-3)^n$$
$$= (-3)^n \left(18(3 + 2n) - 9(5n + 7) + 9(n + 1)\right) = 0,$$

and the initial values $s_0 = 1$ and $s_1 = -3$ are easily checked. This finishes the proof of the statement $s_n = (-3)^n$.

It should be mentioned that—as for the WZ method—Zeilberger's method faces problems if a pole appears for some $n \in \mathbb{N}_{\geq 0}$ anywhere in the intermediate calculations. This happens if the rational certificate of the intermediate application of Gosper's algorithm (see (7.5))

$$\widetilde{R}(n, k) = \frac{G(n, k)}{a_k}$$

has nonnegative integer denominator zeros with respect to n.

We summarize the above in the following algorithm.

Algorithm 7.5 (*Zeilberger*) Given $F(n, k)$, this algorithm searches for a holonomic recurrence equation for $s_n := \sum_{k=-\infty}^{\infty} F(n, k)$:

1. Input: $F(n, k) \not\equiv 0$, a hypergeometric term with respect to both n and k.
2. Set $J := 1$.
3. Set $a_k := F(n, k) + \sum_{j=1}^{J} \sigma_j(n) F(n + j, k)$ with undetermined variables σ_j depending on n, but not depending on k.
4. Apply the adaptation of Gosper's algorithm described to a_k: In the last step, solve the linear system for the coefficients of f_k, and at the same time for the unknowns σ_j $(j = 1, \ldots, J)$. If the procedure is successful, Gosper's algorithm finds $G(n, k)$ with
$$G(n, k + 1) - G(n, k) = a_k.$$

Be aware of possible nonnegative integer denominator zeros, with respect to n, of the rational certificate
$$\widetilde{R}(n, k) = \frac{G(n, k)}{a_k}$$

where the resulting recurrence equation might not be valid.

---

[3] This can fail only if some of the zeros of the highest or lowest coefficient polynomials are integers.

The calculation also determines the functions $\sigma_j \in \mathbb{Q}(n)$ $(j = 1, \ldots, J)$.
If the procedure is not successful then increase J by one and continue with
step 3.

5. Output: By summation, we have

$$s_n + \sum_{j=1}^{J} \sigma_j(n)\, s_{n+j} = 0$$

for s_n, if the right-hand side is telescoping, in particular if $F(n, k)$ has finite
support with respect to k. Multiplication by the common denominator results in
the holonomic recurrence equation sought. $\square$

The next example shows how sums with non-natural bounds can be treated, leading
generally to inhomogeneous recurrence equations.

Example 7.6 (Non-Natural Bounds; compare [PS95]*)* We consider the sum

$$s_n = \sum_{k=0}^{\lfloor n/3 \rfloor} \binom{n - 2k}{k} \left(-\frac{4}{27}\right)^k.$$

For $F(n, k) = \binom{n - 2k}{k} \left(-\frac{4}{27}\right)^k$, the calculation

```
>   F:=binomial(n-2*k,k)*(-4/27)^k:
```

```
>   Sumtohyper(F,k);
```

$$\text{Hypergeom}\left(\left[-\frac{1}{3}n,\ -\frac{1}{3}n + \frac{2}{3},\ -\frac{1}{3}n + \frac{1}{3}\right],\ \left[-\frac{1}{2}n,\ -\frac{1}{2}n + \frac{1}{2}\right],\ 1\right)$$

suggests that the natural bounds are $k = 0, \ldots, \lfloor n/3 \rfloor$. But, be careful! The term
ratio $F(n, k + 1)/F(n, k)$

```
>   ratio(F,k);
```

$$\frac{4}{27} \frac{(-n + 3k + 2)(-n + 3k + 1)(-n + 3k)}{(k + 1)(-n + 2k + 1)(-n + 2k)}$$

shows, on the one hand, that, for $k = \lfloor n/3 \rfloor$, it follows that $F(n, k + 1) = 0$. On the
other hand, since both $n/2$ and $(n + 1)/2$ are zeros of its denominator, it may happen
that the value of $F(n, k)$ begins to differ from zero again for $k > \lfloor n/2 \rfloor$. And it does!
See, for example, what happens for $n = 20$:

```
>   map(expand,subs(n=20,
>   [seq(binomial(n-2*j,j)*(-4/27)^j,j=0..15)]));
```

$$\Big[1, \frac{-8}{3}, \frac{640}{243}, \frac{-23296}{19683}, \frac{14080}{59049}, \frac{-28672}{1594323}, \frac{114688}{387420489}, 0, 0, 0, 0,$$

$$\frac{16777216}{1853020188851841}, \frac{7633633280}{1500946352969999121},$$

$$\frac{63887638528}{450283905890997363}, \frac{3468186091520}{12157665459056928801},$$

$$\frac{1403921729847296}{2954312706550833698643}\Big]$$

Hence, the given bounds are *not* the natural ones, and $F(n, k)$ does not have finite support. Nevertheless, we take $J = 2$ (the choice $J = 1$ is not successful), set $a_k = F(n, k) + \sigma_1 F(n+1, k) + \sigma_2 F(n+2, k)$, and use Zeilberger's procedure to get

$$F(n, k) + \sigma_1 F(n+1, k) + \sigma_2 F(n+2, k) = G(n, k+1) - G(n, k) \quad (7.10)$$

for

$$\sigma_1 = \frac{3(n+4)}{2(n+3)}, \qquad \sigma_2 = -\frac{9(n+2)}{2(n+3)},$$

and $G(n, k) = R(n, k)F(n, k)$ with

$$R(n, k) = \frac{27}{2} \frac{(n-2k+2)\, k\, (n-2k+1)}{(n+3)\, (n-3k+2)\, (n-3k+1)}.$$

Summation of (7.10) from $k = 0$ to $k = \lfloor (n+2)/3 \rfloor$ by telescoping generates the recurrence equation for s_n

$$\sum_{k=0}^{\lfloor \frac{n+2}{3} \rfloor} F(n, k) + \frac{3(n+4)}{2(n+3)} \sum_{k=0}^{\lfloor \frac{n+2}{3} \rfloor} F(n+1, k) - \frac{9(n+2)}{2(n+3)} \sum_{k=0}^{\lfloor \frac{n+2}{3} \rfloor} F(n+2, k)$$

$$= G(n, \lfloor (n+2)/3 \rfloor + 1) - G(n, 0) = 0.$$

We choose the upper summation bound $k = \lfloor (n+2)/3 \rfloor$ since $F(n+2, k)$ is involved in (7.10). On the other hand, since $F(n, k) = 0$ for k immediately after $\lfloor n/3 \rfloor$, we get

$$s_n + \frac{3(n+4)}{2(n+3)} s_{n+1} - \frac{9(n+2)}{2(n+3)} s_{n+2} = G(n, [(n+2)/3] + 1) - G(n, 0) = 0.$$

Even though in general we get an *inhomogeneous recurrence equation* if non-natural bounds are involved, in the current case it turns out that the resulting recurrence equation is homogeneous despite the fact that the bounds are not the natural ones. This is so since $F(n, k) = 0$ for all k between $\lfloor n/3 \rfloor$ and $\lfloor n/2 \rfloor$, and $G(n, k)$ is a multiple of $F(n, k)$. Hence we have determined the holonomic recurrence equation

$$9\,(n+2)\,s_{n+2} - 3\,(n+4)\,s_{n+1} - 2\,(n+3)\,s_n = 0$$

for s_n. Note that the hsum package contains an implementation sumrecursion (F,k=a..b,s(n)) of the inhomogeneous version of Zeilberger's algorithm, which finds the above recurrence equation automatically:

```
>  re:=sumrecursion(F,k=0..n/3,s(n));
```

$$-9\,(2+n)\,s(2+n) + 3\,(n+4)\,s(n+1) + 2\,(n+3)\,s(n) = 0$$

This establishes our result for nonnegative integers n that are divisible by 3. For other values of n we compute sumrecursion(F,k=0..n/3-1/3,s(n)); and sumrecursion(F,k=0..n/3-2/3,s(n)); with the same result.

We will come back to this example in Session 9.12.   △

In the next two examples we show two methods by means of which one can find hypergeometric representations of certain double sums without the need of Wegschaider's algorithm.

Example 7.7 (Clausen's Formula, see e.g. [Rainville60]) Here we deduce *Clausen's formula*

$$\left({}_2F_1\left(\begin{array}{c} a,b \\ a+b+1/2 \end{array} \middle| x \right) \right)^2 = {}_3F_2\left(\begin{array}{c} 2a, 2b, a+b \\ a+b+1/2, 2a+2b \end{array} \middle| x \right) \quad (7.11)$$

which gives all possible squares of ${}_2F_1$ functions generating a ${}_3F_2$. Therefore we equate coefficients using the Cauchy product, and apply Zeilberger's algorithm to the left-hand side with respect to the summation variable.

Rewriting the left-hand side of (7.11) using the Cauchy product yields

$$\left(\sum_{k=0}^{\infty} \frac{(a)_k\,(b)_k}{\left(a+b+\frac{1}{2}\right)_k\,k!}\,x^k \right)^2 = \sum_{k=0}^{\infty}\sum_{j=0}^{k} \frac{(a)_j\,(b)_j}{\left(a+b+\frac{1}{2}\right)_j\,j!}\,\frac{(a)_{k-j}\,(b)_{k-j}}{\left(a+b+\frac{1}{2}\right)_{k-j}\,(k-j)!}\,x^k.$$

The coefficient of x^k on the right is a sum for which Zeilberger's algorithm generates the hypergeometric term[4]

```
>  Closedform(hyperterm([a,b],[a+b+1/2],1,j)*
>  hyperterm([a,b],[a+b+1/2],1,k-j),j,k);
```

$$\text{Hyperterm}\left([2b,\, 2a,\, a+b],\, [2a+2b,\, a+b+\tfrac{1}{2}],\, 1,\, k \right)$$

Hence we have discovered the right-hand side of (7.11)!   △

Next, we give an example of a rather complicated double sum identity that can be discovered by Zeilberger's algorithm. The motivation for dealing with this sum will be apparent soon.

---

[4] Closedform is the same as closedform, with hyperterm replaced by the inert form Hyperterm in the output, preventing evaluation, and hence emphasizing the hypergeometric structure.

Example 7.8 (*Double Sum Identity, see* [Gasper86]) We try to find a hypergeometric representation for the double sum

$$\sum_{j=0}^{\lfloor\frac{n-k}{2}\rfloor} \frac{\left(\frac{1}{2}\right)_j \left(\frac{\alpha}{2}+1\right)_{n-j} \left(\frac{\alpha+3}{2}\right)_{n-2j} (\alpha+1)_{n-2j}}{j! \left(\frac{\alpha+3}{2}\right)_{n-j} \left(\frac{\alpha+1}{2}\right)_{n-2j} (n-2j)!} \, {}_3F_2\left(\begin{array}{c} 2j-n, n-2j+\alpha+1, \frac{\alpha+1}{2} \\ \alpha+1, \frac{\alpha+2}{2} \end{array}\middle| x\right).$$

Application of Zeilberger's algorithm to the summand gives

```
>   zeilberger(pochhammer(1/2,j)*pochhammer(alpha/2+1,n-j)*
>   pochhammer((alpha+3)/2,n-2*j)*pochhammer(alpha+1,n-2*j)/
>   j!/pochhammer((alpha+3)/2,n-j)/pochhammer((alpha+1)/2,n-2*j)/
>   (n-2*j)!*hyperterm([2*j-n,n-2*j+alpha+1,(alpha+1)/2],
>   [alpha+1,(alpha+2)/2],x,k),k,AG(j));
```

```
Error, (in zeilberger) Algorithm finds no recurrence equation
of first order
```

Unfortunately, no recurrence equation of first order is discovered, so that we don't see a hypergeometric term solution. Therefore, did we lose the battle? No, because following Gasper, we can change the order of summation, summing with respect to j first! Using this approach we get for the outer summand the first order recurrence equation with respect to k

```
>   zeilberger(pochhammer(1/2,j)*pochhammer(alpha/2+1,n-j)*
>   pochhammer((alpha+3)/2,n-2*j)*pochhammer(alpha+1,n-2*j)/
>   j!/pochhammer((alpha+3)/2,n-j)/pochhammer((alpha+1)/2,n-2*j)/
>   (n-2*j)!*hyperterm([2*j-n,n-2*j+alpha+1,(alpha+1)/2],
>   [alpha+1,(alpha+2)/2],x,k),j,AG(k));
```

$$x\,(\alpha+1+2k)\,(-n+k)\,(k+\alpha+2+n)\,AG(k)$$
$$-\,(k+1)\,(\alpha+1+k)\,(\alpha+3+2k)\,AG(k+1) = 0$$

Hence we have discovered the representation

$$\sum_{j=0}^{\lfloor\frac{n-k}{2}\rfloor} \frac{\left(\frac{1}{2}\right)_j \left(\frac{\alpha}{2}+1\right)_{n-j} \left(\frac{\alpha+3}{2}\right)_{n-2j} (\alpha+1)_{n-2j}}{j! \left(\frac{\alpha+3}{2}\right)_{n-j} \left(\frac{\alpha+1}{2}\right)_{n-2j} (n-2j)!}$$
$$\times {}_3F_2\left(\begin{array}{c} 2j-n, n-2j+\alpha+1, \frac{\alpha+1}{2} \\ \alpha+1, \frac{\alpha+2}{2} \end{array}\middle| x\right)$$
$$= C(n,\alpha) \cdot {}_3F_2\left(\begin{array}{c} -n, n+\alpha+2, \frac{\alpha+1}{2} \\ \alpha+1, \frac{\alpha+3}{2} \end{array}\middle| x\right)$$

where $C(n,\alpha)$ depends only on n and α. To find $C(n,\alpha)$, we apply Zeilberger's algorithm to the coefficient of the left-hand ${}_3F_2$ function with respect to n with the result

```
>   closedform(pochhammer(1/2,j)*pochhammer(alpha/2+1,n-j)*
>   pochhammer((alpha+3)/2,n-2*j)*pochhammer(alpha+1,n-2*j)/j!/
>   pochhammer((alpha+3)/2,n-j)/pochhammer((alpha+1)/2,n-2*j)/
>   (n-2*j)!,j,n);
```

$$\frac{\mathrm{pochhammer}(\alpha+2,\,n)}{n!}$$

Hence $C(n,\alpha) = \dfrac{(\alpha+2)_n}{n!}$, and therefore

$$\frac{(\alpha+2)_n}{n!}\,{}_3F_2\left(\begin{array}{c} -n,\, n+\alpha+2,\, \frac{\alpha+1}{2} \\ \alpha+1,\, \frac{\alpha+3}{2} \end{array}\,\middle|\,x\right)$$

$$= \sum_{j=0}^{\lfloor\frac{n-k}{2}\rfloor} \frac{\left(\frac{1}{2}\right)_j \left(\frac{\alpha}{2}+1\right)_{n-j} \left(\frac{\alpha+3}{2}\right)_{n-2j} (\alpha+1)_{n-2j}}{j!\left(\frac{\alpha+3}{2}\right)_{n-j} \left(\frac{\alpha+1}{2}\right)_{n-2j} (n-2j)!}$$

$$\times\, {}_3F_2\left(\begin{array}{c} 2j-n,\, n-2j+\alpha+1,\, \frac{\alpha+1}{2} \\ \alpha+1,\, \frac{\alpha+2}{2} \end{array}\,\middle|\,x\right).$$

Applying Clausen's formula with $a = j - n/2$, and $b = (n - 2j + \alpha + 1)/2$ to the right-hand side yields the *Askey-Gasper identity* [AG76]

$$\frac{(\alpha+2)_n}{n!}\,{}_3F_2\left(\begin{array}{c} -n,\, n+\alpha+2,\, \frac{\alpha+1}{2} \\ \alpha+1,\, \frac{\alpha+3}{2} \end{array}\,\middle|\,x\right)$$

$$= \sum_{j=0}^{\lfloor\frac{n-k}{2}\rfloor} \frac{\left(\frac{1}{2}\right)_j \left(\frac{\alpha}{2}+1\right)_{n-j} \left(\frac{\alpha+3}{2}\right)_{n-2j} (\alpha+1)_{n-2j}}{j!\left(\frac{\alpha+3}{2}\right)_{n-j} \left(\frac{\alpha+1}{2}\right)_{n-2j} (n-2j)!}\left({}_2F_1\left(\begin{array}{c} j-\frac{n}{2},\, \frac{n-2j+\alpha+1}{2} \\ \frac{\alpha+2}{2} \end{array}\,\middle|\,x\right)\right)^2,$$

$$\tag{7.12}$$

which implies the *Askey-Gasper inequality*

$${}_3F_2\left(\begin{array}{c} -n,\, n+\alpha+2,\, \frac{\alpha+1}{2} \\ \alpha+1,\, \frac{\alpha+3}{2} \end{array}\,\middle|\,x\right) \geq 0 \qquad (x\in[-1,1],\,\alpha > -2,\,\alpha \neq -1).$$

For $\alpha = 2, 4, 6, \ldots$ (compare Exercise 2.16) this was the key inequality in de Branges' proof [deBranges85] of the *Bieberbach conjecture*. Note that the same argument can be applied to Weinstein's version of the proof [Weinstein91]; compare ([Todorov92, Wilf94, KS96]).

Note that, whereas it turned out to be easy to deduce the left-hand side from the right-hand side of (7.12), there is no simple way to reverse this process. Finding this representation starting with the left-hand side of (7.12) was Askey's and Gasper's brilliant achievement. △

The given method can be used to prove and discover many more hypergeometric representations of double sums, in particular for sums of orthogonal polynomials, for example those in [Feldheim43, AF69], or [AG71]. Some examples of this type are given in the exercises.

Like the Wilf-Zeilberger method, Zeilberger's algorithm is accompanied by a rational certification mechanism. With this rational certificate (see Exercise 7.16) it is again possible to prove the resulting recurrence equation by rational arithmetic without any knowledge of where either the certificate and the recurrence equation come from. This is an important issue as the following example shows:

Example 7.9 Assume that we find an antiderivative using Maple and we would like to prove it using the fundamental theorem of calculus:

```
>  integral:=int(x/(1-x^2),x);
```

$$-\frac{1}{2}\ln(x-1) - \frac{1}{2}\ln(x+1)$$

This calculation is obviously proved if we get the integrand back after differentiation.

```
>  derivative:=diff(integral,x);
```

$$-\frac{1}{2}\frac{1}{x-1} - \frac{1}{2}\frac{1}{x+1}$$

We see, however, that differentiation does *not* generate the integrand. In the given case, this is not a crucial issue since the integrand is rational, and the identity

$$\frac{x}{1-x^2} = -\frac{1}{2(x-1)} - \frac{1}{2(x+1)}$$

is easily checked, for example by

```
>  normal(derivative-x/(1-x^2));
```

$$0$$

This step is much more difficult if nonrational functions are involved. We get, for example

```
>  integral:=int(sqrt(x^2+y^2),x);
```

$$\frac{1}{2}x\sqrt{x^2+y^2} + \frac{1}{2}y^2\ln\left(x+\sqrt{x^2+y^2}\right)$$

```
>  derivative:=diff(integral,x);
```

$$\frac{1}{2}\sqrt{x^2+y^2} + \frac{1}{2}\frac{x^2}{\sqrt{x^2+y^2}} + \frac{1}{2}\frac{y^2\left(1+\dfrac{x}{\sqrt{x^2+y^2}}\right)}{x+\sqrt{x^2+y^2}}$$

```
>  normal(derivative);
```

$$\sqrt{x^2+y^2}$$

Here we are lucky that, although the differentiated antiderivative is quite different from the original integrand, we get the integrand back using `normal`. This is so since the term under investigation is rational in the variables x, y and $\sqrt{x^2 + y^2}$.

We give a final example.

```
>   integral:=int(sin(3*x)/cos(x),x);
```

$$-2\cos(x)^2 + \ln(\cos(x))$$

```
>   derivative:=diff(integral,x);
```

$$4\cos(x)\sin(x) - \frac{\sin(x)}{\cos(x)}$$

```
>   expand(sin(3*x)/cos(x)-derivative);
```

$$0$$

The underlying identity

$$\frac{\sin(3\,x)}{\cos(x)} = 4\,\sin(x)\cos(x) - \frac{\sin(x)}{\cos(x)}$$

cannot be shown by rational arithmetic. To prove this statement, addition formulas for trigonometric functions are utilized, which can be done using the `expand` command.   △

This leaves us with one unanswered question about Zeilberger's algorithm: How can we guarantee that the algorithm terminates? We continue to increase the order J until we find the resulting recurrence equation. Since we saw that Zeilberger's algorithm might miss the lowest order recurrence equation, how can we be sure that it terminates at all? Luckily, one can prove that for *proper hypergeometric terms* Zeilberger's algorithm does actually terminate. Recall that a hypergeometric term $F(n, k)$ is called proper if it has finite support, and is of the form $F(n, k) = P(n, k)\frac{Q(n,k)}{R(n,k)}w^n z^k$ where $P(n, k)$ is a polynomial (*polynomial part*) and $Q(n, k)$, $R(n, k)$ are Γ-term products with integer-linear arguments (*factorial part*).

We have as a first step

Theorem 7.10 (see [WZ92, GKP94]) *Let $F(n, k)$ be a proper hypergeometric term, written in the form*

$$F(n, k) = P(n, k)\frac{\Gamma(\alpha_1 k + \beta_1 n + c_1)\cdots\Gamma(\alpha_p k + \beta_p n + c_p)}{\Gamma(\gamma_1 k + \delta_1 n + d_1)\cdots\Gamma(\gamma_q k + \delta_q n + d_q)}w^n z^k, \qquad (7.13)$$

$P \in \mathbb{Q}[n, k]$, $\alpha_l, \beta_l, \gamma_l, \delta_l \in \mathbb{Z}$, *and* $c_l, d_l, w, z \in \mathbb{Q}$.

Then there exists a k-free holonomic recurrence equation with polynomials $a_{ij}(n)$

$$\sum_{i=0}^{I}\sum_{j=0}^{J} a_{ij}(n)\, F(n+j,k+i) = 0 \tag{7.14}$$

for large enough I and J. To be precise, the condition

$$J \geq J_0 := \sum_{l=1}^{p} |\alpha_l| + \sum_{l=1}^{q} |\gamma_l| \ \text{and}\ I \geq \left(\sum_{l=1}^{p} |\beta_l| + \sum_{l=1}^{q} |\delta_l| - 1 \right) \cdot J_0 + \deg_{k} P(n,k) \tag{7.15}$$

is sufficient (but not necessary) for such a recurrence equation to exist.

Proof We use operator notation that will be especially helpful in the proof of the theorem's corollary.

Let K and N denote the (forward) *shift operators* with respect to k and n, respectively, hence

$$K^i\, N^j\, F(n,k) = F(n+j,k+i).$$

Then (7.14) reads $H(N,K,n)\, F(n,k) = 0$ with the *holonomic operator*

$$H(N,K,n) = \sum_{i=0}^{I}\sum_{j=0}^{J} a_{ij}(n) K^i\, N^j. \tag{7.16}$$

The operator $H(N,K,n)$ is a polynomial in the variables N, K and n since each $a_{ij}(n)$ is a polynomial in n. Note, however, that the polynomial ring in the variables (N,K,n,k) is not commutative since by the calculation $K(ka_k) - kKa_k = (k+1)a_{k+1} - ka_{k+1} = a_{k+1} = Ka_k$ we see that the commutator rule $Kk - kK = K$ is valid. Similarly $Nn - nN = N$, but the other pairs of variables commute with each other.

Now, assume $F(n,k)$ is a proper hypergeometric term. We will show that the application of $H(N,K,n)$ to $F(n,k)$ results in a proper hypergeometric term again. Let $F(n,k)$ and $H(N,K,n)$ be given by (7.13) and (7.16), respectively. Then we set

$$\widetilde{H}(n,k) := \frac{\displaystyle\prod_{l=1}^{p} \Gamma(\alpha_l k + \beta_l n + s(-\alpha_l)\alpha_l I + s(-\beta_l)\beta_l J + c_l)}{\displaystyle\prod_{l=1}^{q} \Gamma(\gamma_l k + \delta_l n + s(\gamma_l)\gamma_l I + s(\delta_l)\delta_l J + d_l)}\, w^n z^k, \tag{7.17}$$

with $s(x)$ denoting the Heaviside function

$$s(x) := \begin{cases} 1 & \text{if } x > 0 \\ 0 & \text{otherwise} \end{cases}.$$

The importance of $\widetilde{H}(n, k)$ comes from the fact that for all $i = 0, \ldots, I$ and $j = 0, \ldots, J$ the expression $a_{ij}(n) K^i N^j F(n, k)$ is a polynomial multiple of $\widetilde{H}(n, k)$. This is a consequence of the fundamental identity (1.4) of the Γ function. Since a finite sum of polynomials forms a polynomial, we see that $H(N, K, n) F(n, k)$ is indeed a proper hypergeometric term.

Moreover, a closer look shows that

$$a_{ij}(n) K^i N^j F(n, k) = p_{ij}(n, k) \widetilde{H}(n, k) \qquad (p_{ij}(n, k) \in \mathbb{Q}[n, k]),$$

where the polynomial factor $p_{ij}(n, k)$ has degree at most

$$D := \deg_k P(n, k) + \left(\sum_{l=1}^{p} |\alpha_l| + \sum_{l=1}^{q} |\gamma_l| \right) I + \left(\sum_{l=1}^{p} |\beta_l| + \sum_{l=1}^{q} |\delta_l| \right) J$$

with respect to k (see Exercise 7.17).

Hence the identity $H(N, K, n) F(n, k) = 0$ to be proved reads

$$\sum_{i=0}^{I} \sum_{j=0}^{J} p_{ij}(n, k) \widetilde{H}(n, k) = 0.$$

This must be valid for all $k \in \mathbb{Z}$, which is only possible if

$$p(n, k) := \sum_{i=0}^{I} \sum_{j=0}^{J} p_{ij}(n, k) \equiv 0$$

for all $k \in \mathbb{Z}$. Equating coefficients yields at most $D + 1$ homogeneous linear equations in the $(I + 1)(J + 1)$ variables. This system undoubtedly has a nontrivial solution if there are more variables than equations, which is the case if (7.15) is valid. $\square$

Note that with more effort better (often sharp) bounds can be obtained [MZ05] using the signs of the coefficients α_l and γ_l. However, the most important applications of Theorem 7.10 are the next two corollaries which rely on the existence of I and J.

Corollary 7.11 (see [WZ92, GKP94]) *Let $F(n, k)$ be a proper hypergeometric term, given by (7.13). Then there exist polynomials $\sigma_j(n) \in \mathbb{Q}[n]$ $(j = 0, \ldots, J)$ (not all zero) and a proper hypergeometric term $G(n, k)$ such that*

$$\sum_{j=0}^{J} \sigma_j(n) F(n + j, k) = G(n, k + 1) - G(n, k) \tag{7.18}$$

for large enough J.

Proof Let I be the smallest possible choice such that Theorem 7.10 applies. We write

$$H(N, K, n) = H(N, 1, n) - (K - 1) h(N, K, n) \qquad (7.19)$$

for some linear difference operator $h(N, K, n)$. Note that this is equivalent to a noncommutative division by $K - 1$ (which is possible since $H(N, 1, n) - H(N, K, n)$ has a zero at $K = 1$) in the polynomial ring considered corresponding to

$$H(N, 1, n) - H(N, K, n) = (K - 1) h(N, K, n).$$

Since $H(N, 1, n)$ is a polynomial of degree J in N, we may write

$$H(N, 1, n) = \sum_{j=0}^{J} \sigma_j(n) N^j.$$

By the proof of Theorem 7.10

$$G(n, k) := h(N, K, n) F(n, k) \qquad (7.20)$$

is a proper hypergeometric term. Applying (7.19) to $F(n, k)$, we get

$$0 = H(N, K, n) F(n, k) = \Big(H(N, 1, n) - (K - 1) h(N, K, n) \Big) F(n, k)$$

$$= \sum_{j=0}^{J} \sigma_j(n) N^j F(n, k) - (K - 1) G(n, k),$$

so that (7.18) is valid.

It remains to prove that at least one of the polynomials σ_j is nonzero. Assume all σ_j are identically zero, so then in particular $G(n, k)$ is independent of k. Since $G(n, k)$ is a hypergeometric term with respect to n, there are polynomials $\tau_0(n)$ and $\tau_1(n)$ such that

$$(\tau_0(n) + \tau_1(n) N) G(n, k) = 0.$$

Therefore, by (7.20)

$$(\tau_0(n) + \tau_1(n) N) h(N, K, n)$$

forms a nonzero linear difference operator of degree $I - 1$ with respect to K which, applied to $F(n, k)$, yields zero. This contradicts the minimality of I. $\square$

Finally we get

Corollary 7.12 (see [WZ92, GKP94]) *For series of proper hypergeometric terms, Zeilberger's algorithm terminates.*

Proof For proper hypergeometric $F(n,k)$, Corollary 7.11 shows that Zeilberger's sum

$$\sum_{j=0}^{J} \sigma_j(n)\, F(n+j,k)$$

has a proper hypergeometric antidifference $G(n,k)$ for suitably chosen $\sigma_j(n)$ ($j = 0,\ldots,J$) and $J \in \mathbb{N}$. Since Gosper's algorithm is a decision procedure for such an output, it will find $G(n,k)$. Summation yields the desired recurrence equation for the sum, if $F(n,k)$ has finite support. $\qquad\qquad\square$

In the past decade many more details concerning Zeilberger's algorithm have been studied.

In Exercise 7.33 it is shown that Zeilberger's algorithm does not terminate for all hypergeometric series. However, Abramov ([Abramov02], see also [AL00]) gave a criterion, by means of which the applicability of Zeilberger's algorithm to hypergeometric terms can be checked a priori. This is implemented in the Maple package SumTools as SumTools[Hypergeometric][IsZApplicable]. Koornwinder [Koornwinder98] and Vidunas [Vidunas02] extended Zeilberger's algorithm to nonterminating series. This is implemented as Maple package infhsum.mpl, see [Vidunas01], and put into the q-context in [CHM08]. Gerhard [Gerhard01] used a modular approach and gave a worst case asymptotic cost analysis for Zeilberger's algorithm.

In [Zeilberger90a] Zeilberger had given a different approach to summation based on non-commutative elimination techniques. This was further developed by Chyzak and Salvy [CS98], Koepf [Koepf97a, Koepf97b] and others. An extension of Zeilberger's fast algorithm to the general holonomic case was given in [Chyzak00]. In [CKS09] it was shown how the methods of both Fasenmyer's and Zeilberger's approach can be generalized to certain non-holonomic summations.

The CAOP project (Computer Algebra and Orthogonal Polynomials) [CAOP] is a web resource which computes recurrence and differential equations online for families of the Askey-Wilson scheme using the Maple programs of this chapter and of Chap. 10.

A multivariate variant of Zeilberger's algorithm was considered in [AZ06].

q-Zeilberger Algorithm

In the previous chapter, the q-analogue of Gosper's algorithm was used to obtain a q-analogue of the WZ method for proving q-identities.

Applying the same adaptation that we met in this chapter to the q-Gosper algorithm by solving the linear system for the coefficients of f_k (which turns out to be a Laurent polynomial here), and at the same time for the unknowns σ_j ($j = 1,\ldots,J$) yields the q-analogue of Zeilberger's algorithm for finding recurrence equations for *q*-hypergeometric sums. This algorithm is much stronger in generating *q*-identities than the *q*-Fasenmyer method that we saw in Chap. 4.

An implementation [BK99] based on Koornwinder's [Koornwinder93] and the implementations of the present book is given in the qsum package containing the Maple procedure qsumrecursion(F,q,k,S(n)) for this purpose. The request[5]

```
>    qsumrecursion(qbinomial(n,k,q),q,k,S(n),recursion=up);
```

$$-S(2+n) + 2\,S(n+1) + (-1+q^{(n+1)})\,S(n) = 0$$

e.g., confirms the recurrence equation that was already computed in Chap. 4. Whereas in Exercise 4.19 the *q*-analogue of Fasenmyer's method failed to generate the *q*-analogues of the Chu-Vandermonde and of the Pfaff-Saalschütz identities (3.7) and (3.6), the *q*-analogue of Zeilberger's algorithm is quickly successful; see Exercise 7.35.

Further Reading

For further reading on Zeilberger's algorithm see [GKP94, PWZ96, Koepf06], and for the *q*-case [Koornwinder93, PR97] and [BK99].

Exercises

Exercise 7.1 *Prove the identities (2.4)–(2.5) using Zeilberger's algorithm in detail (without Maple!).*

Exercise 7.2 *Prove the entries of our hypergeometric database in Chap. 3 using Zeilberger's algorithm. Which entries with respect to which variables cannot be solved? Why?*

Exercise 7.3 *Prove Bailey's hypergeometric identities (Table 6.1, under Algorithm 6.5) using Zeilberger's algorithm.*

Exercise 7.4 *Show that if s_n satisfies a holonomic recurrence equation, then it satisfies a holonomic recurrence equation of lowest order which is unique up to polynomial multiples.*

Exercise 7.5 *Determine for which x the following sums have a hypergeometric term representation:*

---

[5] Without the option recursion=up, the procedure gives the recurrence equation in terms of downward shifts.

(a) $\displaystyle\sum_{k=0}^{n} \binom{n+k}{k} x^k,$

(b) $\displaystyle\sum_{k=0}^{\infty} \binom{n+k}{k} x^k,$

(c) $\displaystyle\sum_{k=0}^{n} \binom{n-k}{k} x^k,$

(d) $\displaystyle\sum_{k=0}^{\infty} \binom{n-k}{k} x^k.$

Exercise 7.6   *Show that the hypergeometric function of Gauss' identity (3.1)*

$$s_m := {}_2F_1\left(\begin{array}{c} a,b \\ c+m \end{array}\Bigg| 1\right)$$

satisfies the recurrence equation

$$(c+m)(a+b-c-m)\,s_m + (b-c-m)(a-m-c)\,s_{m+1} = 0$$

with respect to the parameter m. Therefore one has, for $m \in \mathbb{N}$,

$$ {}_2F_1\left(\begin{array}{c} a,b \\ c+m \end{array}\Bigg| 1\right) = \frac{(c-a)_m\,(c-b)_m}{(c)_m\,(c-a-b)_m}\, {}_2F_1\left(\begin{array}{c} a,b \\ c \end{array}\Bigg| 1\right).$$

Show that letting $m \to \infty$ implies Gauss' identity for arbitrary a, b and c for which we have convergence (see [PS95]).

Exercise 7.7   *Take the left-hand sides of the identities of Exercise 6.7, and deduce the right-hand sides by Zeilberger's algorithm whenever possible.*

Exercise 7.8   *Prove the following identity, proposed in SIAM Review 37, 1995, Problem 95-1 (a) [XT95]:*

$$\sum_{k=0}^{m} \binom{m}{k}\binom{n+k}{k}\binom{n+1}{j-k} = \sum_{k=0}^{n} \binom{n}{k}\binom{m+k}{k}\binom{m+1}{j-k}.$$

Exercise 7.9   *Prove, by Zeilberger's algorithm,*

$$(x+y)_n = \sum_{k=0}^{n} \binom{n}{k} (x)_k\,(y)_{n-k}.$$

Exercise 7.10   (Classical Discrete Orthogonal Polynomials)   *Find three-term recurrence equations with respect to all possible variables for the following families of discrete orthogonal polynomials (see [KLS10, Koepf97a]). The parameters x, N and n denote nonnegative integers.*

Krawtchouk Polynomials

$$K_n(x; p, N) = {}_2F_1\left(\begin{array}{c} -n, -x \\ -N \end{array}\middle| \frac{1}{p}\right).$$

Meixner Polynomials

$$M_n(x; \beta, c) = {}_2F_1\left(\begin{array}{c} -n, -x \\ \beta \end{array}\middle| 1 - \frac{1}{c}\right).$$

Charlier Polynomials

$$C_n(x, a) = {}_2F_0\left(\begin{array}{c} -n, -x \\ - \end{array}\middle| -\frac{1}{a}\right) = \frac{(-1)^n}{a^n}(x - n + 1)_n \cdot {}_1F_1\left(\begin{array}{c} -n \\ x - n + 1 \end{array}\middle| a\right).$$

Hahn Polynomials

$$Q_n(x; \alpha, \beta, N) = \frac{(\alpha + x + 1)_n\,(x - N)_n}{(\alpha + 1)_n\,(-N)_n}\,{}_3F_2\left(\begin{array}{c} -n, -x, \beta + N + 1 - x \\ N + 1 - x - n, -\alpha - x - n \end{array}\middle| 1\right)$$

$$= {}_3F_2\left(\begin{array}{c} -n, -x, \alpha + \beta + n + 1 \\ \alpha + 1, -N \end{array}\middle| 1\right).$$

Prove that the different representations define the same functions whenever applicable.

$\diamond$ **Exercise 7.11** *In the previous exercise we introduced the classical discrete orthogonal polynomials. Their recurrence equations with respect to x can be written in the form*

$$\sigma(x)\,\Delta\nabla p_n(x) + \tau(x)\,\Delta p_n(x) + \lambda_n\,p_n(x) = 0,$$

or, equivalently, in the form

$$\overline{\sigma}(x)\,\Delta p_n(x) + \overline{\tau}(x)\nabla p_n(x) + \lambda_n\,p_n(x) = 0$$

where $\Delta f(x) = f(x+1) - f(x)$ denotes the forward and $\nabla f(x) = f(x) - f(x-1)$ denotes the backward difference operator, with $\overline{\sigma}(x) = \sigma(x + 1) + \tau(x + 1)$ and $\overline{\tau}(x) = -\sigma(x)$. Write two Maple procedures `sumdeltanabla(F,k,s(x))` *and* `'sumdelta+nabla'(F,k,s(x))` *to search for recurrence equations in these forms, respectively, and apply these implementations to the polynomial systems that were defined in the previous exercise. Note that in these forms the results are much simpler than the recurrence equation results of the previous exercise. Furthermore, for the classical discrete orthogonal polynomials, $\sigma(x)$ and $\tau(x)$ are polynomials of degree at most 2 and 1, respectively.*

Exercise 7.12 (Wilson Polynomials) *The Wilson polynomials* [Wilson80] *(see* [KLS10], (10.1.1)) *are given by*

$$W_n(x^2) = W_n(x^2; a, b, c, d)$$
$$= (a+b)_n\,(a+c)_n\,(a+d)_n$$
$$\times\, {}_4F_3\left(\begin{array}{c} -n, a+b+c+d+n-1, a-ix, a+ix \\ a+b, a+c, a+d \end{array}\middle|\, 1\right).$$

Find three-term recurrence equations with respect to all parameters involved, particularly with respect to n and ix.[6]

Furthermore, prove the identity

$$W_n(x^2; a, b, c, d) = W_n(x^2; b, a, c, d).$$

The three-term recurrence equations for $W_n(x^2)$ with respect to n and ix are quite complicated. There are much simpler recurrence equations for

$$\widetilde{W}_n(x^2) = \frac{W_n(x^2)}{(a+b)_n\,(a+c)_n\,(a+d)_n}$$

of the form ([KLS10], (10.1.4))

$$\alpha(x)\,\widetilde{W}_n(x^2) = A_n\,\widetilde{W}_{n+1}(x^2) - (A_n + C_n)\,\widetilde{W}_n(x^2) + C_n\,\widetilde{W}_{n-1}(x^2)$$

where $\alpha(x)$ does not depend on n, and A_n, C_n do not depend on x, and (compare [KLS10], (10.1.6))

$$\beta_n\,\widetilde{W}_n(x^2) = \widetilde{B}(x)\,\widetilde{W}_n(i(x+1)) - (\widetilde{B}(x) + \widetilde{D}(x))\,\widetilde{W}_n(x) + \widetilde{D}(x)\,\widetilde{W}_n(i(x-1)),$$

where β_n does not depend on x, and $\widetilde{B}(x)$, $\widetilde{D}(x)$ do not depend on n, respectively. Find these! Also use the procedure `sumdelta+nabla` from the previous exercise for which the result looks rather simple.

Exercise 7.13 (Whipple Transformation) *Show that*

$$ {}_4F_3\left(\begin{array}{c} -n, b, c, d \\ e, f, g \end{array}\middle|\, 1\right) = \frac{(f-b)_n\,(g-b)_n}{(f)_n\,(g)_n}$$
$$\times\, {}_4F_3\left(\begin{array}{c} -n, b, e-c, e-d \\ e, b-f-n+1, b-g-n+1 \end{array}\middle|\, 1\right)$$

constitutes an identity provided that $e + f + g = -n + b + c + d + 1$.

Exercise 7.14 *Show, for* $d = 2, 3, 4, 5$, *that*

$$\sum_{k=-\infty}^{\infty} (-1)^k \binom{n}{k}\binom{dk}{n} = (-d)^n, \tag{7.21}$$

---

[6] There is no recurrence w.r.t. x, but w.r.t. $y = ix$.

by an application of Zeilberger's algorithm [PS95]. What are the orders of the resulting recurrence equations?

Can one prove (7.21) for arbitrary $d \in \mathbb{N}$ by Zeilberger's algorithm?

Exercise 7.15 *Show that the two results concerning the Watson identity in Session 8.2 are equivalent to Watson's identity.*

Exercise 7.16 *Define an appropriate rational certificate $R(n, k)$ for Zeilberger's algorithm, and give a detailed description of which rational identity must be checked to prove the resulting recurrence equation with the aid of $R(n, k)$.*

Exercise 7.17 *Prove that for all $i = 0, \ldots, I$ and $j = 0, \ldots, J$ the expression $K^i N^j F(n, k)$ is a polynomial multiple of $\widetilde{H}(n, k)$, given by (7.17),*

$$K^i N^j F(n, k) = p(n, k)\widetilde{H}(n, k) \qquad (p(n, k) \in \mathbb{Q}[n, k]),$$

where the polynomial factor $p(n, k)$ has degree at most

$$\deg_k P(n, k) + \left(\sum_{l=1}^{p} |\alpha_l| + \sum_{l=1}^{q} |\gamma_l|\right) I + \left(\sum_{l=1}^{p} |\beta_l| + \sum_{l=1}^{q} |\delta_l|\right) J$$

with respect to k.

Exercise 7.18 *Find hypergeometric term representations for the following sums*

(a) $\displaystyle \sum_{k=0}^{m-1} \frac{4(-1)^k \binom{m-1}{k}\binom{2m-1}{2k}\left(4m^2+16k^2-16km+16k-6m+3\right)}{\binom{4m-1}{4k}\left(4m-4k-3\right)\left(4m-4k-1\right)},$

(b) $\displaystyle \sum_{k=0}^{n-p} \binom{1+2n}{2p+2k+1}\binom{p+k}{k},$

(c) $\displaystyle \sum_{k=0}^{\lfloor n/2 \rfloor} \frac{(-n/2)_k (-n/2+1/2)_k}{k! (b+1/2)_k},$

(d) $\displaystyle \sum_{k=0}^{n} \binom{-1/4}{k}^2 \binom{-1/4}{n-k}^2,$

(e) $\displaystyle {}_3F_2\left(\begin{matrix} -n, 1-a-n, 1-b-n \\ a, b \end{matrix} \;\middle|\; 1\right),$

(f) $\displaystyle {}_3F_2\left(\begin{matrix} -n+k, k+1/2, n+k+2 \\ k+3/2, 2k+1 \end{matrix} \;\middle|\; 1\right).$

Exercise 7.19 *Show that Zeilberger's algorithm applied to*

$$S_n := {}_3F_2\left(\begin{matrix} -n, b, c+m \\ b+1, c \end{matrix} \;\middle|\; 1\right),$$

for any fixed given $m \in \mathbb{N}_{\geq 0}$, generates the recurrence equation of first order

$$(n + 1 + b)\, s_{n+1} - (n + 1)\, s_n = 0,$$

valid for $n \geq m$, the degree bound to find f_k being equal to m. This shows that the complexity of Zeilberger's algorithm can be arbitrarily high despite its success for $J = 1$.

Exercise 7.20 *Prove the identity*

$$\sum_{k=0}^{\lfloor n/2 \rfloor} \binom{n-k}{k} x^k = \frac{1}{2^n} \sum_{k=0}^{\lfloor n/2 \rfloor} \binom{n+1}{2k+1} (1 + 4x)^k.$$

Give two different hypergeometric representations of the Fibonacci numbers, thus extending the result of Exercise 4.6.

Exercise 7.21 *Prove the following identity, proposed in SIAM Review 37, 1995, Problem 95-1 (b) [XT95]:*

$$\sum_{k=0}^{m} \binom{2n-m-1-k}{n-k} \binom{m+k}{k} = \binom{2n-1}{n} \quad (m \leq n-1).$$

What is the result for the corresponding sum

$$\sum_{k=0}^{n} \binom{2n-m-1-k}{n-k} \binom{m+k}{k}$$

with natural bounds?

Exercise 7.22 (Clausen's Product Identity) *Use the method of Example 7.7 to prove the Clausen product identity*

$$\begin{aligned}
{}_2F_1 &\left(\begin{array}{c} 1/4 + a,\ 1/4 + b \\ 1 + a + b \end{array} \middle| x \right) \cdot {}_2F_1 \left(\begin{array}{c} 1/4 - a,\ 1/4 - b \\ 1 - a - b \end{array} \middle| x \right) \\
&= {}_3F_2 \left(\begin{array}{c} 1/2,\ 1/2 + a - b,\ 1/2 - a + b \\ 1 + a + b,\ 1 - a - b \end{array} \middle| x \right).
\end{aligned}$$

Exercise 7.23 *Use the method of Example 7.7 to find hypergeometric representations for the following products*

(a) ${}_0F_1 \left(\begin{array}{c} - \\ a \end{array} \middle| x \right) \cdot {}_0F_1 \left(\begin{array}{c} - \\ b \end{array} \middle| x \right),$

(b) ${}_1F_1 \left(\begin{array}{c} a \\ b \end{array} \middle| x \right) \cdot {}_1F_1 \left(\begin{array}{c} a \\ b \end{array} \middle| -x \right),$

(c) $\ _0F_2 \left(\begin{matrix} - \\ a,b \end{matrix} \middle| x \right) \cdot \ _0F_2 \left(\begin{matrix} - \\ a,b \end{matrix} \middle| -x \right).$

Exercise 7.24 *Which hypergeometric identity of Bailey's list (Table 6.1, under Algorithm 6.5) was proved by the application of Zeilberger's algorithm in Example 7.8; see [Gasper86].*

Exercise 7.25 *Use the method of Example 7.8 to discover a hypergeometric representation of the double sum*

$$\sum_{k=0}^{n} \binom{n}{k} (c)_k \, (m)_{n-k} \cdot \ _3F_2 \left(\begin{matrix} -k,a,b \\ c,d \end{matrix} \middle| 1 \right);$$

(see [Gasper74]).

Exercise 7.26 *Consider the infinite matrix*

$$A_{mn} := \sum_{k=-\infty}^{\infty} \sum_{j=-\infty}^{\infty} \binom{m}{j}^2 \binom{m}{k}^2 \binom{2m+n-j-k}{2m}.$$

Prove that

$$A_{mn} = \sum_{k=-\infty}^{\infty} \binom{m+n-k}{k}^2 \binom{m+n-2k}{m-k}^2 = \sum_{k=-\infty}^{\infty} \binom{m}{k}\binom{n}{k}\binom{m+k}{k}\binom{n+k}{k}.$$

This shows in particular that A_{mn} is symmetric, and its diagonal elements A_{nn} are the Apéry numbers A_n of (7.8).

Exercise 7.27 *The Maple procedure* sumrecursion *of Session 7.3 is not very efficient since* ratio *is applied to the sum a_k, i.e.,* simpcomb *is applied to rather complicated expressions in general. Rewrite* sumrecursion *by a more direct approach using (7.4). Compare the timings of the two different versions of the procedure using the examples of this chapter, in particular Dougall's identity.*

Exercise 7.28 *Prove (10)*

$$S(n) := \sum_{k=0}^{\lfloor n/2 \rfloor} (-1)^k \binom{n}{k} \binom{2n-2k}{n} = 2^n,$$

using Zeilberger's algorithm.

Exercise 7.29 (Parameter Derivative) *Use the method of Example 7.8 to obtain the identity ([AF69], (3.46), see also [Askey68], (8))*

$$L_n^{(\alpha+\mu)}(x) = \sum_{k=0}^{n} \frac{(\mu)_{n-k}}{(n-k)!} L_k^{(\alpha)}(x)$$

for the generalized Laguerre polynomials

$$L_n^{(\alpha)}(x) = \sum_{k=0}^{n} \frac{(-1)^k}{k!} \binom{n+\alpha}{n-k} x^k = \binom{n+\alpha}{n} {}_1F_1\left(\begin{array}{c} -n \\ 1+\alpha \end{array} \middle| x\right).$$

Take the limit as $\mu \to 0$ to get for the derivative with respect to the parameter α ([Koepf97a, KS98b])

$$\frac{\partial}{\partial \alpha} L_n^{(\alpha)}(x) = \sum_{k=0}^{n-1} \frac{1}{n-k} L_k^{(\alpha)}(x).$$

Exercise 7.30 (Jacobi Polynomials) *The Jacobi polynomials $P_n^{(\alpha,\beta)}(x)$ generalize the Legendre polynomials $P_n(x) = P_n^{(0,0)}(x)$, and can be defined by the hypergeometric representation*

$$P_n^{(\alpha,\beta)}(x) = \binom{n+\alpha}{n} {}_2F_1\left(\begin{array}{c} -n,\, n+\alpha+\beta+1 \\ \alpha+1 \end{array} \middle| \frac{1-x}{2}\right).$$

Deduce the identities (see [Askey68, AG71], (2.7)–(2.8))

$$P_n^{(a,\beta)}(x) = \sum_{k=0}^{n}(2k+\alpha+\beta+1)\frac{\Gamma(n+\beta+1)}{\Gamma(k+\beta+1)}\frac{\Gamma(n+k+a+\beta+1)}{\Gamma(n+a+\beta+1)}$$
$$\times \frac{\Gamma(k+\alpha+\beta+1)}{\Gamma(n+k+\alpha+\beta+2)}\frac{(a-\alpha)_{n-k}}{(n-k)!}P_k^{(\alpha,\beta)}(x)$$

and

$$P_n^{(\alpha,b)}(x) = \sum_{k=0}^{n}(-1)^{n-k}(2k+\alpha+\beta+1)\frac{\Gamma(n+\alpha+1)}{\Gamma(k+\alpha+1)}\frac{\Gamma(n+k+\alpha+b+1)}{\Gamma(n+\alpha+b+1)}$$
$$\times \frac{\Gamma(k+\alpha+\beta+1)}{\Gamma(n+k+\alpha+\beta+2)}\frac{(b-\beta)_{n-k}}{(n-k)!}P_k^{(\alpha,\beta)}(x)$$

by applying the method of Example 7.8. From these identities the representations

$$\frac{\partial}{\partial \alpha} P_n^{(\alpha,\beta)}(x) = \sum_{k=0}^{n-1}\frac{1}{\alpha+\beta+1+k+n}\cdot\left(P_n^{(\alpha,\beta)}(x)\right.$$
$$\left. +\frac{\alpha+\beta+1+2k}{n-k}\frac{(\beta+k+1)_{n-k}}{(\alpha+\beta+k+1)_{n-k}}P_k^{(\alpha,\beta)}(x)\right)$$

and

$$\frac{\partial}{\partial\beta} P_n^{(\alpha,\beta)}(x) = \sum_{k=0}^{n-1} \frac{1}{\alpha+\beta+1+k+n} \cdot \Big(P_n^{(\alpha,\beta)}(x)$$

$$+(-1)^{n-k}\frac{\alpha+\beta+1+2k}{n-k}\frac{(\alpha+k+1)_{n-k}}{(\alpha+\beta+k+1)_{n-k}} P_k^{(\alpha,\beta)}(x)\Big)$$

for the parameter derivatives of the Jacobi polynomials can be obtained ([Fröhlich, KS98b]).

Exercise 7.31 (Bessel Functions) *The product of two Bessel functions*

$$J_n(x) = \left(\frac{x}{2}\right)^n \sum_{k=0}^{\infty} \frac{(-1)^k}{4^k\, k!\, \Gamma(k+1+n)} x^{2k}$$

and $J_m(x)$ has a hypergeometric representation. Generate this representation.

Exercise 7.32 *Let $F(n,k)$ and s_n be hypergeometric terms. Prove: The WZ method fails to prove the identity*

$$\sum_{k=-\infty}^{\infty} \frac{F(n,k)}{s_n} = 1$$

if and only if Zeilberger's algorithm fails to discover the first order recurrence equation valid for

$$s_n = \sum_{k=-\infty}^{\infty} F(n,k).$$

Exercise 7.33 *Show by an explicit consideration that Zeilberger's algorithm does not terminate for*

$$F(n,k) = \frac{1}{k^2+n^2+1}\binom{n}{k}$$

([Stölting], *Lemma 13, p. 60).*

Exercise 7.34 *In SIAM Review 38, 1996, Problem 96-16 [IR96], the following question was posed:*
    Define

$$S_n(p) = \sum_{j=0}^{n}\left\{\binom{pn+p+1}{pj+p-1} - \binom{pn+p+1}{pj+p-2}\right\}$$

for integers $n \geq 0$ and $p \geq 1$.
Evaluate $S_n(p)$ for $p = 1, 2, 3, 4, 5, 6. \ldots$
Solve this problem for $p = 3, 4, 5, 6$ using Zeilberger's algorithm. For $p = 1$ and $p = 2$ the problem was solved in Exercise 5.21 by Gosper's algorithm. Hint: For $p = 3$ the bounds are not the natural ones.

Exercise 7.35 *Use the q-analogue of Zeilberger's algorithm to generate the q-analogues of the Chu-Vandermonde and of the Pfaff-Saalschütz identities* (3.7)

$$\,_2\phi_1\left(\begin{matrix} q^{-n}, b \\ c \end{matrix}\,\bigg|\, q, \frac{cq^n}{b}\right) = \frac{(c/b;\, q)_n}{(c;\, q)_n}$$

and (3.6)

$$\,_3\phi_2\left(\begin{matrix} q^{-n}, a, b \\ c, \frac{ab}{cq^{n-1}} \end{matrix}\,\bigg|\, q, q\right) = \frac{(c/a;\, q)_n\,(c/b;\, q)_n}{(c;\, q)_n\,(c/(ab);\, q)_n};$$

compare (4.19).

Exercise 7.36 *Use the q-Zeilberger algorithm to generate the q-analogues of the binomial theorem, and of Kummer's, Dixon's and Dougall's identities that were proved by the q-WZ method in Exercise 6.10 on p. 115.*

Exercise 7.37 (q-Orthogonal Polynomials) *Find three-term recurrence equations for the little q-Legendre polynomials, the big q-Legendre polynomials, the continuous q-Legendre polynomials, and the q-Laguerre polynomials, see pp. 71. Prove the identity* (4.9)

$$P_n(x|q^2) = P_n(x;\, q).$$

For the little and big q-Legendre polynomials, give recurrence equations w.r.t. x. Replace x by q^x if necessary.

Exercise 7.38 (*Stanton's Conjecture*) *Prove Stanton's Conjecture*

$$\sum_{k=0}^{2n}(-1)^k\,q^{4k^2}\left(\begin{matrix} 2n \\ n-4k \end{matrix}\right)_q = \sum_{k=0}^{n} q^{2k^2}\left[\begin{matrix} n \\ 2k \end{matrix}\right]_{q^2}\left(-q;\, q^2\right)_{n-2k}\left(-1;\, q^4\right)_k.$$

This conjecture was first proved by Paule and Riese [PS95] using the q-Zeilberger algorithm.

Exercise 7.39 (q-Hypergeometric Transformations) *Prove the following q-hypergeometric transformations:*
Watson's transformation:

$$\,_8\phi_7\left(\begin{matrix} a, q\sqrt{a}, -q\sqrt{a}, b, c, d, e, q^{-n} \\ \sqrt{a}, -\sqrt{a}, aq/b, aq/c, aq/d, aq/e, aq^{1+n} \end{matrix}\,\bigg|\, q, \frac{a^2q^{2+n}}{bcde}\right)$$
$$= \frac{(aq, aq/(de);\, q)_n}{(aq/d, aq/e;\, q)_n}\,_4\phi_3\left(\begin{matrix} aq/(bc), d, e, q^{-n} \\ aq/b, aq/c, deq^{-n}/a \end{matrix}\,\bigg|\, q, q\right)$$

([GR90], Appendix (III.18)), Bailey's transformation:

$$\,_{10}\phi_9\left(\begin{array}{c} a, q\sqrt{a}, -q\sqrt{a}, b, c, d, e, f, \lambda a q^{1+n}/(ef), q^{-n} \\ \sqrt{a}, -\sqrt{a}, aq/b, aq/c, aq/d, aq/e, aq/f, efq^{-n}/\lambda, aq^{1+n} \end{array}\,\Bigg|\, q, q\right)$$

$$= \frac{(aq, aq/(ef), \lambda q/e, \lambda q/f;\ q)_n}{(aq/e, aq/f, \lambda q/(ef), \lambda q;\ q)_n}\cdot$$

$$\,_{10}\phi_9\left(\begin{array}{c} \lambda, q\sqrt{\lambda}, -q\sqrt{\lambda}, \lambda b/a, \lambda c/a, \lambda d/a, e, f, \lambda a q^{1+n}/(ef), q^{-n} \\ \sqrt{\lambda}, -\sqrt{\lambda}, aq/b, aq/c, aq/d, \lambda q/e, \lambda q/f, efq^{-n}/a, \lambda q^{1+n} \end{array}\,\Bigg|\, q, q\right)$$

([GR90], *Appendix* (III.28)), *where* $c = qa^2/(bd\lambda)$, *and*

$$\,_5\phi_4\left(\begin{array}{c} a, b, c, d, q^{-n} \\ aq/b, aq/c, aq/d, a^2 q^{-n}/\lambda^2 \end{array}\,\Bigg|\, q, q\right) = \frac{(\lambda q/a, \lambda^2 q/a;\, q)_n}{(\lambda q, \lambda^2 q/a^2;\, q)_n}\cdot$$

$$\,_{12}\phi_{11}\left(\begin{array}{c} \lambda, q\sqrt{\lambda}, -q\sqrt{\lambda}, \lambda b/a, \lambda c/a, \lambda d/a, \sqrt{a}, -\sqrt{a}, \sqrt{aq}, -\sqrt{aq}, \lambda^2 q^{1+n}/a, q^{-n} \\ \sqrt{\lambda}, -\sqrt{\lambda}, aq/b, aq/c, aq/d, \lambda q/\sqrt{a}, -\lambda q/\sqrt{a}, \lambda\sqrt{q/a}, -\lambda\sqrt{q/a}, aq^{-n}/\lambda, \lambda q^{1+n} \end{array}\,\Bigg|\, q, q\right)$$

([GR90], *Appendix* (III.25)), *where* $c = qa^2/(bd\lambda)$.

References

Abramov02. Abramov, S.A.: Applicability of Zeilberger's algorithm to hypergeometric terms. In: Proceedings of ISSAC 02, pp. 1–7. ACM Press, New York (2002).

AL00. Abramov, S.A., Le, H.Q.: Applicability of Zeilberger's algorithm to rational functions. In: Proceedings FPSAC'2000, pp. 91–102. Springer LNCS (2000).

AZ06. Apagodu, M., Zeilberger, D.: Multi-variable Zeilberger and Almkvist-Zeilberger algorithms and the sharpening of Wilf-Zeilberger theory. Adv. Appl. Math. **37**, 139–152 (2006)

Apéry79. Apéry, R.: Irrationalité de $\zeta(2)$ et $\zeta(3)$. Astérisque **61**, 11–13 (1979)

Askey68. Askey, R.: Dual equations and classical orthogonal polynomials. J. Math. Anal. Appl. **24**, 677–685 (1968)

AF69. Askey, R., Fitch, J.: Integral representations for Jacobi polynomials and some applications. J. Math. Anal. Appl. **26**, 411–437 (1969)

AG71. Askey, R., Gasper, G.: Jacobi polynomial expansions of Jacobi polynomials with non-negative coefficients. Proc. Camb. Phil. Soc. **70**, 243–255 (1971)

AG76. Askey, R., Gasper, G.: Positive Jacobi polynomial sums II. Amer. J. Math. **98**, 709–737 (1976)

BK99. Böing, H., Koepf, W.: Algorithms for q-hypergeometric summation in computer algebra. J. Symbolic Comput. **28**, 777–799 (1999)

CAOP. CAOP: Computer Algebra and Orthogonal Polynomials. Web resource www.caop.org. Idea by Tom H. Koornwinder, first realization by René Swarttouw, now designed by Torsten Sprenger and maintained by Wolfram Koepf.

CHM08. Chen, W.Y.C., Hou, Q.-H., Mu, Y.-P.: Non-terminating basic hypergeometric series and the q-Zeilberger algorithm. Proc. Edinb. Math. Soc. 2(51), 609–633 (2008)

Chyzak00. Chyzak, F.: An extension of Zeilberger's fast algorithm to general holonomic functions. Discrete Math. **217**, 115–134 (2000)

CKS09. Chyzak, F., Kauers, M., Salvy, B.: A non-holonomic systems approach to special function identities. In: Proceedings of ISSAC 09, pp. 111–118. ACM Press, New York (2009).

CS98. Chyzak, F., Salvy, B.: Non-commutative elimination in Ore algebras proves multivariate identities. J. Symbolic Comput. **26**, 187–227 (1998)

deBranges85. De Branges, L.: A proof of the Bieberbach conjecture. Acta Math. **154**, 137–152 (1985)

Feldheim43. Feldheim, E.: Relations entre les polynomes de Jacobi. Laguerre et Hermite. Acta Math. **75**, 117–138 (1943)

Fröhlich. Fröhlich, J.: Parameter derivatives of the Jacobi polynomials and the Gaussian hypergeometric function. Integr. Transform. Spec. Funct. **2**, 252–266 (1994)

Gasper74. Gasper, G.: Projection formulas for orthogonal polynomials of a discrete variable. J. Math. Anal. Appl. **45**, 176–198 (1974)

Gasper86. Gasper, G.: A short proof of an inequality used by de Branges in his proof of the Bieberbach. Robertson and Milin conjectures. Complex Variables **7**, 45–50 (1986)

GR90. Gasper, G., Rahman, M.: Basic Hypergeometric Series. Encyclopedia of Mathematics and its Applications, vol. 35. Cambridge University Press, London and New York (1990) (2nd edn 2004).

Gerhard01. Gerhard, J.: Modular algorithms in symbolic summation and symbolic integration. PhD thesis, Universität Paderborn (2001).

GKP94. Graham, R.L., Knuth, D.E., Patashnik, O.: Concrete Mathematics, 2nd edn. A Foundation for Computer Science. Addison-Wesley, Reading, Massachussets (1994)

IR96. Ierley, G.R., Ruehr, O.G.: Problem 96–16. SIAM Rev. **38**, 668 (1996)

KLS10. Koekoek, R., Lesky, P., Swarttouw, R.F.: Hypergeometric Orthogonal Polynomials and their q-Analogues. Springer Monographs in Mathematics, Springer, Berlin (2010)

KS96. Koepf, W., Schmersau, D.: On the de Branges theorem. Complex Variables **31**, 213–230 (1996)

Koepf97a. Koepf, W.: Identities for families of orthogonal polynomials and special functions. Integr. Transform. Spec. Funct. **5**, 69–102 (1997)

Koepf97b. Koepf, W.: The algebra of holonomic equations. Math. Semesterberichte **44**, 173–194 (1997)

Koepf06. Koepf, W.: Computeralgebra. Springer, Berlin (2006)

KS98b. Koepf, W., Schmersau, D.: Representations of orthogonal polynomials. J. Comput. Appl. Math. **90**, 57–94 (1998)

Koornwinder93. Koornwinder, T.H.: On Zeilberger's algorithm and its q-analogue. J. Comput. Appl. Math. **48**, 91–111 (1993)

Koornwinder98. Koornwinder, T.H.: Identities of nonterminating series by Zeilberger's algorithm. J. Comput. Appl. Math. **99**, 449–461 (1998)

MZ05. Mohammed, M., Zeilberger, D.: Sharp upper bounds for the orders of the recurrences outputted by the Zeilberger and q-Zeilberger algorithms. J. Symbolic Comput. **39**, 201–207 (2005)

PR97. Paule, P., Riese, A.: A Mathematica q-analogue of Zeilberger's algorithm based on an algebraically motivated approach to q-hypergeometric telescoping. In: Ismail, M.E.H., et al. (eds.) Fields Institute Communications, vol. 14, pp. 179–210. American Mathematical Society, Rhode Island (1997)

PS95. Paule, P., Schorn, M.: A Mathematica version of Zeilberger's algorithm for proving binomial coefficient identities. J. Symbolic Comput. **20**, 673–698 (1995)

PWZ96. Petkovšek, M., Wilf, H., Zeilberger, D.: $A = B$. AK Peters, Wellesley (1996)

Rainville60. Rainville, E.D.: Special Functions. The MacMillan Co., New York (1960)

Stölting. Stölting, G.: Algorithmische Berechnung von Summen. Diploma thesis, Freie Universität Berlin (1996).

Todorov92. Todorov, P.: A simple proof of the Bieberbach conjecture. Bull. Cl. Sci., VI. Sér., Acad. R. Belg. 3 12, 335–356 (1992).

vanderPoorten78. van der Poorten, A.: A proof that Euler missed. Apéry's proof of the irrationality of $\zeta(3)$. Math. Intelligencer **1**, 195–203 (1978)

Vidunas01. Vidunas, R.: Maple package infhsum. http://staff.science.uva.nl/thk/specfun/infhsum.mpl

Vidunas02. Vidunas, R.: A Generalization of Kummer's Identity. Rocky Mountain J. Math. **32**, 919–936 (2002)

Weinstein91. Weinstein, L.: The Bieberbach conjecture. Int. Math. Res. Not. **5**, 61–64 (1991)

Wilf94. Wilf, H.S.: A footnote on two proofs of the Bieberbach-de Branges theorem. Bull. London Math. Soc. **26**, 61–63 (1994)

WZ92. Wilf, H.S., Zeilberger, D.: An algorithmic proof theory for hypergeometric (ordinary and "q") multisum/integral identities. Invent. Math. **108**, 575–633 (1992)

Wilson80. Wilson, J.A.: Some hypergeometric orthogonal polynomials. Siam J. Math. Anal. **11**, 690–701 (1980)

XT95. Xin-Rong, M., Tian-Ming, W.: Problem 95–1. SIAM Rev. **37**, 98 (1995)

Zeilberger90a. Zeilberger, D.: A holonomic systems approach to special functions identities. J. Comput. Appl. Math. **32**, 321–368 (1990)

Zeilberger90b. Zeilberger, D.: A fast algorithm for proving terminating hypergeometric identities. Discrete Math. **80**, 207–211 (1990)

Zeilberger91a. Zeilberger, D.: The method of creative telescoping. J. Symbolic Comput. **11**, 195–204 (1991)

Zeilberger91b. Zeilberger, D.: A Maple program for proving hypergeometric identities. SIGSAM Bull. **25**, 1–13 (1991)

Chapter 8
Extensions of the Algorithms

In this chapter, we extend Gosper's, Wilf-Zeilberger's and Zeilberger's methods to accept rational-linear Γ inputs rather than only integer-linear ones [Koepf95a]. For such an input a_{k+1}/a_k is not always rational, so that Gosper's algorithm may not apply. Therefore, we raise a different question: Given a nonnegative integer m, can we find a sequence s_k for given a_k satisfying

$$a_k = s_{k+m} - s_k \tag{8.1}$$

in the particular case that s_k is an *m-fold hypergeometric term*, i.e.,

$$\frac{s_{k+m}}{s_k} \in \mathbb{Q}(k) \ ? \tag{8.2}$$

Note that in this case the input function a_k is itself an m-fold hypergeometric term since by (8.1) and (8.2)

$$\frac{a_{k+m}}{a_k} = \frac{s_{k+2m} - s_{k+m}}{s_{k+m} - s_k} = \frac{s_{k+m}}{s_k} \frac{\frac{s_{k+2m}}{s_{k+m}} - 1}{\frac{s_{k+m}}{s_k} - 1} = \frac{u_k}{v_k}$$

is rational, i.e., u_k and v_k can be chosen to be polynomials, $u_k, v_k \in \mathbb{Q}[k]$.

Assume that, given a_k, we have found an *m-fold antidifference*, namely s_k with the property $s_{k+m} - s_k = a_k$. Then we can easily construct an antidifference $\tilde{s}_k$ of a_k by the simple definition

$$\tilde{s}_k := s_k + s_{k+1} + \cdots + s_{k+(m-1)} \tag{8.3}$$

since then, by telescoping,

$$\tilde{s}_{k+1} - \tilde{s}_k = (s_{k+1} + \cdots + s_{k+m}) - (s_k + \cdots + s_{k+(m-1)}) = s_{k+m} - s_k = a_k.$$

W. Koepf, *Hypergeometric Summation*, Universitext,
DOI: 10.1007/978-1-4471-6464-7_8, © Springer-Verlag London 2014

We summarize

Lemma 8.1 and Algorithm *If s_k is an m-fold antidifference of a_k then $\tilde{s}_k$ given by* (8.3) *forms an antidifference of a_k.* $\square$

Assume next that an m-fold antidifference s_k of a_k forms a *hypergeometric term.* Then obviously

$$\frac{s_{k+m}}{s_k} = \frac{s_{k+m}}{s_{k+(m-1)}} \cdot \frac{s_{k+(m-1)}}{s_{k+(m-2))}} \cdots \frac{s_{k+1}}{s_k}$$

is also rational, and therefore our algorithm below will find s_k.

An m-fold hypergeometric m-fold antidifference can always be constructed by an application of Gosper's original algorithm in the following way:

Algorithm 8.2 (extended_gosper)

The following steps generate an m-fold hypergeometric m-fold antidifference:

1. Input: $m \in \mathbb{N}$, and a_k in terms of rational functions, powers, factorials, Γ function terms, binomial coefficients and Pochhammer symbols that are rational-linear in their arguments.
2. Define $b_k := a_{km}$.
3. Apply Gosper's algorithm to b_k with respect to k. Get the antidifference t_k of b_k, or the statement: "No hypergeometric term antidifference of b_k, and therefore no m-fold hypergeometric m-fold term antidifference of a_k exists."
4. The output $s_k := t_{k/m}$ is a solution of (8.1) with the property (8.2).

Proof The existence of an m-fold hypergeometric solution s_k of

$$s_{k+m} - s_k = a_k \tag{8.4}$$

is equivalent to the existence of a rational solution $S(k)$ of

$$r(k)\, S(k+m) - S(k) = 1 \tag{8.5}$$

where $r(k) = a_{k+m}/a_k$ and $S(k) = s_k/a_k$. The existence of a hypergeometric solution t_k of

$$t_{k+1} - t_k = a_{km} = b_k \tag{8.6}$$

is equivalent to the existence of a rational solution $T(k)$ of

$$r(km)\, T(k+1) - T(k) = 1 \tag{8.7}$$

where $T(k) = t_k/a_{km}$. Clearly (8.5) and (8.7) are either both solvable and have solutions such that $T(k) = S(km)$, or are both unsolvable. So either (8.6) has no hypergeometric solution and (8.4) has no m-fold hypergeometric solution, or (8.6) has a hypergeometric solution $t_k = T(k)a_{km}$ and (8.4) has an m-fold hypergeometric solution $s_k = S(k)a_k = T(k/m)a_k$, hence $s_k = t_{k/m}$. $\square$

Example 8.3 As an example, we consider $a_k := k\left(\frac{k}{2}\right)!$, and $m = 2$. Then $b_k = a_{2k} = 2k\,k!$, and Gosper's algorithm yields $t_k = 2k!$. Therefore $s_k = t_{k/2} = 2\left(\frac{k}{2}\right)!$ has the property that

$$s_{k+2} - s_k = a_k.$$

By (8.3), we find the antidifference

$$\tilde{s}_k = s_k + s_{k+1} = 2\left(\frac{k}{2}\right)! + 2\left(\frac{k+1}{2}\right)!$$

of a_k.

We consider two more examples: If $a_k = \binom{k/3}{n}$ then our algorithm generates the antidifference

$$\tilde{s}_k = -\frac{\left(n - \frac{k}{3}\right)}{n+1}\binom{\frac{k}{3}}{n} - \frac{\left(n - \frac{k}{3} - 1/3\right)}{n+1}\binom{\frac{k}{3} + 1/3}{n} - \frac{\left(n - \frac{k}{3} - 2/3\right)}{n+1}\binom{\frac{k}{3} + 2/3}{n},$$

and if $a_k = \binom{n}{k/2+1} - \binom{n}{k/2}$ then

$$\tilde{s}_k = \frac{\left(\frac{k}{2}+1\right)}{n-k-1}\left(\binom{n}{\frac{k}{2}+1} - \binom{n}{\frac{k}{2}}\right) + \frac{\left(\frac{k}{2}+3/2\right)}{n-k-2}\left(\binom{n}{\frac{k}{2}+3/2} - \binom{n}{\frac{k}{2}+1/2}\right).$$

For more examples, see Exercise 8.5. $\triangle$

Next, we give an algorithm that finds an appropriate nonnegative integer m for suitable input.

Algorithm 8.4 (`find_mfold`)

The following is an algorithm generating a successful choice of m for an application of Algorithm 8.2.

1. Input: a_k as a ratio of products of rational functions, powers, factorials, Γ function terms, binomial coefficients, and Pochhammer symbols that are rational-linear in their arguments.
2. Build the list of all arguments. They are of the form $p_j/q_j\,k + \alpha_j$ with integer p_j and q_j, p_j/q_j in lowest terms, q_j positive.
3. Calculate $m := \mathrm{lcm}\{q_j\}$.

Proof It is clear that the procedure generates a representation for $b_k = a_{km}$ with the given choice of m which is integer-linear in the arguments involved. Since in this case b_{k+1}/b_k is rational, Algorithm 8.2 is applicable. $\square$

We mention that in our examples above, the given procedure yields the desired values $m = 2$ for $a_k := k\left(\frac{k}{2}\right)!$, $m = 3$ for $a_k = \binom{k/3}{n}$, and $m = 2$ for $a_k = \binom{n}{k/2+1} - \binom{n}{k/2}$.

Session 8.5 In Exercise 8.2 you are asked to write an implementation of the above extended version of Gosper's algorithm. Such an implementation is available in the hsum package .

The function extended_gosper(a,k,m) returns the m-fold antidifference of a_k if applicable, whereas with extended_gosper(a,k), we get an antidifference of a_k using Lemma 8.1 and the number m obtained by using Algorithm 8.4.

The above calculations are done by

```
> extended_gosper(k*(k/2)!,k,2);
```

$$2\left(\frac{1}{2}k\right)!$$

```
> extended_gosper(k*(k/2)!,k);
```

$$2\left(\frac{1}{2}k\right)!+2\left(\frac{1}{2}k+\frac{1}{2}\right)!$$

```
> extended_gosper(binomial(k/3,n),k,3);
```

$$\frac{\left(\frac{1}{3}k-n\right)\text{binomial}\left(\frac{1}{3}k,\ n\right)}{n+1}$$

```
> extended_gosper(binomial(k/3,n),k);
```

$$\frac{\left(\frac{1}{3}k-n\right)\text{binomial}\left(\frac{1}{3}k,\ n\right)}{n+1}+\frac{\left(\frac{1}{3}k+\frac{1}{3}-n\right)\text{binomial}\left(\frac{1}{3}k+\frac{1}{3},\ n\right)}{n+1}$$
$$+\frac{\left(\frac{1}{3}k+\frac{2}{3}-n\right)\text{binomial}\left(\frac{1}{3}k+\frac{2}{3},\ n\right)}{n+1}$$

```
> extended_gosper(binomial(n,k/2+1)-binomial(n,k/2),k,2);
```

$$-\frac{\left(\frac{1}{2}k+1\right)\left(\text{binomial}\left(n,\ \frac{1}{2}k+1\right)-\text{binomial}\left(n,\ \frac{1}{2}k\right)\right)}{k+1-n}$$

```
> extended_gosper(binomial(n,k/2+1)-binomial(n,k/2),k);
```

$$-\frac{\left(\frac{1}{2}k+1\right)\left(\text{binomial}\left(n,\ \frac{1}{2}k+1\right)-\text{binomial}\left(n,\ \frac{1}{2}k\right)\right)}{k+1-n}$$
$$-\frac{\left(\frac{1}{2}k+\frac{3}{2}\right)\left(\text{binomial}\left(n,\ \frac{1}{2}k+\frac{3}{2}\right)-\text{binomial}\left(n,\ \frac{1}{2}k+\frac{1}{2}\right)\right)}{k+2-n}$$

Note that Algorithm 8.2 is also implemented in the Maple package `SumTools` under the name `SumTools[Hypergeometric][KoepfGosper]`:

```
>    SumTools[Hypergeometric][KoepfGosper](
>    binomial(n,k/2+1)-binomial(n,k/2),k);
```

$$-\frac{\left(\frac{1}{2}k+1\right)\left(\text{binomial}\left(n,\frac{1}{2}k+1\right)-\text{binomial}\left(n,\frac{1}{2}k\right)\right)}{k+1-n}$$

$$-\frac{\left(\frac{1}{2}k+\frac{3}{2}\right)\left(\text{binomial}\left(n,\frac{1}{2}k+\frac{3}{2}\right)-\text{binomial}\left(n,\frac{1}{2}k+\frac{1}{2}\right)\right)}{k+2-n}$$

Next, we will give an extended version of the WZ method which resolves some questions that remained unanswered in Chap. 6 so that finally Bailey's complete list (Table 6.1) can be settled using a unified approach.

Assume that for a hypergeometric identity the WZ method fails. This may happen either because a_{k+1}/a_k is not rational, or because there is no single formula for the result as in *Andrews' statement*

$$_3F_2\left(\begin{matrix}-n,n+3a,a\\3a/2,(3a+1)/2\end{matrix}\middle|\frac{3}{4}\right)=\begin{cases}0 & \text{if } n\neq 0 \text{ (mod 3)}\\\dfrac{n!\,(a+1)_{n/3}}{(n/3)!\,(3a+1)_n} & \text{otherwise}\end{cases}\tag{8.8}$$

which—together with many similar identities listed in Table 8.1—can be found in a paper of Gessel and Stanton [GS82].

In such cases, we proceed as follows. Assume we want to prove an identity of the form

$$S_n:=\sum_{k=-\infty}^{\infty}F(n,k)=\text{constant}\qquad(n\bmod m\text{ constant}),\tag{8.9}$$

such as, e.g., (8.8) divided by $\frac{n!\,(a+1)_{n/3}}{(n/3)!\,(3a+1)_n}$, m denoting a certain positive integer and $F(n,k)$ being an (m,l)-*fold hypergeometric term* with respect to (n,k), i.e.

$$\frac{F(n+m,k)}{F(n,k)},\quad\frac{F(n,k+l)}{F(n,k)}\quad\in\quad\mathbb{Q}(n,k),$$

with finite support with respect to k. Then we apply the extended version of Gosper's algorithm to find an l-fold antidifference of the expression

$$a_k:=F(n+m,k)-F(n,k)$$

with respect to the variable k.[1] If successful, this generates $G(n,k)$ with

---

[1] In most cases $l=1$, so that Gosper's original algorithm is applied.

Table 8.1 Gessel and Stanton's hypergeometric identities

Equation	Identity

(1.1)
$${}_3F_2\left(\begin{array}{c} -n\,,\,n+3a\,,\,a \\ 3a/2\,,\,(3a+1)/2 \end{array}\middle|\,\frac{3}{4}\right)=\begin{cases} 0 & \text{if } n\neq 0 \ (\mathrm{mod}\ 3) \\[2mm] \dfrac{n!\,(a+1)_{n/3}}{(n/3)!\,(3a+1)_n} & \text{otherwise}\end{cases}$$

(1.2)
$${}_5F_4\left(\begin{array}{c} 2a\,,\,2b\,,\,1-2b\,,\,1+2a/3\,,\,-n \\ a-b+1\,,\,a+b+1/2\,,\,2a/3\,,\,1+2a+2n \end{array}\middle|\,\frac{1}{4}\right)=\frac{(a+1/2)_n\,(a+1)_n}{(a+b+1/2)_n\,(a-b+1)_n}$$

(1.3)
$${}_5F_4\left(\begin{array}{c} a,\,b,\,a+1/2-b,\,1+2a/3,\,-n \\ 2a+1-2b,\,2b,\,2a/3,\,1+a+n/2 \end{array}\middle|\,4\right)=\begin{cases} 0 & \text{if } n \text{ odd} \\[2mm] \dfrac{n!\,(a+1)_{n/2}\,2^{-n}}{(\frac{n}{2})!\,(a-b+1)_{n/2}(b+\frac{1}{2})_{n/2}} & \text{otherwise}\end{cases}$$

(1.4)
$${}_3F_2\left(\begin{array}{c} 1/2+3a\,,\,1/2-3a\,,\,-n \\ 1/2\,,\,-3n \end{array}\middle|\,\frac{3}{4}\right)=\frac{(1/2-a)_n\,(1/2+a)_n}{(1/3)_n\,(2/3)_n}$$

(1.5)
$${}_3F_2\left(\begin{array}{c} 1+3a\,,\,1-3a\,,\,-n \\ 3/2\,,\,-1-3n \end{array}\middle|\,\frac{3}{4}\right)=\frac{(1+a)_n\,(1-a)_n}{(2/3)_n\,(4/3)_n}$$

(1.6)
$${}_3F_2\left(\begin{array}{c} 2a\,,\,1-a\,,\,-n \\ 2a+2\,,\,-a-1/2-3n/2 \end{array}\middle|\,1\right)=\frac{((n+3)/2)_n\,(n+1)(2a+1)}{(1+(n+2a+1)/2)_n\,(2a+n+1)}$$

(1.7)
$${}_7F_6\left(\begin{array}{c} 2a\,,\,2b\,,\,1-2b\,,\,1+2a/3\,,\,a+d+n+1/2\,,\,a-d\,,\,-n \\ a-b+1\,,\,a+b+1/2\,,\,2a/3\,,\,-2d-2n\,,\,2d+1\,,\,1+2a+2n \end{array}\middle|\,1\right)$$
$$=\frac{(2a+1)_{2n}\,(b+d+1/2)_n\,(d-b+1)_n}{(2d+1)_{2n}\,(a+b+1/2)_n\,(a-b+1)_n}=\frac{(a+1/2)_n\,(a+1)_n\,(b+d+1/2)_n\,(d-b+1)_n}{(a+b+1/2)_n\,(a-b+1)_n\,(d+1/2)_n\,(d+1)_n}$$

(1.8)
$${}_7F_6\left(\begin{array}{c} a\,,\,b\,,\,a+1/2-b\,,\,1+2a/3\,,\,1-2d\,,\,2a+2d+n\,,\,-n \\ 2a-2b+1\,,\,2b\,,\,2a/3\,,\,a+d+1/2\,,\,1-d-n/2\,,\,1+a+n/2 \end{array}\middle|\,1\right)$$
$$=\begin{cases} 0 & \text{if } n \text{ odd} \\[2mm] \dfrac{(b+d)_{n/2}\,(d-b+a+1/2)_{n/2}\,n!\,(a+1)_{n/2}\,2^{-n}}{(b+1/2)_{n/2}\,(a+d+1/2)_{n/2}\,(d)_{n/2}\,(n/2)!\,(a-b+1)_{n/2}} & \text{otherwise}\end{cases}$$

(3.7)
$${}_2F_1\left(\begin{array}{c} -n\,,\,-2n-2/3 \\ 4/3 \end{array}\middle|\,-8\right)=\frac{(5/6)_n}{(3/2)_n}\,(-27)^n$$

(5.21)
$${}_3F_2\left(\begin{array}{c} 3a+1/2\,,\,3a+1\,,\,-n \\ 6a+1\,,\,-n/3+2a+1 \end{array}\middle|\,\frac{4}{3}\right)=\begin{cases} 0 & \text{if } n\neq 0 \ (\mathrm{mod}\ 3) \\[2mm] \dfrac{(1/3)_{n/3}\,(2/3)_{n/3}}{(1+2a)_{n/3}\,(-2a)_{n/3}} & \text{otherwise}\end{cases}$$

(5.22)
$${}_2F_1\left(\begin{array}{c} -n\,,\,1/2 \\ 2n+3/2 \end{array}\middle|\,\frac{1}{4}\right)=\frac{(1/2)_n}{(2n+3/2)_n}\left(\frac{27}{4}\right)^n$$

(5.23)
$${}_2F_1\left(\begin{array}{c} -n\,,\,-1/3-2n \\ 2/3 \end{array}\middle|\,-8\right)=(-27)^n$$

(5.24)
$${}_2F_1\left(\begin{array}{c} -n\,,\,n/2+1 \\ 4/3 \end{array}\middle|\,\frac{8}{9}\right)=\begin{cases} 0 & \text{if } n \text{ odd} \\[2mm] \dfrac{(1/2)_{n/2}}{(7/6)_{n/2}}\,(-3)^{-(n/2)} & \text{otherwise}\end{cases}$$

(5.25)
$${}_2F_1\left(\begin{array}{c} -n\,,\,1/2 \\ (n+3)/2 \end{array}\middle|\,4\right)=\begin{cases} 0 & \text{if } n \text{ odd} \\[2mm] \dfrac{(1/2)_{n/2}\,(3/2)_{n/2}}{(5/6)_{n/2}\,(7/6)_{n/2}} & \text{otherwise}\end{cases}$$

(5.27)
$${}_4F_3\left(\begin{array}{c} 1/3-n\,,\,-n/2\,,\,(1-n)/2\,,\,22/21-3n/7 \\ 5/6\,,\,4/3\,,\,1/21-3n/7 \end{array}\middle|\,-27\right)=\frac{(-8)^n}{1-9n}$$

$$a_k = F(n+m, k) - F(n, k) = G(n, k+l) - G(n, k) , \qquad (8.10)$$

and summation over all k leads to

$$s_{n+m} - s_n = \sum_{k=-\infty}^{\infty} \Big(F(n+m, k) - F(n, k) \Big) = \sum_{k=-\infty}^{\infty} \Big(G(n, k+l) - G(n, k) \Big) = 0$$

since the right-hand side can be viewed as the sum of l telescoping series. Therefore s_n is constant mod m, and these m constants can be calculated using suitable initial values. This can be accomplished if the series considered is terminating, i.e., if $F(n, k)$ has finite support. Note that again the function

$$R(n, k) = \frac{G(n, k)}{F(n, k)} \qquad (8.11)$$

acts as a rational certificate function, the *extended WZ certificate.*

Once the extended WZ certificate is known, it is a matter of pure rational arithmetic to decide the validity of (8.9) since the only thing that one has to show is (8.10) which after division by $F(n, k)$ is equivalent to the purely rational identity

$$\frac{F(n+m, k)}{F(n, k)} - 1 - R(n, k+l) \frac{F(n, k+l)}{F(n, k)} + R(n, k) = 0$$

Example 8.6 (Andrews' Identity) As an example, we prove (8.8): In the given case, we set $m = 3, l = 1$, and further

$$F(n, k) := \frac{(-n)_k \, (n+3a)_k \, (a)_k}{k! \, (3a/2)_k \, ((3a+1)/2)_k} \, \frac{(n/3)! \, (3a+1)_n}{n! \, (a+1)_{n/3}} \left(\frac{3}{4}\right)^k .$$

We notice that

$$\frac{F(n, k+1)}{F(n, k)} \quad \text{and} \quad \frac{F(n+3, k)}{F(n, k)}$$

are (complicated) rational functions:

```
>   F:=hyperterm([-n,n+3*a,a],[3*a/2,(3*a+1)/2],3/4,k)*(n/3)!*
>   pochhammer(3*a+1,n)/(n!*pochhammer(a+1,n/3)):

>   ratio(F,k);
```

$$\frac{3 \, (k-n) \, (n+3\,a+k) \, (a+k)}{(3\,a+2\,k) \, (3\,a+1+2\,k) \, (k+1)}$$

```
>   simpcomb(subs(n=n+3,F)/F);
```

$$- \frac{(n+3) \, (n+3\,a+k) \, (n+3\,a+k+1) \, (n+3\,a+k+2)}{(k-n-1) \, (-n-2+k) \, (-n-3+k) \, (n+3\,a)}$$

An application of Gosper's algorithm to $a_k = F(n+3, k) - F(n, k)$ is successful, and leads to the rational certificate

$$R(n, k) = \frac{(2n+3+3a)\,k\,(3a-1+2k)\,(3a+2k-2)}{(n+3a)\,(k-n-1)\,(-n-2+k)\,(-n-3+k)}$$

by the calculation

```
>  A:=subs(n=n+3,F)-F: gos:=gosper(A,k):
>  simpcomb(gos/F);
```

$$\frac{(2n+3+3a)\,k\,(3a-1+2k)\,(3a+2k-2)}{(n+3a)\,(k-n-1)\,(-n-2+k)\,(-n-3+k)}$$

Therefore

$$\sum_{k=-\infty}^{\infty} F(n, k) = \sum_{k=0}^{n} F(n, k) = \text{constant} \qquad (n \bmod 3 \text{ constant}),$$

and statement (8.8) follows using three trivial initial values (check those!).

Session 8.7 We can automate the calculation of the extended WZ certificate by the procedure

```
WZcertificate:=proc(F,k,n)
local a,gos,m,l;
if nargs>3 then m:=args[4] else m:=1 end if;
if nargs>4 then l:=args[5] else l:=1 end if;
a:=subs(n=n+m,F)-F;
try
   gos:=extended_gosper(a,k,l);
catch:
   error `Extended WZ method fails`
end try;
return simpcomb(gos/F);
end proc:
```

with optional fourth argument m and fifth argument l, extending the procedure given in Session 6.7. For Andrews' example, we get

```
>  WZcertificate(F,k,n,3);
```

$$\frac{(2n+3+3a)\,k\,(3a-1+2k)\,(3a+2k-2)}{(n+3a)\,(-n-1+k)\,(-n-2+k)\,(-n-3+k)}$$

Table 8.1 lists the hypergeometric identities of the Gessel-Stanton paper (Eq. (1.4) corrects a misprint in [GS82]), and Table 8.2 contains their rational certificates (8.11), calculated by WZcertificate, together with the certificates of Bailey's

Table 8.2 The extended WZ method

Bailey p.	n	m	$R(n,k)$
11, Gauss	$-a$	2	$-\dfrac{k\,(1+n-b-2\,k)}{(n+1-k)\,(n+2-k)}$
11, Bailey	$-a$	2	$\dfrac{2\,(c-1+k)\,k\,(2\,n+3)}{(n+2-k)\,(n+1-k)\,(n+2-c)}$
16, Watson	$-a$	2	$-\dfrac{k\,(2\,c-1+k)\,(1+n-b-2\,k)}{(n+1-k)\,(n+2-k)\,(1+n-b+2\,c)}$
16, Whipple	$-a$	2	$\dfrac{2\,k\,(-2\,c+e-k)\,(e-1+k)\,(2\,n+3)}{(n+1-k)\,(n+2-k)\,(n+1-2\,c+e)\,(n+2-e)}$

Equation	m	$R(n,k)$
(1.1)	3	$-\dfrac{(3\,a+2\,k-2)\,k\,(3\,a-1+2\,k)\,(2\,n+3+3\,a)}{(n+3\,a)\,(n+1-k)\,(n+2-k)\,(n+3-k)}$
(1.2)	1	$-\dfrac{2\,(2\,a+2\,b-1+2\,k)\,k\,(a-b+k)}{(n+1-k)\,(1+2\,a+2\,n+k)\,(2\,a+3\,k)}$
(1.3)	2	$-\dfrac{(2\,a-2\,b+k)\,(2\,b+k-1)\,k}{(n+2-k)\,(n+1-k)\,(2\,a+3\,k)}$
(1.4)	1	$-\dfrac{8\,(2\,k-1)\,k\,(3\,n-k+1)}{27\,(n+1-k)\,(1-2\,a+2\,n)\,(1+2\,a+2\,n)}$
(1.5)	1	$-\dfrac{2\,(2\,k+1)\,k\,(3\,n+2-k)}{27\,(n+1-k)\,(1-a+n)\,(a+1+n)}$
(1.6)	2	$-\dfrac{(2a+k+1)\,(2a+3+3n-2k)\,k\,\left(9n^2+2na+34n+2a+33-8nk-16k\right)}{3\,(n+2-k)\,(n+1-k)\,(2\,a+n+1)\,(3\,n+5)\,(3\,n+7)}$
(1.7)	1	$-\dfrac{(2\,d+2\,n-k+1)\,(2\,d+k)\,(2\,a+2\,b-1+2\,k)\,k\,(a-b+k)\,(4\,n+2\,a+3+2\,d)}{(n+1-k)\,(1+2a+2n+k)\,(2a+3k)\,(2a+2d+2n+1)\,(2b+2d+1+2n)\,(d-b+1+n)}$
(1.8)	2	$-\dfrac{2\,(2\,a+2\,d-1+2\,k)\,k\,(2\,b+k-1)\,(2\,a-2\,b+k)\,(2\,d+n-2\,k)\,(n+1+a+d)}{(n+2-k)\,(n+1-k)\,(2d-2b+2a+1+n)\,(2a+3k)\,(2b+2d+n)\,(2a+2d+n)}$
(3.7)	1	$\dfrac{(7\,n+9-3\,k)\,(3\,k+1)\,k}{9\,(6\,n+8-3\,k)\,(6\,n+5-3\,k)\,(n+1-k)}$
(5.21)	3	$-\dfrac{k\,(6\,a+k)\,(n-6\,a-3\,k)}{(n+1-k)\,(n+2-k)\,(n+3-k)}$
(5.22)	1	$-\dfrac{4\,(-1+6\,k)\,k}{9\,(n+1-k)\,(4\,n+3+2\,k)}$
(5.23)	1	$\dfrac{k\,(-1+3\,k)\,(21\,n+23-9\,k)}{27\,(n+1-k)\,(4+6\,n-3\,k)\,(7+6\,n-3\,k)}$
(5.24)	2	$\dfrac{3\,(1+3\,k)\,k}{(n+1-k)\,(n+2-k)}$
(5.25)	2	$-\dfrac{(-1+3\,k)\,k}{9\,(n+2-k)\,(n+1-k)}$
(5.27)	1	$-\dfrac{(-1+6\,k)\,(1+3\,k)\,k}{4\,(1+n-2\,k)\,(2+3\,n-3\,k)\,(-1+9\,n-21\,k)}$

list (Table 6.1) to which the WZ method did not apply. In all cases considered we
have $l = 1$, so that Gosper's original algorithm is applied.

Table 8.3 Gessel and Stanton's open problems

Equation	Identity
(6.2)	$${}_7F_6\left(\begin{matrix} a+1/2\,,a\,,b\,,1-b\,,-n\,,(2a+1)/3+n\,,a/2+1 \\ 1/2\,,(2a-b+3)/3\,,(2a+b+2)/3\,,-3n\,,2a+1+3n\,,a/2 \end{matrix}\;\middle|\;1\right)$$ $$=\frac{((2a+2)/3)_n\,(2a/3+1)_n\,((1+b)/3)_n\,((2-b)/3)_n}{((2a-b)/3+1)_n\,((2a+b+2)/3)_n\,(2/3)_n\,(1/3)_n}$$
(6.3)	$${}_5F_4\left(\begin{matrix} a+1/2,\,a,\,-n,\,(2a+1)/3+n,\,a/2+1 \\ 1/2\,,-3n\,,2a+1+3n\,,a/2 \end{matrix}\;\middle|\;9\right)=\frac{((2a+2)/3)_n\,(2a/3+1)_n}{(2/3)_n\,(1/3)_n}$$
(6.5)	$${}_2F_1\left(\begin{matrix} -n\,,-n+1/4 \\ 2n+5/4 \end{matrix}\;\middle|\;\frac{1}{9}\right)=\frac{(5/4)_{2n}}{(2/3)_n\,(13/12)_n}\left(\frac{2^6}{3^5}\right)^n$$
(6.6)	$${}_2F_1\left(\begin{matrix} -n\,,-n+1/4 \\ 2n+9/4 \end{matrix}\;\middle|\;\frac{1}{9}\right)=\frac{(9/4)_{2n}}{(4/3)_n\,(17/12)_n}\left(\frac{2^6}{3^5}\right)^n$$

Rational certificates

Equation	m	$R(n,k)$
(6.2)	1	$-\dfrac{2\,(a+2+3\,n)\,(3\,n-k+1)\,(2\,a+b-1+3\,k)\,(2\,a-b+3\,k)\,(2\,k-1)\,k}{3\,(a+2k)\,(2-b+3n)\,(1+b+3n)\,(2a+1+3n+k)\,(2a+2+3n+k)\,(n+1-k)}$
(6.3)	1	$-\dfrac{(2\,a+4+6\,n)\,(3\,n-k+1)\,(2\,k-1)\,k}{(3\,a+6\,k)\,(2\,a+1+3\,n+k)\,(2\,a+2+3\,n+k)\,(n+1-k)}$
(6.5)	1	$-\dfrac{9\,k\,\left(52\,n^2+75\,n+26+16\,kn+24\,k-32\,k^2\right)}{16\,(8\,n+5+4\,k)\,(4\,n+3-4\,k)\,(n+1-k)}$
(6.6)	1	$-\dfrac{9\,k\,\left(52\,n^2+127\,n+72+16\,kn-4\,k-32\,k^2\right)}{16\,(8\,n+9+4\,k)\,(4\,n+3-4\,k)\,(n+1-k)}$

Note, that Gessel and Stanton were not able to present proofs for their statements (6.2), (6.3), (6.5), and (6.6)[2]: Table 8.3 contains proofs (see [Koepf95a]).

As with the original WZ approach, this method is not capable of proving Gessel-Stanton's (6.1), which is a non-terminating version of (6.2). Also, Gessel-Stanton's result (1.9)

$$_3F_2\left(\begin{matrix} -sb+s+1\,,b-1\,,-n \\ b+1\,,s(-n-b)-n \end{matrix}\;\middle|\;1\right)=\frac{(1+s+sn)_n\,b\,(n+1)}{(1+s(b+n))_n\,(b+n)} \tag{8.12}$$

is beyond the capabilities of the given method since in this case the summand is an (m,l)-fold hypergeometric term only for *fixed* (rational), but not *arbitrary s*; cf. Exercise 8.8.

Next, we give examples of an application for which $l\neq 1$. To prove the identity $(n\in\mathbb{N})$

---

[2] Obviously these were proved subsequently by Zeilberger's algorithm.

$$-\sum_{k=0}^{n}(-2)^n\binom{n}{k}\cdot\binom{k/2}{n} = 1 \; , \tag{8.13}$$

we apply the extended WZ method with $l = 2$, $m = 1$, and get the rational certificate

$$R(n, k) = -\frac{(k - 1)\,(2\,n - k)}{n\,(n + 1 - k)} \; ,$$

which proves (8.13). Note that (8.13) is not valid for $n = 0$ where the denominator of the extended WZ certificate is zero.

Finally, we consider an extension of Zeilberger's algorithm dealing with the question of determining a holonomic recurrence equation (7.1) for sums (7.2) for which $F(n, k)$ is an (m, l)-fold hypergeometric term with respect to (n, k).

In particular, this applies to all cases when the input function $F(n, k)$ is given as a ratio of products of rational functions, powers, factorials, Γ function terms, binomial coefficients, and Pochhammer symbols that are *rational-linear* in their arguments with respect to both n and k.

We mention that Zeilberger's original algorithm may be applicable although this is generally the case only if the arguments are integer-linear. An example of that type is given by the function

$$s_n := {}_2F_1\left(\begin{array}{c}-n/2\,,\,-n/2 + 1/2\\ b + 1/2\end{array}\middle|\,1\right) = \sum_{k=0}^{\infty}\frac{(-n/2)_k\,(-n/2 + 1/2)_k}{k!\,(b + 1/2)_k} \; ,$$

for which an application of Zeilberger's algorithm yields the recurrence equation

$$(2b + n)\,s_{n+1} - 2(b + n)\,s_n = 0 \; ,$$

and therefore the explicit representation

$$s_n = \frac{2^n\,(b)_n}{(2b)_n}.$$

Zeilberger's algorithm applies since $F(n+1, k)/F(n, k)$ and $F(n, k+1)/F(n, k)$ are rational although the expression for $F(n, k)$ is not integer-linear in its arguments.

On the other hand, Zeilberger's algorithm is not directly applicable to every $F(n, k)$ with rational-linear Γ-arguments. An example of this situation is the left-hand side of Watson's theorem (Table 6.1) with respect to variable a.

We present now an algorithm which can be applied for arbitrary rational-linear input [Koepf95a].

Algorithm 8.8 (`extended_sumrecursion`)
The following steps constitute an algorithm to determine a holonomic recurrence equation (7.1) for sums (7.2).

1. Input: $F(n, k)$, given as a ratio of products of rational functions, powers, factorials, Γ function terms, binomial coefficients, and Pochhammer symbols with rational-linear arguments in n and k.
2. Form the list of all arguments. They are of the form $p_j/q_j\, n + s_j/t_j\, k + \alpha_j$ with integers p_j, q_j, s_j, t_j, p_j/q_j and s_j/t_j in lowest terms, q_j and t_j positive.
3. Calculate $m := \operatorname{lcm}\{q_j\}$ and $l := \operatorname{lcm}\{t_j\}$.
4. Define $\tilde{F}(n, k) := F(mn, kl)$. Then $\tilde{F}(n, k)$ is integer-linear in its arguments.
5. Apply Zeilberger's algorithm to $\tilde{F}(n, k)$ to get the recurrence equation

$$\sum_{j=0}^{J} P_j(n)\, \tilde{s}_{n+j} = 0 \tag{8.14}$$

with polynomials P_j in n, for the sum

$$\tilde{s}_n := \sum_{k=-\infty}^{\infty} \tilde{F}(n, k).$$

6. The output is the recurrence equation

$$\sum_{j=0}^{J} P_j(n/m)\, s_{n+mj} = 0$$

for the sum

$$s_n := \sum_{k=-\infty}^{\infty} F(n, k).$$

Proof Our construction provides us with $\tilde{F}(n, k)$ integer-linear in the arguments involved. Therefore Zeilberger's algorithm can be applied, and yields the recurrence equation (8.14), say. Assume first that $l = 1$. Then, by definition, we have $\tilde{s}_n = s_{mn}$ so that we get

$$0 = \sum_{j=0}^{J} P_j(n)\, \tilde{s}_{n+j} = \sum_{j=0}^{J} P_j(n)\, s_{mn+mj} \, ,$$

and the substitution $n = n/m$ gives the result.

If $l > 1$, then Zeilberger's algorithm is an application of Gosper's to

$$a_k = \tilde{F}(n, k) + \sum_{j=1}^{J} \sigma_j(n)\, \tilde{F}(n + j, k).$$

If successful, we get $G(n, k)$ with

$$a_k = \tilde{F}(n, k) + \sum_{j=1}^{J} \sigma_j(n)\, \tilde{F}(n + j, k) = G(n, k + 1) - G(n, k).$$

We sum this equation with respect to all $k = \tilde{k}/l$ ($\tilde{k} \in \mathbb{Z}$). Then the right-hand side can be viewed as a sum of l telescoping sums, again, and we have

$$
\begin{aligned}
0 &= \sum_{k=\tilde{k}/l} \tilde{F}(n, k) + \sum_{j=1}^{J} \sigma_j(n) \sum_{k=\tilde{k}/l} \tilde{F}(n + j, k) \\
&= \sum_{k=\tilde{k}/l} F(mn, kl) + \sum_{j=1}^{J} \sigma_j(n) \sum_{k=\tilde{k}/l} F(mn + mj, kl) \\
&= s_{mn} + \sum_{j=1}^{J} \sigma_j(n)\, s_{mn+mj} \,,
\end{aligned}
$$

which yields the result. $\qquad\square$

Example 8.9 As a first example, we apply the algorithm to Watson's function

$$s_n = {}_3F_2\left(\begin{matrix} -n\,,b\,,c \\ (-n + b + 1)/2\,, 2c \end{matrix} \,\middle|\, 1 \right)$$

with respect to the variable n (Watson's original integer variable) to which Zeilberger's algorithm does not apply directly. In this case, the algorithm determines $m = 2$ and $l = 1$, and leads to the two-fold recurrence equation

$$(b - 2c - n - 1)\,(n + 1)\,s_n - (b - n - 1)\,(2c + n + 1)\,s_{n+2} = 0$$

from which the explicit right-hand representation listed in Table 6.1 can be deduced for integer n since $s_0 = 1$ and

$$s_1 = 1 + \frac{-1\,b\,c}{1\,(b/2)\,(2c)} = 0.$$

Example 8.10 As another example, we consider one of the identities of the paper of Gessel and Stanton [GS82]: The evaluation of ([GS82], (1.8))

$$s_n := {}_7F_6\left(\begin{matrix} a\,,b\,,a + 1/2 - b\,,1 + 2a/3\,,1 - 2d\,,2a + 2d + n\,,-n \\ 2a - 2b + 1\,,2b\,,2a/3\,,a + d + 1/2\,,1 - d - n/2\,,1 + a + n/2 \end{matrix} \,\middle|\, 1 \right)$$

$$= \begin{cases} 0 & \text{if } n \text{ odd} \\[2ex] \dfrac{(b + d)_{n/2}\,(d - b + a + 1/2)_{n/2}\,n!\,(a + 1)_{n/2}\,2^{-n}}{(b + 1/2)_{n/2}\,(a + d + 1/2)_{n/2}\,(d)_{n/2}\,(n/2)!\,(a - b + 1)_{n/2}} & \text{otherwise} \end{cases}$$

cannot be handled with respect to n using Zeilberger's algorithm. The extended version, however, leads to the equivalent 2-fold recurrence equation

$$0 = (n+1+2d+2a)(2b-n-2a-2)(n+1+2b)(n+2d)s_{n+2}$$
$$+(n+1+2d-2b+2a)(n+2d+2b)(2a+n+2)(n+1)s_n.$$

Session 8.11 The procedure `sumrecursion` included in the `hsum` package contains an implementation of the extended Zeilberger algorithm. Let's deal with the Watson example (Example 8.9) again. Set

```
> summand:=hyperterm([-n,b,c],[(-n+b+1)/2,2*c],1,k);
```

$$summand := \frac{\text{pochhammer}(-n,\ k)\,\text{pochhammer}(b,\ k)\,\text{pochhammer}(c,\ k)}{\text{pochhammer}\left(-\dfrac{n}{2}+\dfrac{b}{2}+\dfrac{1}{2},\ k\right)\text{pochhammer}(2\,c,\ k)\,k!}$$

Then this summand is not a hypergeometric term w.r.t. the variable n as the computation

```
> ratio(summand,n);
```

$$-\frac{\Gamma\left(-\dfrac{n}{2}+\dfrac{b}{2}\right)(n+1)\,\Gamma\left(-\dfrac{n}{2}+\dfrac{b}{2}+\dfrac{1}{2}+k\right)}{\Gamma\left(-\dfrac{n}{2}+\dfrac{b}{2}+k\right)(-n-1+k)\,\Gamma\left(-\dfrac{n}{2}+\dfrac{b}{2}+\dfrac{1}{2}\right)}$$

shows. Therefore the implementation `sumrecursion` of Session 7.3 is not applicable. However, for the given example Algorithm 8.8 `extended_sumrecursion` is applied with the result

```
> sumrecursion(summand,k,s(n));
```

$$(b-2\,c-n-1)\,(n+1)\,s(n) - (2\,c+n+1)\,(-n+b-1)\,s(n+2) = 0$$

Furthermore, the procedure `hyperrecursion(upper,lower,x,s(n))` yields a recurrence equation with respect to n for the hypergeometric function with upper parameters `upper`, lower parameters `lower`, and point x. We get for example

```
> hyperrecursion([-n,n+3*a,a],[3/2*a,(3*a+1)/2],3/4,s(n));
```

$$-(n+1+3\,a)\,(n+2+3\,a)\,s(n+3) + (n+1)\,(n+2)\,s(n) = 0$$

Gessel-Stanton (1.6)

```
> hyperrecursion([2*a,1-a,-n],[2*a+2,-a-1/2-3/2*n],1,s(n));
```

$$3\,(3\,n+7)\,(3\,n+5)\,(n+2\,a+1)\,s(n)$$
$$-\,(2\,a+7+3\,n)\,(2\,a+5+3\,n)\,(2\,a+3+3\,n)\,s(n+2) = 0$$

Gessel-Stanton (1.7)

```
> hyperrecursion([2*a,2*b,1-2*b,1+2/3*a,a+d+n+1/2,a-d,-n],
> [a-b+1,a+b+1/2,2/3*a,-2*d-2*n, 2*d+1,1+2*a+2*n],1,s(n));
```

$$(d+n+1)\,(2\,d+2\,n+1)\,(a+1+n-b)\,(2\,a+2\,b+1+2\,n)\,s(n+1)+$$
$$(2\,d+2\,n+1+2\,b)\,(-d-n-1+b)\,(a+1+n)\,(2\,a+1+2\,n)\,s(n) = 0$$

Gessel-Stanton (6.2)

```
>    hyperrecursion([a+1/2,a,b,1-b,-n,(2*a+1)/3+n,a/2+1],
>    [1/2,(2*a-b+3)/3,(2*a+b+2)/3,-3*n,2*a+1+3*n,a/2],1,s(n));
```

$$(1+3\,n)\,(3\,n+2)\,(2\,a-b+3+3\,n)\,(2\,a+b+2+3\,n)\,\mathrm{s}(n+1)$$
$$+\,(b+1+3\,n)\,(-2+b-3\,n)\,(2\,a+2+3\,n)\,(2\,a+3+3\,n)\,\mathrm{s}(n)=0$$

Note that Algorithm 8.8 is also implemented in the Maple package `SumTools` under the name `SumTools[Hypergeometric][KoepfZeilberger]`.

Finally, we characterize the input to which Algorithm 8.8 can be safely applied. If $F(n,k)$ is a hypergeometric term with respect to both n and k that has finite support, and is of the form $F(n,k)=P(n,k)\,\frac{Q(n,k)}{R(n,k)}\,w^n z^k$ where $P(n,k)$ is a polynomial and $Q(n,k),R(n,k)$ are Γ-term products with rational-linear arguments we call it an *admissible hypergeometric term*. Note that in particular, any proper hypergeometric term is admissible. For admissible terms, however, rational-linear rather than only integer-linear Γ terms are allowed. We have

Theorem 8.12 *(Applicability of Extended Algorithm) For sums of admissible hypergeometric terms Algorithm 8.8 terminates.*

Proof Given any admissible hypergeometric term $F(n,k)$, it is easily seen that $\tilde{F}(n,k):=F(mn,kl)$, which is constructed in Step 4 of Algorithm 8.8, forms a proper hypergeometric term. Therefore, by Corollary 7.12, the application of Zeilberger's algorithm terminates, and results in a holonomic recurrence equation for $\tilde{s}_n$. Hence, the holonomic recurrence equation for s_n, generated by the algorithm, is valid. $\square$

Further Reading

For further reading on the extensions of this chapter see [GS82, Koepf95a, HKS12].

Exercises

Exercise 8.1 *Show that Algorithm 8.2, the extension of Gosper's algorithm, carries a rational certificate s_k/a_k. Describe the certification procedure.*

⋄ **Exercise 8.2** *Write a Maple procedure* `extended_gosper(a,k,m)` *corresponding to Algorithm 8.2, Session 8.11.*

⋄ **Exercise 8.3** *Write a Maple procedure* `find_mfold` *corresponding to Algorithm 8.4, Session 8.11.*

⋄ **Exercise 8.4** *Give an algorithm to find the hypergeometric term solution for s_n, given by the recurrence equation ($m \in \mathbb{N}$)*

$$Q(n)\, s_{n+m} = P(n)\, s_n \qquad (P, Q \in \mathbb{Q}[n])$$

and implement it in Maple.

Exercise 8.5 *Find the antidifferences of*

(a) $\frac{3k+2}{k+2}\binom{k}{k/2}$,

(b) $\frac{3k+4}{k+4}\binom{k/2}{k/4}$,

(c) $\frac{23k^2+42k+16}{4(k+1)(k+2)}\binom{3k/2}{k/2}$.

$\diamond$ **Exercise 8.6** Write a Maple procedure `extended_sumrecursion` corresponding to Algorithm 8.8.

Exercise 8.6 *Generate the right-hand sides of the hypergeometric identities of Table 8.1 by an application of the extended Zeilberger procedure to the left-hand sides.*

Exercise 8.7 *Prove (8.12) for $s = 1, \ldots, 5$ as well as for $s = 1/2, 1/3, 1/4$.*

Exercise 8.8 *Apply both Zeilberger's algorithm and its extended version to the sum*

$$_2F_1\left(\begin{array}{c} -n/2,\, -n/2 + 1/2 \\ b + 1/2 \end{array} \middle| \, 1\right).$$

How can one construct the recurrence equation generated by the extended algorithm from the (simpler) recurrence equation generated by Zeilberger's algorithm? Write the result as a factorization in operator notation.

Exercise 8.9 *The numbers*

$$f_n := \sum_{k=0}^{\lfloor n/2 \rfloor} \binom{n-k}{k} = \sum_{k=0}^{\lfloor n/2 \rfloor} \frac{(-\frac{n}{2})_k\,(-\frac{n-1}{2})_k}{(-n)_k}(-4)^k$$

represent the Fibonacci numbers, cf. Exercise 7.20.

Apply both Zeilberger's algorithm and its extended version to both representations of f_n and describe what happens.

References

GS82. Gessel, I., Stanton, D.: Strange evaluations of hypergeometric series. Siam J. Math. Anal.
        13, 295–308 (1982)
HKS12. Horn, P., Koepf, W., Sprenger, T.: m-fold hypergeometric solutions of linear recurrence
        equations revisited. Math Comput Sci **6**, 61–77 (2012)
Koepf95a. Koepf, W.: Algorithms for m-fold hypergeometric summation. J. Symbolic Comput.
        20, 399–417 (1995)

Chapter 9
Petkovšek's and van Hoeij's Algorithm

We saw that in many cases Zeilberger's algorithm obtains the holonomic recurrence equation of lowest order for a given definite sum s_n. In particular, if the order of the resulting recurrence equation is one, or if the latter contains only two shifts s_n and s_{n+m} for some $m \in \mathbb{N}$, then one finds a hypergeometric term representation for the sum under consideration using m initial values.

In this chapter we show what to do if s_n is a hypergeometric term, but Zeilberger's algorithm fails to find the corresponding recurrence equation of order one. Indeed, we study the more general situation of finding *all* hypergeometric term solutions of any given holonomic recurrence equation. This situation had been investigated by Petkovšek [Petkovšek92] who found an algorithm that solves the above problem.

Note that this algorithm is completely independent of Zeilberger's, and is itself of interest. Like Gosper's algorithm it is a decision procedure to find hypergeometric terms, this time as solutions of arbitrary holonomic recurrence equations rather than the fairly specific recurrence equation $s_{n+1} - s_n = a_n$.[1] In particular, this algorithm can be used to find the hypergeometric term solutions of recurrence equations returned by Zeilberger's algorithm.

Petkovšek's algorithm comes in two parts. In a first step, Petkovšek gives an algorithm to find all *polynomial solutions* of a given holonomic recurrence equation. In a second part, this subalgorithm is used to determine the *hypergeometric term solutions* of a given holonomic recurrence equation.

To omit lengthy notation, we describe the details for the situation where the given recurrence equation has order two, which indeed is the most interesting case.

Example 9.1 (*Polynomial Solutions of Holonomic Recurrence Equations*) Here, we consider the generic second order example. Our considerations will lead us to a generic algorithm for this case [Petkovšek92].

---

[1] If $a_{n+1}/a_n = u_n/v_n$ ($u_n, v_n \in \mathbb{Q}[n]$), then the difference equation $s_{n+1} - s_n = a_n$ implies the holonomic recurrence equation $v_n \, s_{n+2} - (u_n + v_n) \, s_{n+1} + u_n \, s_n = 0$.

W. Koepf, *Hypergeometric Summation*, Universitext,
DOI: 10.1007/978-1-4471-6464-7_9, © Springer-Verlag London 2014

Assume that a second order holonomic recurrence equation

$$P_n\, s_{n+2} + Q_n\, s_{n+1} + R_n\, s_n = 0 \qquad (P_n, Q_n, R_n \in \mathbb{Q}[n]) \tag{9.1}$$

for s_n is given. We wish to find *all* polynomial solutions $s_n \not\equiv 0$ of this equation.

Note that the only thing we need is an *upper bound* for the degree of any polynomial solution. As soon as we have such an upper bound, we can substitute a generic polynomial for the solution s_n into (9.1), equate coefficients, and solve the corresponding linear system for the unknown coefficients of s_n. If the linear system possesses a nontrivial solution, we have found s_n, and if not, no such polynomial solution exists.

Now we show how an upper bound for the unknown degree N of the nontrivial solution

$$s_n = n^N + \delta_1\, n^{N-1} + \cdots + \delta_N$$

with the unknown coefficients δ_l $(l = 1, \ldots, N)$ can be determined. Since the recurrence equation is linear and homogeneous, any multiple of a solution is also a solution so that it is sufficient to consider solution polynomials whose leading coefficients equal to one. Those polynomials are called *monic*.

Assume we have the representations

$$P_n = \alpha_0\, n^M + \alpha_1\, n^{M-1} + \cdots + \alpha_M,$$
$$Q_n = \beta_0\, n^M + \beta_1\, n^{M-1} + \cdots + \beta_M$$

and

$$R_n = \gamma_0\, n^M + \gamma_1\, n^{M-1} + \cdots + \gamma_M$$

for the given polynomials, where M is the maximal degree of the triple (P_n, Q_n, R_n).

Note that for any $j \in \mathbb{N}$ the shift s_{n+j} can be expanded by the binomial theorem as

$$
\begin{aligned}
s_{n+j} &= (n + j)^N + \delta_1\, (n + j)^{N-1} + \delta_2\, (n + j)^{N-2} + \cdots + \delta_N \\
&= n^N + (\delta_1 + jN)\, n^{N-1} + \left(\delta_2 + j(N - 1)\delta_1 + j^2 \frac{N(N - 1)}{2} \right) n^{N-2} + \cdots .
\end{aligned}
$$

Substituting all polynomials into (9.1) yields

$$
\begin{aligned}
0 = &\left(\alpha_0\, n^M + \alpha_1\, n^{M-1} + \cdots \right) \left(n^N + (\delta_1 + 2N)\, n^{N-1} + \cdots \right) \\
&+ \left(\beta_0\, n^M + \beta_1\, n^{M-1} + \cdots \right) \left(n^N + (\delta_1 + N)\, n^{N-1} + \cdots \right) \\
&+ \left(\gamma_0\, n^M + \gamma_1\, n^{M-1} + \cdots \right) \left(n^N + \delta_1 n^{N-1} + \cdots \right) .
\end{aligned} \tag{9.2}
$$

Equating the coefficients of n^{M+N}, we get, in particular,

$$\alpha_0 + \beta_0 + \gamma_0 = 0. \tag{9.3}$$

Therefore, for any nontrivial polynomial solution, this equation must be valid. If (9.3) is not satisfied, no polynomial solution exists, and we can quit. Therefore, assume that (9.3) is valid. Then we equate the coefficients of n^{M+N-1} in (9.2), use (9.3) to replace α_0 in terms of β_0 and γ_0, and get the condition

$$\alpha_1 + \beta_1 + \gamma_1 - (\beta_0 + 2\gamma_0) N = 0. \tag{9.4}$$

Now two things can happen. Either $\beta_0 + 2\gamma_0 \neq 0$, in which case (9.4) gives a unique choice for the exact degree N of s_n. If this is not a nonnegative integer, we quit. Otherwise we have found the degree bound searched for.

Assume finally, that

$$\beta_0 + 2\gamma_0 = 0. \tag{9.5}$$

Then, furthermore, by (9.4),
$$\alpha_1 + \beta_1 + \gamma_1 = 0 \tag{9.6}$$

must be valid. We equate the coefficients of n^{M+N-2} in (9.2), use (9.3), (9.5) and (9.6) as replacement rules, and get the condition

$$N^2\gamma_0 - (\beta_1 + \gamma_0 + 2\gamma_1) N + \alpha_2 + \beta_2 + \gamma_2 = 0.$$

To prove that this final condition yields only two possible choices for N, we will check that $\gamma_0 \neq 0$. Assume, to the contrary, $\gamma_0 = 0$. Then, by (9.5), $\beta_0 = 0$, and, by (9.3), $\alpha_0 = 0$, a contradiction to the choice of M. This finishes the search for the degree bound, and the proof of the algorithm to find all polynomial solutions of (9.1).

To convince ourselves that the calculations that were hidden above are correct, we repeat them with Maple:

```
>   P:=add(alpha[l]*n^(M-l),l=0..2):
>   Q:=add(beta[l]*n^(M-l),l=0..2):
>   R:=add(gamma[l]*n^(M-l),l=0..2):
>   s:=add(delta[l]*n^(N-l),l=0..2):
>   term:=expand(P*subs(n=n+2,s)+Q*subs(n=n+1,s)+R*s):
>   term:=expand(
>   subs((n+1)^N=n^N+N*n^(N-1)+N*(N-1)/2*n^(N-2),term)):
>   term:=expand(
>   subs((n+2)^N=n^N+2*N*n^(N-1)+4*N*(N-1)/2*n^(N-2),term)):
>   term:=numer(normal(term/(n^M*n^N))):
>   term:=collect(term,n):
>   deg:=degree(term,n):
```

```
>   eq1:=factor(coeff(term,n,deg));
```
$$2\,\delta_0\,(\gamma_0 + \alpha_0 + \beta_0)$$
```
>   eq1:=normal(eq1/(2*delta[0]));
```
$$\gamma_0 + \alpha_0 + \beta_0$$
```
>   alpha[0]:=solve(eq1,alpha[0]);
```
$$-\gamma_0 - \beta_0$$
```
>   eq2:=factor(coeff(term,n,deg-1));
```
$$-2\,\delta_0\,(-\beta_1 + 2\,N\,\gamma_0 + \beta_0\,N - \gamma_1 - \alpha_1)$$
```
>   eq2:=collect(normal(eq2/(2*delta[0])),N);
```
$$(-2\,\gamma_0 - \beta_0)\,N + \beta_1 + \gamma_1 + \alpha_1$$
```
>   alpha[1]:=solve(coeff(eq2,N,0),alpha[1]);
```
$$-\beta_1 - \gamma_1$$
```
>   beta[0]:=solve(coeff(eq2,N,1),beta[0]);
```
$$-2\,\gamma_0$$
```
>   eq3:=factor(coeff(term,n,deg-2));
```
$$2\,\delta_0\,(-\beta_1\,N - 2\,\gamma_1\,N - N\,\gamma_0 + N^2\,\gamma_0 + \gamma_2 + \alpha_2 + \beta_2)$$
```
>   eq3:=collect(normal(eq3/(2*delta[0])),N);
```
$$N^2\,\gamma_0 + (-\beta_1 - 2\,\gamma_1 - \gamma_0)\,N + \gamma_2 + \alpha_2 + \beta_2$$

Example 9.2 Now we consider a less general example. We search for the polynomial solutions $s_n \not\equiv 0$ of the recurrence equation

$$n(n+1)s_{n+2} - 2n(n+100)s_{n+1} + (n+99)(n+100)s_n = 0. \tag{9.7}$$

Here $P_n = n^2 + n$, $Q_n = -2n^2 - 200n$ and $R_n = n^2 + 199n + 9900$. Therefore, we have $M = 2$. We get

$$\alpha_0 + \beta_0 + \gamma_0 = 1 - 2 + 1 = 0,$$

so that condition (9.3) is satisfied. Since

$$\beta_0 + 2\,\gamma_0 = -2 + 2 = 0,$$

Equation (9.5) is also valid. Therefore we calculate

$$\alpha_1 + \beta_1 + \gamma_1 = 1 - 200 + 199 = 0,$$

and we see that (9.6) holds too. Hence we know that the degree N must satisfy the quadratic equation

$$N^2 - 199N + 9900 = (N - 99)(N - 100) = 0$$

with the solutions $N = 99, 100$. Since both are nonnegative integers, these two cases may actually appear. They lead to the polynomial solutions

$$(n)_{99} \quad \text{and} \quad (n)_{100} \tag{9.8}$$

of (9.7) (check!). With Maple, the general polynomial solution (of degree 100) is generated as follows.

```
>   P:=n*(n+1):
>   Q:=-2*n*(n+100):
>   R:=(n+99)*(n+100):
>   s:=add(delta[100-l]*n^l,l=0..100):
>   rec:=collect(P*subs(n=n+2,s)+Q*subs(n=n+1,s)+R*s,n):
>   sol:=solve({coeffs(rec,n)},{seq(delta[l],l=0..100)}):
>   factor(subs(sol,s));
```

$$n\,(n+98)\,(n+97)\,(n+96)\,(n+95)\,(n+94)\,(n+93)\,(n+92)\,(n+91)$$
$$(n+90)\,(n+89)\,(n+88)\,(n+87)\,(n+86)\,(n+85)\,(n+84)$$
$$(n+83)\,(n+82)\,(n+81)\,(n+80)\,(n+79)\,(n+78)\,(n+77)$$
$$(n+76)\,(n+75)\,(n+74)\,(n+73)\,(n+72)\,(n+71)\,(n+70)$$
$$(n+69)\,(n+68)\,(n+67)\,(n+66)\,(n+65)\,(n+64)\,(n+63)$$
$$(n+62)\,(n+61)\,(n+60)\,(n+59)\,(n+58)\,(n+57)\,(n+56)$$
$$(n+55)\,(n+54)\,(n+53)\,(n+52)\,(n+51)\,(n+50)\,(n+49)$$
$$(n+48)\,(n+47)\,(n+46)\,(n+45)\,(n+44)\,(n+43)\,(n+42)$$
$$(n+41)\,(n+40)\,(n+39)\,(n+38)\,(n+37)\,(n+36)\,(n+35)$$
$$(n+34)\,(n+33)\,(n+32)\,(n+31)\,(n+30)\,(n+29)\,(n+28)$$
$$(n+27)\,(n+26)\,(n+25)\,(n+24)\,(n+23)\,(n+22)\,(n+21)$$
$$(n+20)\,(n+19)\,(n+18)\,(n+17)\,(n+16)\,(n+15)\,(n+14)$$
$$(n+13)\,(n+12)\,(n+11)\,(n+10)\,(n+9)\,(n+8)\,(n+7)\,(n+6)$$
$$(n+5)\,(n+4)\,(n+3)\,(n+2)\,(n+1)\,(\delta_1 - 4851\,\delta_0 + n\,\delta_0)$$

In this example, the final linear system is quite complicated, and takes a second to solve.[2] Have a look at these equations! Finally, the factorization takes some time.

Example 9.3 Let us consider the same example from a different perspective. Using the shift operator N, as we did in the previous chapters, (9.7) can be rewritten in the form

$$\left(n(n+1)N^2 - 2n(n+100)N + (n+99)(n+100)\right)s_n = 0. \tag{9.9}$$

Now observe that the operator

$$T(N,n) = n(n+1)N^2 - 2n(n+100)N + (n+99)(n+100)$$

has the following three different polynomial factorizations

---

[2] In the article [ABP95] a method is introduced to obtain the resulting polynomial essentially by solving a linear system not depending on its degree, and iterative computations. In cases like the given one this method is obviously advantageous.

$$T(N, n) = \Big(nN - (n + 100)\Big)\Big(nN - (n + 99)\Big)$$
$$= \Big(nN - (n + 99)\Big)\Big(nN - (n + 100)\Big) \tag{9.10}$$
$$= \Big((n + 1)N - (n + 100)\Big)\Big((n - 1)N - (n + 99)\Big).$$

You can check these results by writing the operator equations out, or by using the commutator rule $Nn - nN = N$ and polynomial arithmetic. Note that the existence of three essentially different factorizations is a consequence of the noncommutativity of the polynomial ring we are dealing with.

In view of (9.10), we see that if any of the right factors applied to s_n is zero, then $T(N, n)s_n$ is zero too, and hence (9.9) is satisfied. Since the right factors correspond to hypergeometric terms satisfying the first order recurrence equations

$$ns_{n+1} - (n + 99)s_n = 0, \quad ns_{n+1} - (n + 100)s_n = 0,$$
$$\text{and} \quad (n - 1)s_{n+1} - (n + 99)s_n = 0, \tag{9.11}$$

respectively, these are hypergeometric term solutions of (9.9). Let us check which of these hypergeometric term solutions are polynomials. The first term satisfies

$$\frac{s_{n+1}}{s_n} = \frac{n + 99}{n},$$

and one therefore realizes that up to a constant factor

$$s_n = \frac{(100)_{n-1}}{(n - 1)!} = \frac{(98 + n)!}{99!\,(n - 1)!} = \frac{(n)_{99}}{99!}.$$

This is one of the polynomial solutions that we met in (9.8). The second term of (9.11) yields similarly up to a constant factor

$$s_n = \frac{(101)_{n-1}}{(n - 1)!} = \frac{(99 + n)!}{100!\,(n - 1)!} = \frac{(n)_{100}}{100!},$$

the second polynomial solution that we met in (9.8). Finally, the third term of (9.11) gives

$$s_n = \frac{(101)_{n-2}}{(n - 2)!} = \frac{(98 + n)!}{100!\,(n - 2)!} = \frac{(n - 1)_{100}}{100!}.$$

This is a third (linearly dependent) polynomial solution of (9.7). Check that all three solutions that we have generated are covered by Maple's output.

The remaining question is: Can factorizations like (9.10) of operators $T(N, n)$ be generated automatically? The answer is twofold:

- Using the algorithm of this chapter and further techniques leads to an algorithm to find factorizations of operators $T(N, n)$ that are polynomials w.r.t. N and have rational coefficients w.r.t. n. This will be discussed later.
- Noncommutative polynomial factorizations of operator polynomials $T(N, n)$ can be found using advanced commutative and noncommutative Gröbner basis techniques. We mention that there is an implementation in the computer algebra system REDUCE [Hearn95] for the calculation of noncommutative Gröbner bases and polynomial factorization in noncommutative polynomial rings, given by commutator rules [MA94]. This package `ncpoly` easily generates the factorizations (9.10).

Note that, for the given example, the REDUCE factorization algorithm generates the factorizations (9.10) much faster than Maple generated the polynomial solutions by Petkovšek's algorithm. This is not very surprising: Whereas with Petkovšek's algorithm the complete polynomial solution has to be calculated, in particular all the coefficients of the polynomials of degree 100 have to be found (Note, however, Footnote 2), each of the operator factorizations requires only the calculation of the much simpler recurrence equation that is valid for the corresponding solution. The advantages and disadvantages of both approaches will be discussed in more detail later.

Example 9.4 Let's check whether there are polynomial solutions of the Apéry recurrence equation

$$(n + 2)^3 s_{n+2} - (2n + 3) \left(17n^2 + 51n + 39\right) s_{n+1} + (n + 1)^3 s_n = 0$$

of the sum

$$s_n = \sum_{k=0}^{n} \binom{n}{k}^2 \binom{n+k}{k}^2.$$

We see at a glance that Apéry's recurrence equation does not possess any polynomial solution since $\alpha_0 + \beta_0 + \gamma_0 = 1 - 34 + 1 \neq 0$.

Session 9.5 The Maple procedure

```
rec2poly:=proc()
local rec,s,n,P,Q,R,M,N,alpha,beta,gamma,delta,sol,tmp,l,S,REC;
rec:=expand(args[1]):
if type(rec,`equation`) then rec:=op(1,rec)-op(2,rec) end if;
s:=op(0,args[2]);
n:=op(1,args[2]);
P:=collect(coeff(rec,s(n+2)),n);
Q:=collect(coeff(rec,s(n+1)),n);
R:=collect(coeff(rec,s(n)),n);
M:=max(degree(P,n),degree(Q,n),degree(R,n));
alpha[0]:=coeff(P,n,M);
beta[0]:=coeff(Q,n,M);
gamma[0]:=coeff(R,n,M);
# check first condition
```

```
if not(simplify(alpha[0]+beta[0]+gamma[0])=0) then
  error 'No polynomial solution exists';
end if;
alpha[1]:=coeff(P,n,M-1);
beta[1]:=coeff(Q,n,M-1);
gamma[1]:=coeff(R,n,M-1);
# check second condition
if not(simplify(beta[0]+2*gamma[0])=0) then
  N:=normal((alpha[1]+beta[1]+gamma[1])/(beta[0]+2*gamma[0]));
# check third condition
elif not(simplify(alpha[1]+beta[1]+gamma[1])=0) then
  error 'No polynomial solution exists';
else
  alpha[2]:=coeff(P,n,M-2);
  beta[2]:=coeff(Q,n,M-2);
  gamma[2]:=coeff(R,n,M-2);
  sol:={solve(N^2*gamma[0]-(beta[1]+gamma[0]+
    2*gamma[1])*N+alpha[2]+beta[2]+gamma[2],N)};
  N:=max(op(select(type,sol,nonnegint)));
end if;
if type(N,negint) then
  error 'No polynomial solution exists';
end if;
S:=add(delta[N-l]*n^l,l=0..N);
REC:=collect(P*subs(n=n+2,S)+Q*subs(n=n+1,S)+R*S,n);
sol:={solve(normal({coeffs(REC,n)}),{seq(delta[l],l=0..N)})};
if sol={} or {seq(op(2,op(l,op(1,sol))),l=1..nops(op(1,sol)))}={0}
  then error 'No polynomial solution exists'
end if;
return factor(subs(op(1,sol),S));
end proc:
```

is an implementation of the above algorithm to find all monic polynomial solutions
of a second order holonomic recurrence equation. We get for example

```
>   rec2poly(
>   n*(n+1)*s(n+2)-2*n*(n+10)*s(n+1)+(n+9)*(n+10)*s(n),s(n));
```

$$n\,(n+8)\,(n+7)\,(n+6)\,(n+5)\,(n+4)\,(n+3)\,(n+2)\,(n+1)$$
$$(n\,\delta_0 + \delta_1 - 36\,\delta_0)$$

```
>   rec2poly(
>   n*(n+1)*s(n+2)-2*n*(n+20)*s(n+1)+(n+19)*(n+20)*s(n),s(n));
```

$$n\,(n+18)\,(n+17)\,(n+16)\,(n+15)\,(n+14)\,(n+13)\,(n+12)\,(n+11)$$
$$(n+10)\,(n+9)\,(n+8)\,(n+7)\,(n+6)\,(n+5)\,(n+4)\,(n+3)\,(n+2)$$
$$(n+1)\,(n\,\delta_0 + \delta_1 - 171\,\delta_0)$$

Now, having described and proved how the procedure works for second order recurrence equations, we state the general algorithm without proof; see Exercise 9.4.

Algorithm 9.6 (*Polynomial Solutions of Holonomic Recurrence Equations*) The following algorithm finds all polynomial solutions of a given holonomic recurrence equation.

1. Input: A holonomic recurrence equation

$$\sum_{j=0}^{J} P_j(n)\, s_{n+j} = 0 \tag{9.12}$$

   with polynomials

$$P_j(n) = \sum_{l=0}^{M} \alpha_{jl}\, n^{M-l} \in \mathbb{Q}[n],$$

   such that one of $\alpha_{j0} \neq 0\ (j = 0, \ldots, J)$.
2. Set $m := 0$.
3. Compute for all $l = 0, \ldots, m$

$$b_{lm} := \sum_{j=0}^{J} j^{l}\, \alpha_{j,m-l}.$$

   If $b_{lm} = 0$ for all $l = 0, \ldots, m$, increase m by one, and repeat Step 3.
4. Let $\mathcal{N}$ be the set of nonnegative integer roots $N \in \mathbb{N}_{\geq 0}$ of the polynomial

$$\sum_{l=0}^{m} \binom{N}{l} b_{lm}. \tag{9.13}$$

5. If $\mathcal{N} = \emptyset$ then return "no polynomial solution exists"; exit.
6. Set $N := \max \mathcal{N}$. Find the general polynomial solution s_n of (9.12) by substituting the generic polynomial of degree N, equating coefficients, and solving the corresponding linear system.
7. Output: The polynomial solution s_n, determined in the previous step. $\qquad\square$

Note that the main part in proving the algorithm is to show that the iteration in Step 3 stops (it turns out that it does not exceed J steps), and that formula (9.13) is valid; see [Petkovšek92, PWZ96].

We move on to our main problem of finding the hypergeometric term solutions of a given holonomic recurrence equation. For this purpose we will need a refined version of Gosper's representation lemma for rational functions (Lemma 5.1) which is due to Petkovšek.

Lemma 9.7 and Algorithm (*Gosper-Petkovšek Representation of Rational Functions*) Any rational function $t_k \in \mathbb{Q}(k) \backslash \{0\}$ has a representation of the form

$$t_k = C\, \frac{p_{k+1}}{p_k}\, \frac{q_{k+1}}{r_{k+1}} \tag{9.14}$$

where $p_k, q_k, r_k \in \mathbb{Q}[k]$ are monic polynomials, $C \in \mathbb{Q}$, and the following properties are valid:

(a) $\gcd(q_k, r_{k+j}) = 1$ for all $j \in \mathbb{N}_{\geq 0}$;
(b) $\gcd(p_k, q_{k+1}) = 1$;
(c) $\gcd(p_k, r_k) = 1$.

Proof Let

$$t_k = C \frac{u_k}{v_k},$$

where $u_k, v_k \in \mathbb{Q}[k]$ are in lowest terms, and are monic.

Gosper's rewriting procedure, given in Lemma 5.1 and applied in a specific way, see below, generates the above representation if we start with the initialization $p_k := 1$, $q_k := u_{k-1}$, and $r_k := v_{k-1}$.

The validity of (a) is then a consequence of Lemma 5.1. It remains to prove (b) and (c).

We apply Gosper's rewriting procedure in such a way that for each rewrite step we take the *minimal* $j \in \mathbb{N}$ for which

$$\gcd(q_k, r_{k+j}) = g_k \not\equiv 1, \tag{9.15}$$

successively increasing this value.

In particular, we have then

$$\gcd(q_k, r_{k+i}) = 1 \quad \text{for all } i < j.$$

Since, by (9.15), g_{k+i-j} is a divisor of r_{k+i}, it follows that

$$\gcd(q_k, g_{k+i-j}) = 1 \quad \text{for all } i < j, \tag{9.16}$$

and since g_{k-i} is a divisor of q_{k-i}, we have also

$$\gcd(g_{k-i}, r_k) = 1 \quad \text{for all } i < j. \tag{9.17}$$

At the beginning of the rewriting, we start with $\gcd(p_k, q_{k+1}) = 1$ since $p_k \equiv 1$. With every rewriting step, we set $p'_k = p_k g_k g_{k-1} \cdots g_{k-j+1}$ and $q'_{k+1} = \frac{q_{k+1}}{g_{k+1}}$ so that by (9.16) p'_k and q'_{k+1} cannot have a common factor either. This proves (b).

Similarly, at the beginning of the rewriting, we have $\gcd(p_k, r_k) = 1$. With every rewriting step, we set $p'_k = p_k g_k g_{k-1} \cdots g_{k-j+1}$ and $r'_k = \frac{r_k}{g_{k-j}}$ so that by (9.17) p'_k and r'_k do not have a common factor, and (c) is proved. $\square$

Note that parts (b) and (c) of the lemma state in particular that in the Gosper-Petkovšek representation (9.14) no canceling between the factors q_{k+1} and p_k, or p_{k+1} and r_{k+1} occurs. It turns out that the Gosper-Petkovšek representation is unique, and for this it is essential to take the minimal $j \in \mathbb{N}$ in each rewrite step; see Exercise 9.15.

We are now ready to state and prove Petkovšek's main algorithm for second order recurrence equations.

Example 9.8 (Hypergeometric Term Solutions of Holonomic Recurrence Equations)
Here, we consider the generic second order example. Our considerations will lead us to a generic algorithm for this case [Petkovšek92].

Assume that a second order holonomic recurrence equation

$$P_n\, s_{n+2} + Q_n\, s_{n+1} + R_n\, s_n = 0 \qquad (P_n,\, Q_n,\, R_n \in \mathbb{Q}[n]) \qquad (9.18)$$

for s_n is given. We would like to find *all* hypergeometric term solutions $s_n \not\equiv 0$ of this equation.

Since s_n is assumed to be a hypergeometric term, the term ratio

$$\frac{s_{n+1}}{s_n} = t_n \in \mathbb{Q}(n) \qquad (9.19)$$

is rational. Hence, by Lemma 9.7, there exist $C \in \mathbb{Q}$ and $p_n, q_n, r_n \in \mathbb{Q}[n]$ such that

$$t_n = C\, \frac{p_{n+1}}{p_n}\, \frac{q_{n+1}}{r_{n+1}}, \qquad (9.20)$$

and the gcd conditions (a)–(c) of Lemma 9.7 are valid.

Dividing (9.18) by s_n and substituting (9.19), we can therefore write

$$P_n\, t_{n+1}\, t_n + Q_n\, t_n + R_n = 0.$$

Using (9.20), we obtain (after multiplication by $p_n\, r_{n+1}\, r_{n+2}$) the equation

$$C^2\, P_n\, p_{n+2}\, q_{n+2}\, q_{n+1} + C\, Q_n\, p_{n+1}\, q_{n+1}\, r_{n+2} + R_n\, p_n\, r_{n+1}\, r_{n+2} = 0. \qquad (9.21)$$

Now, we make extensive use of the gcd conditions of Lemma 9.7. Since the first and second summands of (9.21) have the common factor q_{n+1}, division by this term shows that the third summand $R_n\, p_n\, r_{n+1}\, r_{n+2}$ must be divisible by q_{n+1}. By Lemma 9.7, q_{n+1} is relatively prime to p_n, r_{n+1} and r_{n+2} so that it follows that $R_n/q_{n+1} \in \mathbb{Q}[n]$. In particular, we obtain the information that q_n must be a monic factor of R_{n-1}. There are only finitely many such choices. But note that there might be many, if the degree of R_n is large; see Exercise 9.11.

Similarly, the second and third summands of (9.21) have the common factor r_{n+2}. Division by this term shows therefore that the first summand $P_n\, p_{n+2}\, q_{n+2}\, q_{n+1}$ must be divisible by r_{n+2}. By Lemma 9.7, r_{n+2} is relatively prime to q_{n+1}, q_{n+2} and p_{n+2} so that $P_n/r_{n+2} \in \mathbb{Q}[n]$. In particular, we obtain the information that r_n must be one of the finitely many monic factors of P_{n-2}.

For any chosen pair (q_n, r_n) of monic factors of (R_{n-1}, P_{n-2}) we can cancel $q_{n+1}\, r_{n+2}$ in (9.21) and obtain the polynomial equation

$$C^2 \frac{P_n}{r_{n+2}}\, q_{n+2}\, p_{n+2} + C\, Q_n\, p_{n+1} + \frac{R_n}{q_{n+1}}\, r_{n+1}\, p_n = 0. \tag{9.22}$$

Next, we determine the constant C. Therefore, we consider the leading coefficient of the left-hand side of (9.22), and realize that this generates a quadratic equation for C since p_n, p_{n+1} and p_{n+2} have the same degree. So, for each choice of q_n and r_n (as factors of R_{n-1} and P_{n-2}, respectively) there are at most two possible choices for $C \in \mathbb{Q}$.[3]

For any fixed choice for q_n, r_n and C, we can use Algorithm 9.6 (which for the present situation is described in Example 9.1 and Session 9.5) to determine whether there are any nonzero polynomial solutions p_n of (9.22). Any such solution provides us with a hypergeometric term solution of (9.18). On the other hand, no other hypergeometric term solutions exist.

Example 9.9 Consider the recurrence equation

$$(n + 4)\, s_{n+2} + s_{n+1} - (n + 1)\, s_n = 0 \tag{9.23}$$

with $P_n = n + 4$, $Q_n = 1$, and $R_n = -n - 1$. The only possible choices for q_n (monic factors of R_{n-1}) are $q_n = 1$ or $q_n = n$ and for r_n (monic factors of P_{n-2}) are $r_n = 1$ or $r_n = n + 2$.

The following Maple session generates the possible values for $t_n = \frac{s_{n+1}}{s_n}$:

```
>    P:=n+4:  Q:=1:  R:=-n-1:  q:=1:  r:=1:
>    {solve(coeff(collect(C^2*normal(P/subs(n=n+2,r))*subs(n=n+2,q)
>    +C*Q+normal(R/subs(n=n+1,q))*subs(n=n+1,r),n),n),C)};
```

$$\{-1,\ 1\}$$

```
>    C:=-1:
>    pol:=rec2poly(C^2*normal(P/subs(n=n+2,r))*p(n+2)*subs(n=n+2,q)+
>    C*Q*p(n+1)+normal(R/subs(n=n+1,q))*p(n)*subs(n=n+1,r),p(n));
```

```
Error, (in rec2poly) No polynomial solution exists
```

```
>    C:=1:
>    pol:=rec2poly(C^2*normal(P/subs(n=n+2,r))*p(n+2)*subs(n=n+2,q)+
>    C*Q*p(n+1)+normal(R/subs(n=n+1,q))*p(n)*subs(n=n+1,r),p(n));
```

```
Error, (in rec2poly) No polynomial solution exists
```

```
>    q:=n:  C:='C':
>    {solve(lcoeff(collect(C^2*normal(P/subs(n=n+2,r))*subs(n=n+2,q)+
>    C*Q+normal(R/subs(n=n+1,q))*subs(n=n+1,r),n),n),C)};
```

$$\{0\}$$

```
>    q:=1:  r:=n+2:  C:='C':
```

---

[3] Since the degree of the polynomial w.r.t. C is two, we are not confined to $\mathbb{Q}$ and can find all solutions $C \in \mathbb{C}$ if we want. These computations may take place in an extension field of $\mathbb{Q}$.

```
>    {solve(lcoeff(collect(C^2*normal(P/subs(n=n+2,r))*subs(n=n+2,q)+
>    C*Q+normal(R/subs(n=n+1,q))*subs(n=n+1,r),n),n),C)};
```

$$\{\}$$

```
>    q:=n:
>    {solve(lcoeff(collect(C^2*normal(P/subs(n=n+2,r))*subs(n=n+2,q)+
>    C*Q+normal(R/subs(n=n+1,q))*subs(n=n+1,r),n),n),C)};
```

$$\{-1,\ 1\}$$

```
>    C:=-1:
>    pol:=rec2poly(C^2*normal(P/subs(n=n+2,r))*p(n+2)*subs(n=n+2,q)+
>    C*Q*p(n+1)+normal(R/subs(n=n+1,q))*p(n)*subs(n=n+1,r),p(n)));
```

$$\frac{1}{2}\,\delta_0\,(3+2n)$$

```
>    t:=normal(C*subs(n=n+1,pol)*subs(n=n+1,q)/(pol*subs(n=n+1,r)));
```

$$-\frac{(5+2n)\,(n+1)}{(3+2n)\,(n+3)}$$

```
>    C:=1:
>    pol:=rec2poly(C^2*normal(P/subs(n=n+2,r))*p(n+2)*subs(n=n+2,q)+
>    C*Q*p(n+1)+normal(R/subs(n=n+1,q))*p(n)*subs(n=n+1,r),p(n)));
```

$$\delta_0$$

```
>    t:=normal(C*subs(n=n+1,pol)*subs(n=n+1,q)/(pol*subs(n=n+1,r)));
```

$$\frac{n+1}{n+3}$$

Therefore, we see that there are exactly two linearly independent hypergeometric term solutions (over $\mathbb{Q}(n)$) of (9.23) with

$$\frac{s_{n+1}}{s_n}=-\frac{(5+2n)\,(n+1)}{(3+2n)\,(n+3)}\quad\text{and}\quad\frac{s_{n+1}}{s_n}=\frac{n+1}{n+3},$$

respectively, i.e. (up to a constant factor),

$$s_n=\frac{(1)_n\,(5/2)_n}{(3)_n\,(3/2)_n}\,(-1)^n=\frac{2}{3}\frac{(-1)^n\,(2n+3)}{(n+1)(n+2)}\quad\text{and}\quad s_n=\frac{(1)_n}{(3)_n}=\frac{2}{(n+1)(n+2)}.$$

Example 9.10 Again, we might use a factorization of the operator polynomial

$$T(N,n)=(n+4)\,N^2+N-(n+1)\tag{9.24}$$

corresponding to the recurrence equation (9.23) to check the results of the previous example. As we mentioned earlier, any existing polynomial right factor of $T(N,n)$ of order one in N generates one hypergeometric term solution of the corresponding holonomic recurrence equation, and vice versa.

The implementation [MA94]—which is based on polynomial arithmetic and Gröbner basis computations—shows, however, that there is only *one single* polynomial factorization of $T(N,n)$ with *polynomial coefficients* in $\mathbb{Q}[n]$, namely

$$T(N,n)=(n+4)\,N^2+N-(n+1)=(N+1)\left((n+3)N-(n+1)\right)$$

which corresponds to the hypergeometric solution

$$s_n = \frac{(1)_n}{(3)_n} = \frac{2}{(n+1)(n+2)}$$

of (9.23) that we also found using Petkovšek's algorithm.

Why, however, is the second solution

$$s_n = \frac{(1)_n \, (5/2)_n}{(3)_n \, (3/2)_n} \, (-1)^n = \frac{2}{3} \frac{(-1)^n \, (2n+3)}{(n+1)(n+2)} \tag{9.25}$$

not generated? The operator polynomial

$$(n+1)(2n+5)N - (n+3)(2n+3)$$

corresponding to (9.25) cannot be a right factor of (9.24) since any polynomial right factor of (9.24) has degree at most one in n. Therefore this term is hidden! Note that Petkovšek's algorithm finds factorizations with *rational coefficients*, hence elements of $\mathbb{Q}(n)$ rather than $\mathbb{Q}[n]$ (this is its advantage), but is restricted to first order right factors (this is its disadvantage).

However, Petkovšek's algorithm is the starting point of a general factorization algorithm for recurrence operators $T(N, n)$ over $\mathbb{Q}(n)$. It finds all first order right factors of T. The so-called *adjoint operator* can be used to find all first order left factors of T as well. To find second order right factors of T, one generates the so-called *symmetric product* of T with itself, i.e., the recurrence operator whose solutions are the products of the solutions of T, and applies Petkovšek's algorithm. This result can be used to find the second order right factors of T. Higher order right and left factors of T are dealt with in a similar way. Details can be found, e.g., in [Bronstein94, vdPS03, Horn08]. For the computation of the symmetric product using linear algebra see e.g. [Stanley80, SZ94].[4]

Note that noncommutative polynomial factorization can also be applied to find right factors of operator polynomials corresponding to differential equations or even mixed recurrence-differential equations (see Example 12.5).

Example 9.11 Using Zeilberger's algorithm, we obtain the recurrence equation

$$2(2n+3)s_{n+2} + 3(5n+7)s_{n+1} + 9(n+1)s_n = 0 \tag{9.26}$$

for

$$s_n = \sum_{k=0}^{n} (-1)^k \binom{n}{k} \binom{3k}{n}$$

by the calculation

---

[4] The symmetric product of two recurrence equations can be computed using the function `rec*rec` from the `gfun` package.

```
> rec:=sumrecursion((-1)^k*binomial(n,k)*binomial(3*k,n),k,s(n));
```
$$2\,(3+2\,n)\,s(n+2) + 3\,(5\,n+7)\,s(n+1) + 9\,(n+1)\,s(n) = 0$$

Let's derive the hypergeometric term solutions of this recurrence equation! We have $P_n = 2(2n+3)$, $Q_n = 3(5n+7)$ and $R_n = 9(n+1)$. It turns out that the only choice that generates a hypergeometric term solution is given by $q_n = 1$, $r_n = 1$ and $C = -3$. This leads to the polynomial solution $p_n = 1$ by Algorithm 9.6, and therefore to the rational term ratio

$$\frac{s_{n+1}}{s_n} = t_n = -3,$$

so that $S_n := (-3)^n$ is a solution of (9.26). Since $S_0 = s_0$ and $S_1 = s_1$, and since the solution of (9.26) with two initial values is unique, it turns out that $s_n = S_n = (-3)^n$.

Session 9.12 We implement Petkovšek's algorithm for recurrence equations of order two. The Maple procedure

```
generateproducts:=proc(a)
local f,r,n;
r:=[a];
f:=proc(a,x)
   local p,t;
   if member(x,a,p) then
      t:=subsop(p=NULL,a); procname(t,x),t
   end if;
end;
n:=nops(a);
while (0<n) do
   r:=[op(map(f,r,a[n])),op(r)];
   n:=n-1;
   while (0<n) and (a[n+1]=a[n]) do
      n:=n-1;
   end do;
end do;
return map(convert,r,`*`);
end proc:
```

recursively generates the set of products with factors in the given list r. Hence

```
rec2hyper:=proc()
local rec,s,n,P,Q,R,i,j,Qfactors,Qchoices,Rfactors,Rchoices,
   p,q,r,c,sol,C,tmp,t,cchoices;
rec:=expand(args[1]):
if type(rec,`equation`) then rec:=op(1,rec)-op(2,rec) end if;
s:=op(0,args[2]);
n:=op(1,args[2]);
P:=coeff(rec,s(n+2));
Q:=coeff(rec,s(n+1));
```

```
    R:=coeff(rec,s(n));
    if (P=0) then
      return {-factor_over_Q(R,n,0)/factor_over_Q(Q,n,0)};
    end if;
    R:=factor_over_Q(R,n,-1,Qfactors);
    Qchoices:=generateproducts(Qfactors);
    P:=factor_over_Q(P,n,-2,Rfactors);
    Rchoices:=generateproducts(Rfactors);
    sol:={};
    for q in Qchoices do
      for r in Rchoices do
        cchoices:=(C^2*(P/subs(n=n+2,r))*subs(n=n+2,q)+C*Q+
          (R/subs(n=n+1,q))*subs(n=n+1,r));
        try
          cchoices:=normal(
            {solve(lcoeff(expand(normal(cchoices)),n),C)});
          for c in cchoices do
            try
              tmp:=normal((c^2*(P/subs(n=n+2,r))*
                subs(n=n+2,q)*p(n+2)+c*Q*p(n+1)+
                (R/subs(n=n+1,q))*subs(n=n+1,r)*p(n)));
              tmp:=rec2poly(eval(tmp),p(n));
              t:=normal(c*subs(n=n+1,tmp)*subs(n=n+1,q)/
                (tmp*subs(n=n+1,r)));
              sol:={op(sol),t};
            catch:
            end try;
          end do;
        catch:
        end try;
      end do;
    end do;
    return sol;
    end proc:
```

is an implementation of Petkovšek's algorithm (over $\mathbb{Q}$) for recurrence equations of
order 2, where

```
factor_over_Q:=proc(poly,n,shift,rootlist) local p,i,j,l,lf,f;
f:=factors(subs(n=n+shift,poly)); l:=f[2]; if (nargs=4) then
  lf:=select(x->degree(x[1],n)=1,l);
  rootlist:=[1];
  for i from 1 to nops(lf) do
    for j from 1 to lf[i][2] do
      rootlist:=[op(op(rootlist)),lf[i][1]];
    end do;
  end do;
end if;
return f[1]*mul(factor(subs(n=n-shift,(l[i][1])))^l[i][2],
  i=1..nops(l));
end proc:
```

is an auxiliary function. The procedure `rec2hyper` results in the set of term ratios $t_n = s_{n+1}/s_n$ of all possible hypergeometric term solutions. Let's use the procedure on the above examples:

```
>    rec2hyper((n+4)*s(n+2)+s(n+1)-(n+1)*s(n),s(n));
```

$$\left\{ \frac{n+1}{n+3}, \; -\frac{(5+2n)(n+1)}{(3+2n)(n+3)} \right\}$$

```
>    rec2hyper(
>    2*(2*n+3)*s(n+2)+3*(5*n+7)*s(n+1)+9*s(n)*(n+1),s(n));
```

$$\{-3\}$$

What about the Apéry numbers?

```
>    rec2hyper(
>    (n+1)^3*s(n)-(2*n+3)*(17*n^2+51*n+39)*s(n+1)+s(n+2)*(n+2)^3
>    ,s(n));
```

$$\{\}$$

Hence, Petkovšek's algorithm has *proved* that the Apéry numbers are not hypergeometric terms!

Here are some more results:

```
>    rec2hyper(
>    2*n*(1+n)*s(n)+(2-3*n-n^2)*s(n+1)+(n-1)*s(n+2),s(n));
```

$$\{2, \, n+1\}$$

```
>    rec2hyper(s(n+2)-(n+1)*s(n+1)-(n+1)*s(n),s(n));
```

$$\{n+1\}$$

```
>    rec2hyper(81*(2+3*n)*(4+3*n)*s(n)-
>    6*(70+99*n+36*n^2)*s(n+1)+8*(2+n)*(3+2*n)*s(n+2),s(n));
```

$$\left\{ \frac{27}{4} \right\}$$

In Example 7.6, we had considered the sum

$$s_n = \sum_{k=0}^{\lfloor n/3 \rfloor} \binom{n-2k}{k} \left(-\frac{4}{27} \right)^k$$

and we were able to find the recurrence equation

$$9(n+2)s_{n+2} - 3(n+4)s_{n+1} - 2(n+3)s_n = 0 \tag{9.27}$$

for s_n. Let us use Petkovšek's algorithm to find hypergeometric term solutions.

```
>    rec2hyper(9*(n+2)*s(n+2)-3*(n+4)*s(n+1)-2*(n+3)*s(n),s(n));
```

$$\left\{ -\frac{1}{3}, \; \frac{2(7+3n)}{3(4+3n)} \right\}$$

Hence both

$$u_n = \left(-\frac{1}{3}\right)^n \quad \text{and} \quad v_n = \frac{(7/3)_n}{(4/3)_n}\left(\frac{2}{3}\right)^n = \frac{3n+4}{4}\left(\frac{2}{3}\right)^n$$

are hypergeometric term solutions of (9.27), and by the linear structure every solution of (9.27) must be a linear combination of these. Therefore we try to find a representation for s_n of the form

$$s_n = \alpha u_n + \beta v_n.$$

The values $s_0 = s_1 = 1$ generate the linear system

$$\alpha + \beta = 1 \quad \text{and} \quad -\frac{\alpha}{3} + \frac{7\beta}{6} = 1$$

for (α, β) with the solution

$$\alpha = \frac{1}{9} \quad \text{and} \quad \beta = \frac{8}{9}.$$

Hence we have found that

$$s_n = \frac{1}{9}\left(-\frac{1}{3}\right)^n + \frac{2(3n+4)}{9}\left(\frac{2}{3}\right)^n.$$

Finally, we state without proof Petkovšek's general result on hypergeometric term solutions of higher order recurrence equations [Petkovšek92]. It is a straightforward generalization of Example 9.8; see Exercise 9.8.

Algorithm 9.13 (*Hypergeometric Term Solutions of Holonomic Recurrence Equations*) The following algorithm finds all hypergeometric term solutions of a given holonomic recurrence equation.

1. Input: A holonomic recurrence equation

$$\sum_{j=0}^{J} P_j(n)\, s_{n+j} = 0 \tag{9.28}$$

with polynomials $P_j \in \mathbb{Q}[n]$.
2. Set $L := \{\}$.

3. For all monic factors q_n of $P_0(n-1)$ and r_n of $P_J(n-J)$ do:

   (a) For $j = 0, \ldots, J$ set

   $$h_j(n) := P_j(n) \cdot \prod_{l=1}^{j} q_{n+l} \cdot \prod_{l=j+1}^{J} r_{n+l}.$$

   (b) Let $M := \max_{0 \le j \le J} \deg h_j(n)$, and for $j = 0, \ldots, J$ let α_j denote the coefficient of n^M in $h_j(n)$.

   (c) For any solution $C \in \mathbb{Q}$ (or $C \in \mathbb{C}$) of the polynomial equation

   $$\sum_{j=0}^{J} \alpha_j C^j = 0 \tag{9.29}$$

   do: Apply Algorithm 9.6 to the recurrence equation

   $$\sum_{j=0}^{J} C^j h_j(n) p_{n+j} = 0 \tag{9.30}$$

   to find all polynomial solutions p_n of (9.30). If there is a polynomial solution p_n, then add the term ratio

   $$t_n = C \, \frac{p_{n+1}}{p_n} \frac{q_{n+1}}{r_{n+1}}$$

   to the set L.

4. Output: Return the set L of term ratios of all hypergeometric term solutions of (9.28). $\square$

Note that in Petkovšek's algorithm rational factorization (in (9.29) and (9.13)) plays an essential role again. If factorizations are done over $\mathbb{C}$, then the necessary algebraic extensions make the algorithm very slow. In practice, it turns out that algebraic extensions are most often unnecessary. The hsum package contains an implementation rechyper of Algorithm 9.13 over $\mathbb{Q}$, as described. Most often this suffices. The program algebraicrechyper works in extension fields.

   Using the notion of *similarity* under hypergeometric terms, see p. 94, not only is Petkovšek's algorithm shown to find the set of all hypergeometric term solutions, but this set forms a basis for all linear combinations of hypergeometric term solutions ([PWZ96], Sect. 8.7). Hence if Petkovšek's algorithm does not find any solution, it has proved that no linear combination of hypergeometric terms is a solution.

The algorithmic derivation of hypergeometric term representations of sums

$$s_n = \sum_{k=-\infty}^{\infty} F(n, k)$$

of admissible hypergeometric terms $F(n, k)$ is now complete: An application of Zeilberger's algorithm or its extension generates a holonomic recurrence equation for s_n, and an application of Petkovšek's algorithm finds the hypergeometric term solutions of the resulting recurrence equation whenever those hypergeometric terms exist. Finally, by checking enough initial values, we can determine whether or not one of these hypergeometric terms (or a linear combination of them like in Session 9.12) represents s_n.

In Exercise 9.3, we present a list of hypergeometric sums with hypergeometric term representations for which the Wilf-Zeilberger approach fails, and Zeilberger's algorithm generates a recurrence equation of order two. Obviously in all these cases Petkovšek's algorithm finds the hypergeometric term representations.

Note, however, that the complexity of Petkovšek's algorithm in practice can be quite high, particularly if the recurrence equation has high degree polynomial coefficients, since in this case quite a few choices have to be checked. We will look at this in more detail in the following example.

Session 9.14 To check the efficiency of Petkovšek's algorithm we would like to generate a recurrence equation with "sufficiently difficult" hypergeometric term solutions. The Maple program

```
HolonomicRE:=proc(term,sk)
local s,k,r;
s:=op(0,sk): k:=op(1,sk):
r:=ratio(term,k);
denom(r)*s(k+1)-numer(r)*s(k)=0;
end proc:
```

computes the first order recurrence equation for a hypergeometric term. Hence the computations

```
>   term1:=GAMMA(n+1/2)^5*GAMMA(n+1)/GAMMA(n+3/4)^3/GAMMA(n+1/3);
```

$$term1 := \frac{\Gamma(n + \frac{1}{2})^5 \, \Gamma(n + 1)}{\Gamma(n + \frac{3}{4})^3 \, \Gamma(n + \frac{1}{3})}$$

```
>   RE1:=HolonomicRE(term1,s(n));
```

$$RE1 := (4n + 3)^3 \, (3n + 1) \, s(n + 1) - 6 \, (2n + 1)^5 \, (n + 1) \, s(n) = 0$$

```
>   term2:=GAMMA(n+1/4)/GAMMA(n+1)^3/GAMMA(n+1/3)^4;
```

$$term2 := \frac{\Gamma(n + \frac{1}{4})}{\Gamma(n + 1)^3 \, \Gamma(n + \frac{1}{3})^4}$$

```
>   RE2:=HolonomicRE(term2,s(n));
```

$$RE2 := 4\,(n+1)^3\,(3\,n+1)^4\,s(n+1) - (324\,n+81)\,s(n) = 0$$

yield the recurrences RE1 and RE2 that are valid for $a_n = \dfrac{\Gamma(n+1/2)^5\,\Gamma(n+1)}{\Gamma(n+3/4)^3\,\Gamma(n+1/3)}$ and $b_n = \dfrac{\Gamma(n+1/4)}{\Gamma(n+1)^3\,\Gamma(n+1/3)^4}$, respectively.

Using linear algebra, it is now easy to find a recurrence equation which is valid for the sum $a_n + b_n$ (and for any linear combination of a_n and b_n) (see e.g. [Stanley80, SZ94]). With the gfun package, we get[5]

```
>   tmp:=gfun['rec+rec'](RE1,RE2,s(n)):
>   if type(tmp,set) then tmp:=select(has,tmp,n)[1] end if:
>   RE:=map(factor,tmp);
```

$$\begin{aligned}
RE :={}& 486\,(4\,n+1)\,(n+1)(6912\,n^{12} + 134784\,n^{11} + 1201536\,n^{10}\\
&+ 6475072\,n^9 + 23494256\,n^8 + 60469320\,n^7 + 113205728\,n^6\\
&+ 155330368\,n^5 + 155030016\,n^4 + 109737216\,n^3 + 52242624\,n^2\\
&+ 14979384\,n + 1944351)(2\,n+1)^5\,s(n) - 3\,(3\,n+1)(\\
&47775744\,n^{24} + 1289945088\,n^{23} + 16502538240\,n^{22}\\
&+ 133078892544\,n^{21} + 759170949120\,n^{20} + 3259238326272\,n^{19}\\
&+ 10937158309888\,n^{18} + 29413823444992\,n^{17}\\
&+ 64488148739328\,n^{16} + 116634920972032\,n^{15}\\
&+ 175425566746048\,n^{14} + 220546391942592\,n^{13}\\
&+ 232395627484608\,n^{12} + 205363007155392\,n^{11}\\
&+ 151961459800128\,n^{10} + 93817776256832\,n^9\\
&+ 48035486487104\,n^8 + 20215755160896\,n^7 + 69040265116616\,n^6\\
&+ 1877330292224\,n^5 + 393755684352\,n^4 + 59712834816\,n^3\\
&+ 5524092864\,n^2 + 132123312\,n - 17831097)s(n+1) + 4(\\
&6912\,n^{12} + 51840\,n^{11} + 175104\,n^{10} + 352192\,n^9 + 469808\,n^8\\
&+ 437960\,n^7 + 292648\,n^6 + 141288\,n^5 + 42024\,n^4 - 5416\,n^3\\
&- 13640\,n^2 - 5648\,n - 721)(3\,n+1)\,(n+2)^3\,(4+3\,n)^4\,(4\,n+7)^3\\
&s(n+2)
\end{aligned}$$

This recurrence has leading coefficient P_n

```
>   leading:=coeff(RE,s(n+2));
```

$$\begin{aligned}
leading :={}& 4(6912\,n^{12} + 51840\,n^{11} + 175104\,n^{10} + 352192\,n^9 + 469808\,n^8\\
&+ 437960\,n^7 + 292648\,n^6 + 141288\,n^5 + 42024\,n^4 - 5416\,n^3\\
&- 13640\,n^2 - 5648\,n - 721)(3\,n+1)\,(n+2)^3\,(4+3\,n)^4\,(4\,n+7)^3
\end{aligned}$$

and trailing coefficient R_n

---

[5] gfun treats the *algebra of holonomic functions*. Note that `rec+rec` sometimes returns a recurrence and sometimes a set which also contains some initial values. Therefore we check the output.

```
>  trailing:=coeff(RE,s(n));
```

$$trailing := 486\,(4\,n+1)\,(n+1)(6912\,n^{12}+134784\,n^{11}+1201536\,n^{10}$$
$$+\,6475072\,n^9+23494256\,n^8+60469320\,n^7+113205728\,n^6$$
$$+\,155330368\,n^5+155030016\,n^4+109737216\,n^3+52242624\,n^2$$
$$+\,14979384\,n+1944351)(2\,n+1)^5$$

that already have quite a few factors, hence $2^5 \cdot 4^2 \cdot 5 \cdot 6 = 15{,}360$ cases have to be checked. It turns out that Petkovšek's algorithm takes about one minute to find the solutions, as one can see from the computation

```
>  TIME:=time(): rec2hyper(RE,s(n)); time()-TIME;
```

$$\left\{ \frac{81\,(4\,n+1)}{4\,(n+1)^3\,(3\,n+1)^4}, \frac{6\,(n+1)\,(2\,n+1)^5}{(4\,n+3)^3\,(3\,n+1)} \right\}$$

$$76.612$$

In the sequel we will consider how such an example can be dealt with more efficiently since much fewer cases have to be treated.

Mark van Hoeij [vanHoeij98, CvH06] used a different, and much more efficient, approach to finding hypergeometric term solutions of holonomic recurrence equations by considering the local behavior of the solution terms. He implemented his algorithm in Maple as `LREtools[hypergeomsols]`. This is the state-of the-art algorithm for finding hypergeometric term solutions of holonomic recurrence equations.

In the sequel I would like to present the main ideas of van Hoeij's approach. How does a general hypergeometric term solution s_n of (9.28) look? In the first place, it must be a hypergeometric term, therefore by (2.8) having a representation

$$s_n = \frac{(\alpha_1)_n \cdot (\alpha_2)_n \cdots (\alpha_p)_n}{(\beta_1)_n \cdot (\beta_2)_n \cdots (\beta_q)_n} \frac{x^n}{n!} = C \cdot \frac{\Gamma(n+\alpha_1)\ldots\Gamma(n+\alpha_p)}{\Gamma(n+\beta_1)\ldots\Gamma(n+\beta_q) \cdot \Gamma(n+1)}\,x^n,$$

where we used (1.5) to rewrite the Pochhammer symbols in terms of Gamma functions. Next, by (1.4) every Gamma factor occurring here can be replaced by a rational function times another Gamma factor whose argument differs only by an integer.[6] Therefore, for example, we can choose Re α_k, Re $\beta_k \in (0, 1]$. Then, we get

$$s_n = R(n) \cdot \frac{\Gamma(n+\alpha_1)\ldots\Gamma(n+\alpha_p)}{\Gamma(n+\beta_1)\ldots\Gamma(n+\beta_q) \cdot \Gamma(n+1)}\,x^n$$

for some rational function $R(n) \in \mathbb{Q}(n)$, this time α_k, β_k being *uniquely determined*. Since some of the α_k or β_k now might agree, we finally end up with a representation of the form

---

[6] For the algebraist: We look at these αs and βs not as elements in $\mathbb{C}$, but in $\mathbb{C}/\mathbb{Z}$.

$$s_n = R(n) \cdot x^n \cdot \prod_{j=1}^{J} \Gamma(n - a_j)^{e_j} \quad (a_j \in \mathbb{C}, e_j \in \mathbb{Z}\backslash\{0\}), \qquad (9.31)$$

where all these data are uniquely determined. The numbers $a_j (j = 1, \ldots, J)$ are the finite Gamma singularities of order e_j. By

$$r(s_n) = \frac{s_{n+1}}{s_n} = \frac{R(n+1)}{R(n)} \cdot x \cdot \prod_{j=1}^{J} (n - a_j)^{e_j} \in \mathbb{Q}(n) \qquad (9.32)$$

we denote the *rational certificate* of the hypergeometric term s_n (9.31) as usual. Here the part stemming from the Gamma functions is completely factorized.

The crucial idea behind van Hoeij's approach is now to check the zeros and poles of a possible solution s_n of (9.28). Note that the rational function $R(n)$ has only finitely many zeros and poles, whereas every Gamma term in representation (9.31) creates an infinite number of zeros or poles since $\Gamma(z)$ has poles exactly at the nonnegative integers. We call this the *singularity structure* of s_n. If we can find this singularity structure, then we can find the solutions.

Next, we consider a simple example which shows how the singularity structure of a solution of (9.28) can be observed.

Example 9.15 The easiest example is the recurrence equation

$$s_{n+1} - (n+1)\, s_n = 0. \qquad (9.33)$$

We can use the recurrence in two directions. When we use it to compute—starting with s_0, say—the values of s_n from the previous ones in the forward direction, then all these can be computed without any problem, encountering no zeros or poles. This is because for no value $n \in \mathbb{Z}$ is the coefficient of s_{n+1} zero. However, if we use (9.33) in the backward direction, this is different. Since the coefficient of s_n in (9.33) is zero at $n = -1$, the recurrence cannot be used to compute s_{-1} from s_0. This leads to a pole of s_n for $n = -1$. This pole repeatedly appears for all negative $n \in \mathbb{Z}$. That is exactly how the Gamma function behaves, see Chap. 1. Therefore the Gamma structure of a solution of (9.33) contains $\Gamma(n+1)$, and this can be read off directly from the leading and trailing coefficients of (9.33).

The representation (9.31) gives us the complete information about the finite local singularity structure of a hypergeometric term. We call the set of pairs

$$\text{Sing}(s_n) := \{(a_j, e_j) \mid j = 1, \ldots, J\}$$

the set of *local types* of the hypergeometric term s_n at its finite singularities a_j. Note that—as we saw—the singularities $a_j \in \mathbb{C}$ are known modulo $\mathbb{Z}$, and $e_j \in \mathbb{Z}\backslash\{0\}$, $(j = 1, \ldots, J)$.

When two hypergeometric terms s_n and t_n are similar, i.e. their quotient is a rational function, then their representations (9.31) have the same singularity structure

Sing (s_n), and we must just find the corresponding data $J \in \mathbb{N}$, $a_j \in \mathbb{C}$, $e_j \in \mathbb{Z}\backslash\{0\}$ for $j = 1, \ldots, J$.

One can also study the singularity structure at infinity. Let's discuss the *local type* of s_n at ∞. For this purpose we look at the behavior of the rational certificate at ∞. Substituting $t = \frac{1}{n}$, we can write the asymptotic expansion

$$r(s_n)\left(\frac{1}{t}\right) = c\, t^{-\nu}(1 + d\,t + O(t^2)) = c\, n^{\nu}\left(1 + \frac{d}{n} + O\left(\frac{1}{n^2}\right)\right).$$

The uniquely determined data triple (c, ν, d) is called the *local type* of s_n at ∞. Note that d again is known only modulo $\mathbb{Z}$ since an integer shift in a Gamma term leads to a shift of the same size in d. Note moreover that for two hypergeometric terms s_n and t_n with data (c_1, ν_1, d_1) and (c_2, ν_2, d_2), the product $s_n \cdot t_n$ has the data $(c_1 \cdot c_2, \nu_1 + \nu_2, d_1 + d_2)$, see Exercise 9.16. Therefore, we get

Theorem 9.16 (Fuchs Relations) *Let be* $R(n) = \frac{p(n)}{q(n)}$ *with* $p(n), q(n) \in \mathbb{Q}[n]$. *The following relations between the local types of a hypergeometric term* s_n *given by* (9.31) *are valid:*

(a) $\nu = \displaystyle\sum_{j=1}^{J} e_j,$

(b) $d = -\displaystyle\sum_{j=1}^{J} a_j\, e_j + \deg(p(n)) - \deg(q(n)),$

(c) $c = x,$

where (c, ν, d) *denotes the local type of* s_n *at* ∞.

Proof By expanding the product in (9.32), we get for $n \to \infty$

$$\frac{s_{n+1}}{s_n} = \frac{p(n+1)q(n)}{p(n)q(n+1)} \times \prod_{j=1}^{J}(n - a_j)^{e_j}$$

$$= x\left(n^{\sum_{j=1}^{J} e_j} - \sum_{j=1}^{J} a_j\, e_j\, n^{\sum_{j=1}^{J} e_j - 1} + \left(\deg(p(n)) - \deg(q(n))\right)n^{\sum_{j=1}^{J} e_j - 1} \cdots\right)$$

from which one can read off the data of the theorem. $\square$

Note that—modulo an integer—(b) reads as

$$d \equiv -\sum_{j=1}^{J} a_j\, e_j \quad (\mathrm{mod}\ \mathbb{Z}). \tag{9.34}$$

The Fuchs relations can be used for a fast check of whether a solution term is possible or not.

Session 9.17 We continue with Session 9.14. The rational certificates of the hypergeometric terms `term1` and `term2` are given by

```
>   rat1:=ratio(term1,n);
```

$$rat1 := \frac{6\,(2\,n+1)^5\,(n+1)}{(4\,n+3)^3\,(3\,n+1)}$$

```
>   rat2:=ratio(term2,n);
```

$$rat2 := \frac{81\,(4\,n+1)}{4\,(n+1)^3\,(3\,n+1)^4}$$

The Maple command `asympt` can compute the *asymptotic series* of a function, and for `rat1` and `rat2` we get

```
>   asympt(rat1,n,10);
```

$$n^2 + \frac{11\,n}{12} + \frac{7}{36} + \frac{25}{864\,n} - \frac{605}{20736\,n^2} + \frac{7037}{248832\,n^3} - \frac{39589}{1492992\,n^4} + O\!\left(\frac{1}{n^5}\right)$$

```
>   asympt(rat2,n,10);
```

$$\frac{1}{n^6} - \frac{49}{12\,n^7} + \frac{361}{36\,n^8} - \frac{521}{27\,n^9} + O\!\left(\frac{1}{n^{10}}\right)$$

from which one can read off the local types at ∞, namely $(1, 2, 11/12)$ and $(1, -6, -49/12)$ for `rat1` and `rat2`, respectively. The procedure `infinitetype` computes these data for a hypergeometric term `term`:

```
infinitetype:=proc(term,n)
local rat,t,tay,c,nu,d;
rat:=ratio(term,n);
nu:=degree(numer(rat),n)-degree(denom(rat),n);
rat:=normal(subs(n=1/t,rat/n^nu));
tay:=convert(series(rat,t=0,2),polynom);
[coeff(tay,t,0),nu,coeff(tay,t,1)/coeff(tay,t,0)];
end proc:
```

Using `infinitetype`, we recover the above data

```
>   infinitetype(term1,n);
```

$$\left[1,\ 2,\ \frac{11}{12}\right]$$

```
>   infinitetype(term2,n);
```

$$\left[1,\ -6,\ -\frac{49}{12}\right]$$

Now we continue with the recurrence equation RE from Session 9.14, for which we get, using van Hoeij's algorithm

```
>   TIME:=time():
>   sol:=LREtools[hypergeomsols](RE,s(n),{},output=basis);
>   time()-TIME;
```

$$sol := \left[\frac{\Gamma(n+\frac{1}{2})^5\,\Gamma(n+1)}{\Gamma(n+\frac{1}{3})\,\Gamma(n+\frac{3}{4})^3}, \frac{\Gamma(n+\frac{1}{4})}{\Gamma(n+\frac{1}{3})^4\,\Gamma(n+1)^3} \right]$$

$$0.343$$

which yields of course the same solutions as before

```
>   map(x->ratio(x,n),sol);
```

$$\left[\frac{6\,(n+1)\,(2\,n+1)^5}{(4\,n+3)^3\,(3\,n+1)}, \frac{81\,(4\,n+1)}{4\,(n+1)^3\,(3\,n+1)^4} \right]$$

but in a fraction of the computing time of Petkovšek's algorithm. Let's try to explain how this works.

We recall the leading and trailing coefficients of the recurrence equation for this example:

```
>   leading;
```

$$\begin{aligned}
leading := {} & 4(6912\,n^{12} + 51840\,n^{11} + 175104\,n^{10} + 352192\,n^{9} + 469808\,n^{8} \\
& + 437960\,n^{7} + 292648\,n^{6} + 141288\,n^{5} + 42024\,n^{4} - 5416\,n^{3} \\
& - 13640\,n^{2} - 5648\,n - 721)(3\,n+1)\,(n+2)^3\,(4+3\,n)^4\,(4\,n+7)^3
\end{aligned}$$

```
>   trailing;
```

$$\begin{aligned}
trailing := {} & 486\,(4\,n+1)\,(n+1)(6912\,n^{12} + 134784\,n^{11} + 1201536\,n^{10} \\
& + 6475072\,n^{9} + 23494256\,n^{8} + 60469320\,n^{7} + 113205728\,n^{6} \\
& + 155330368\,n^{5} + 155030016\,n^{4} + 109737216\,n^{3} + 52242624\,n^{2} \\
& + 14979384\,n + 1944351)(2\,n+1)^5
\end{aligned}$$

We can easily read off the zeros (that will lead us to the Gamma singularities) of the leading coefficient (namely, $n = -\frac{1}{3}, -2, -\frac{4}{3}, -\frac{7}{4}$), and the zeros of the trailing coefficient (namely $n = -\frac{1}{4}, -1, -\frac{1}{2}$), omitting the zeros of the two polynomials of 12th degree[7] that almost always constitute so-called apparent singularities, since they don't show up in the solution and therefore only look like singularities. Recall that the Gamma terms of a possible solution of the recurrence equation are only known modulo $\mathbb{Z}$. However, it will also be important to check how often such a zero occurs. Therefore let's take the zeros from the interval $[-1, 0)$ and add their degrees. Then we arrive at the (zero, degree) list

---

[7] If we take the zeros of such polynomials under consideration, we have to work in a complicated algebraic extension field. Van Hoeij showed how this can be avoided, but we will not consider this case here.

$$\left\{\left(-\frac{1}{3},5\right),(-1,3),\left(-\frac{3}{4},3\right)\right\} \text{ corresponding to the factors}$$

$$\left\{\left(n+\frac{1}{3},5\right),(n+1,3),\left(n+\frac{3}{4},3\right)\right\} \tag{9.35}$$

for the leading coefficient and at

$$\left\{\left(-\frac{1}{4},1\right),(-1,1),\left(-\frac{1}{2},5\right)\right\} \text{ corresponding to the factors}$$

$$\left\{\left(n+\frac{1}{4},1\right),(n+1,1),\left(n+\frac{1}{2},5\right)\right\} \tag{9.36}$$

for the trailing coefficient, respectively. By the argument given in Example 9.15, only Gamma terms of arguments given in (9.36) up to the given orders can occur in the numerator of a possible solution of the form (9.31) of RE. In a similar way, by using the recurrence equation in the backward direction, one sees that only Gamma terms of arguments given in (9.35) up to the given orders can occur in the denominator of a possible solution of the form (9.31) of RE.

Let us count how many Gamma products are possible that are in the ranges (9.36) (for the numerator) and (9.35) (for the denominator). There are $6 \cdot 4 \cdot 4 \cdot 2 \cdot 2 \cdot 6 = 2{,}304$ possible solutions. Petkovšek's algorithm had to consider 15,360 cases. Therefore we already have a considerable saving. But we can do much better if we additionally use the Fuchs relations of Theorem 9.16. For this step, we use the *Newton polygon* of the difference operator (see [CvH06], Algorithm Newton τ-polygon, see also [Nörlund24], pp. 312–313, and [Birkhoff30], pp. 210–21) as a black box algorithm which directly reads off from the recurrence equation possible choices for v as well as possible choices for c. This algorithm is included in the hsum package. Applying it to the recurrence equation RE yields

```
>    NewtonpolygonRE(RE,s(n));
```

$$\{[-6, \ 429981696 - 429981696\,n, \ \{1\}],$$
$$[2, \ -429981696 + 429981696\,n, \ \{1\}]\}$$

This fast computation shows that there are only two possible values for v, namely $v = -6$ or $v = 2$. In both cases there is only one possible value for c, namely a zero of the polynomial $429981696 - 429981696\,n$, hence $c = 1$.

Now we can check easily how many Gamma combinations additionally satisfy the first Fuchs relation (Theorem 9.16 (a)). For $v = -2$ or $v = 6$, we compute

```
>    number:=0:
>    for e1 from 0 to 5 do
>       for e2 from 0 to 3 do
>          for e3 from 0 to 3 do
>             for e4 from -1 to 0 do
>                for e5 from -1 to 0 do
```

```
>                        for e6 from -5 to 0 do
>                            if element(e1+e2+e3+e4+e5+e6,{2,-6})
>                                then number:=number+1;
>                            end if;
>                        end do;
>                    end do;
>                end do;
>            end do;
>        end do;
>    end do;
>    number;
```

$$302$$

hence there are only 302 such combinations. That's clearly a relevant reduction from
the original 15,360 cases, and this is the main reason why van Hoeij's algorithm is
so fast.

Van Hoeij's complete algorithm goes one step further and reads off in another
step the possible values for d directly from the recurrence equation. Therefore the
second Fuchs relation (9.34) can be checked, too, and leads to a further reduction.
In our example case, only 22 choices survive and have to be checked, two of which
are the solutions sought.

Finally, for each possible Gamma combination (9.31) the rational function $R(n)$
has to be computed. However, this step can be reduced to the search for a polynomial
as in Petkovšek's algorithm where the *symmetric product* mentioned previously is
utilized again.[8]

q-Petkovšek-van Hoeij Algorithm

There is a q-analogue of Petkovšek's algorithm which was published by Abramov,
Paule, and Petkovšek [APP98] (see also [BK99]).

An implementation of the algorithm of [BK99] is given in the qsum package.
The Maple procedure `qrecsolve(rec,s(n))` contains this implementation.[9]

The q-analogue of Fasenmyer's method generates a second order recurrence equation for the q-Chu-Vandermonde sum

```
>    RE:=qfasenmyer(
>    qphihyperterm([q^(-n),b],[c],q,c/q^(-n)/b,k),q,k,s(n),1,2);
```

$$(-b c q^{(n+1)} + b q + b - b q^{(n+2)} - c q^{(n+1)} + q^{(2+2n)} b c) s(n+1)$$
$$- q (q^{(n+1)} - 1) (q^n c - b) S(n) + b (c q^{(n+1)} - 1) s(n+2) = 0$$

for which the q-Petkovšek algorithm generates the q-hypergeometric solution

---

[8] Moreover, the computation of d in the last step already leads to the degree bound for the polynomial
sought.

[9] `qrecsolve` has enhanced capabilities, which is the reason for its name.

```
>    qrecsolve(RE,q,s(n),output=qhypergeometric);
```

$$\left[\!\left[\left[\frac{\mathrm{qpochhammer}\left(\frac{c}{b},\,q,\,n\right)}{\mathrm{qpochhammer}(c,\,q,\,n)},\,0\le n\right]\right]\!\right]$$

This is equivalent to the first order recurrence equation

```
>    qrecsolve(RE,q,s(n),output=uprec);
```

$$[[b\,(q^n\,c - 1)\,s(n+1) + (-q^n\,c + b)\,s(n) = 0]]$$

Note, however, that in this case an application of the q-analogue of Zeilberger's algorithm determines this recurrence equation in one step (and faster):

```
>    qsumrecursion(qphihyperterm(
>    [q^(-n),b],[c],q,c/q^(-n)/b,k),q,k,s(n),recursion=up);
```

$$b\,(q^n\,c - 1)\,s(n+1) + (-q^n\,c + b)\,s(n) = 0$$

Note that there is also a q-variant of van Hoeij's algorithm. Again, the number of calls of subroutines in the q-Petkovšek algorithm can be drastically lowered by this approach ([CvH06], see also [Horn08, HKS12]). We give an example.

Session 9.18 For this session, we load Sprenger's qFPS package ([KS12, Sprenger09], see also [HKS12]).[10]

```
>    read "qFPS.mpl";
>    with(qFPS):
```

As an example we want to compute a rather complicated recurrence equation which, by construction, has a very simple solution.

```
>    m:=14:
>    RE1:=qshift(F(x),[x],q)-x*F(x):
>    RE2:=mul(x+i*q^i,i=1..floor((m-1)/2))*
>    qshift(F(x),[x$2],q)+mul(x-i*q^i,i=1..floor(m/2))*F(x):
>    RE:=qMultiplyRE(RE2,RE1,F(x))=0;
```

$$\begin{aligned}
RE := {}& (x + q)\,(2\,q^2 + x)\,(3\,q^3 + x)\,(4\,q^4 + x)\,(5\,q^5 + x)\,(6\,q^6 + x)\\
& Sq_{x,\,x,\,x}(F(x)) - (x + q)\,(2\,q^2 + x)\,(3\,q^3 + x)\,(4\,q^4 + x)\,(5\,q^5 + x)\\
& (6\,q^6 + x)\,q^2\,x\,Sq_{x,\,x}(F(x)) - (-x + q)\,(2\,q^2 - x)\,(3\,q^3 - x)\,(4\,q^4 - x)\\
& (5\,q^5 - x)\,(6\,q^6 - x)\,(7\,q^7 - x)\,Sq_x(F(x)) + (-x + q)\,(2\,q^2 - x)\\
& (3\,q^3 - x)\,(4\,q^4 - x)\,(5\,q^5 - x)\,(6\,q^6 - x)\,(7\,q^7 - x)\,x\,F(x) = 0
\end{aligned}$$

which uses the q-shift operator Sq_x, where $x = q^n$, and convert it towards a regular q-recurrence equation

```
>    RE3:=qREtoqRE(RE,F(x),s(n));
```

---

[10] Note that the packages qsum and qFPS are not completely compatible with each other so that it may be useful not to use them in parallel.

$$RE3 := (q^n + q)(2q^2 + q^n)(3q^3 + q^n)(4q^4 + q^n)(5q^5 + q^n)(6q^6 + q^n)$$
$$s(n+3) - (q^n + q)(2q^2 + q^n)(3q^3 + q^n)(4q^4 + q^n)(5q^5 + q^n)$$
$$(6q^6 + q^n)q^2 q^n s(n+2) - (-q^n + q)(2q^2 - q^n)(3q^3 - q^n)$$
$$(4q^4 - q^n)(5q^5 - q^n)(6q^6 - q^n)(7q^7 - q^n)s(n+1) + (-q^n + q)$$
$$(2q^2 - q^n)(3q^3 - q^n)(4q^4 - q^n)(5q^5 - q^n)(6q^6 - q^n)(7q^7 - q^n)$$
$$q^n s(n) = 0$$

The latter equation RE3 can be solved using the q-Petkovšek algorithm

```
>    TIME:=time(): qrecsolve(RE3,q,s(n)); time()-TIME;
```
$$[[q^{\text{binomial}(n,\,2)},\ 0 \le n]]$$
$$252.441$$

which takes several minutes. On the other hand, the q-van Hoeij algorithm, implemented in the qHypergeomSolveRE command, applied to RE3

```
>    TIME:=time(): qHypergeomSolveRE(RE3,s(n)); time()-TIME;
```
$$A_1\, q^{\frac{(n-1)n}{2}}$$
$$2.449$$

is much faster. With the command

```
>    TIME:=time(): qHypergeomSolveRE(RE,F(x)); time()-TIME;
```
$$[x]$$
$$2.293$$

one can directly factorize RE w.r.t. the variable x.

The results in [FKTN12] depend heavily on an efficient implementation of the q-Petkovšek-van Hoeij algorithm and were computed using the qHypergeomSolve RE command of Sprenger's qFPS package.

Further Reading

For further reading on the algorithms of this chapter see [Nörlund24, CvH06, HKS12].

Exercises

Exercise 9.1 *In particular the Maple procedure* rec2poly *of Session 9.5 can be used to find polynomial solutions of first order recurrence equations. Describe this in detail. Find the polynomial solutions of*

$$\text{(a) } n^2\, s_{n+1} - (n+20)^2\, s_n = 0, \qquad \text{(b) } n\, s_{n+1} - (n+40)\, s_n = 0.$$

Write any s_n as hypergeometric term and check the results.

Exercise 9.2 *The fixed point free permutations of n items are called derangements. Their number D_n is given by*

$$D_n = n! \sum_{k=0}^{n} \frac{(-1)^k}{k!}.$$

Show that D_n satisfies the recurrence equation

$$(n+2)\, D_{n+2} - (n+1)\, D_{n+1} - (n+1)\, D_n = 0.$$

Use both Gosper's and Petkovšek's algorithms to prove that D_n does not constitute a hypergeometric term.

Exercise 9.3 *The following is a list of hypergeometric series for which hypergeometric term representations are given (7.5.3.32)–(7.5.3.40) on p. 556 in the extensive bibliography [PBM90]. For all of them, Zeilberger's algorithm does not find the recurrence equation of lowest order and hence the WZ method does not apply (see Exercise 7.32).*

Find hypergeometric term representations for the following hypergeometric sums:

$$\text{(a) } {}_4F_3\left(\begin{array}{c} -n, a, a+1/2, b \\ 2a, (b-n+1)/2, (b-n)/2+1 \end{array} \middle| 1\right),$$

$$\text{(b) } {}_4F_3\left(\begin{array}{c} -n, a, a+1/2, b \\ 2a+1, (b-n)/2, (b-n+1)/2 \end{array} \middle| 1\right),$$

$$\text{(c) } {}_4F_3\left(\begin{array}{c} -n, a, a+1/2, b \\ 2a+1, (b-n+1)/2, (b-n)/2+1 \end{array} \middle| 1\right),$$

$$\text{(d) } {}_4F_3\left(\begin{array}{c} -n, a, b, -1/2-a-b-n \\ -a-n, -b-n, a+b+1/2 \end{array} \middle| 1\right),$$

$$\text{(e) } {}_4F_3\left(\begin{array}{c} -n, a, b, 1/2-a-b-n \\ -a-n, 1-b-n, a+b+1/2 \end{array} \middle| 1\right),$$

$$\text{(f) } {}_4F_3\left(\begin{array}{c} -n, a, b, 1/2-a-b-n \\ 1-a-n, 1-b-n, a+b-1/2 \end{array} \middle| 1\right),$$

$$\text{(g) } {}_4F_3\left(\begin{array}{c} -n, a, b, 3/2-a-b-n \\ 1-a-n, 1-b-n, a+b+1/2 \end{array} \middle| 1\right),$$

$$\text{(h) } {}_4F_3\left(\begin{array}{c} -n, a, b, 3/2-a-b-n \\ 1-a-n, 2-b-n, a+b-1/2 \end{array} \middle| 1\right),$$

$$\text{(i) } {}_4F_3\left(\begin{array}{c} -n, a, b, 5/2-a-b-n \\ 2-a-n, 2-b-n, a+b-1/2 \end{array} \middle| 1\right).$$

Note that the complexity of Petkovšek's algorithm is so high that some of these results take up to about half a minute of computation time or more. Therefore also apply van

Hoeij's algorithm via the `LREtools[hypergeomsols]` command and compare timings and results.

Exercise 9.4 Prove Algorithm 9.6 on the polynomial solutions of holonomic recurrence equations of arbitrary order.

◇ **Exercise 9.5** Write a Maple procedure `recpoly(rec,s(n))` that implements Algorithm 9.6.

Exercise 9.6 In Exercise 8.9 the recurrence equation

$$(n + 1 + 2b)\,(n + 2b)\,s_{n+2} - 4(n + b)\,(n + 1 + b)\,s_n = 0$$

for

$$s_n = {}_2F_1\left(\begin{array}{c} -n/2,\, -n/2 + 1/2 \\ b + 1/2 \end{array} \middle|\, 1\right)$$

was determined by the extended version of Zeilberger's algorithm with $m = 2$. Similarly, one gets the recurrence equations

$$(n + 1 + 2b)\,(n + 2 + 2b)\,(n + 2b)\,s_{n+3} - 8(n+2+b)\,(n + 1 + b)\,(n + b)\,s_n = 0$$

and

$$\begin{aligned} 0 = {} & (n + 2 + 2b)\,(n + 2b)\,(n + 3 + 2b)\,(n + 1 + 2b)\,s_{n+4} \\ & - 16(n + 3 + b)\,(n + 2 + b)\,(n + 1 + b)\,(n + b)\,s_n \end{aligned}$$

when using $m = 3$ and $m = 4$. In a first step, find the hypergeometric term solutions of these recurrence equations by a direct argument, and deduce the closed form of s_n. Then use Petkovšek's algorithm for the same purpose.

Exercise 9.7 Show that the differential equation

$$\begin{aligned} 0 = {} & a(1 - z)\,z\,F'''(x) + \left(a + a^2 + ab + z - 3az - 3a^2z - 2z^2 + z^3\right)F''(z) \\ & + \left(-3a^2 - 2a^3 + b - z - 4az - bz + z^2 + 3az^2\right)F'(z) \\ & + 2a^2(z - 1)\,F(z) \end{aligned} \tag{9.37}$$

leads to the recurrence equation

$$\begin{aligned} 0 = {} & a(2 + n)\,(3 + n)\,(2 + a + b + n)\,u_{n+3} \\ & + (2 + n)\left(1 - 3a - 6a^2 - 2a^3 + b + n - 4an - 3a^2n - an^2\right)u_{n+2} \\ & - \left(1 + 4a + 2a^2 + b + 3n + 4an + bn + 2n^2\right)u_{n+1} \\ & + (a + n)\,(2a + n)\,u_n \end{aligned}$$

for the coefficients of a power series solution

$$F(z) = \sum_{n=0}^{\infty} u_n z^n$$

of (9.37); see e.g. [GK95] or [SZ94]. You can use the `diffeqtorec` *command of the* `gfun` *package [SZ94] for this purpose. Find the hypergeometric function solutions of (9.37).*

Exercise 9.8 *Prove Algorithm 9.13 on the hypergeometric term solutions of holonomic recurrence equations of arbitrary order.*

Exercise 9.9 *Write a Maple procedure* `rechyper(rec,s(n))` *that implements Algorithm 9.13, and apply your implementation to the recurrence equations of Exercises 9.6–9.7.*

Exercise 9.10 (m-fold Hypergeometric Terms) [PS95, HKS12] *Modify Algorithm 9.13 such that one finds, given* $m \in \mathbb{N}$, *all* m-*fold hypergeometric terms* s_n *having a term ratio*

$$\frac{s_{n+m}}{s_n} \in \mathbb{Q}(n)$$

which satisfy the recurrence equation (9.28).

Exercise 9.11 (Complexity) *Assume that two polynomials* P_n *and* R_n *of degree* p *and* r, *respectively, are given that are factorized in linear factors over* $\mathbb{Q}$. *How often does Petkovšek's algorithm, applied to the recurrence equation*

$$P_n\, s_{n+2} + Q_n\, s_{n+1} + R_n\, s_n = 0,$$

have to call Algorithm 9.6, described in Example 9.1?

Exercise 9.12 *Find a hypergeometric term representation of the sum*

$$S_n = \sum_{k=0}^{n} \frac{(-1)^k \binom{r-s-k}{k}\binom{r-2k}{n-k}}{r-n-k+1}.$$

Exercise 9.13 *Derive a second order recurrence equation for the coefficient* A_k *of the double sum*

$$\sum_{k=0}^{n} A_k = \sum_{k=0}^{n} \binom{n}{k} (c)_k\, (m)_{n-k} \cdot {}_3F_2\left(\begin{matrix} -k, a, b \\ c, d \end{matrix} \,\middle|\, 1\right),$$

and show that A_k *does not satisfy any holonomic recurrence equation of first order; compare Exercise 7.25.*

Exercise 9.14 *Show again that the* harmonic numbers

$$H_n = \sum_{k=1}^{n} \frac{1}{k}$$

do not constitute a hypergeometric term. For this purpose find a holonomic recurrence equation for H_n, and apply Petkovšek's algorithm.

Exercise 9.15 *Show that the Gosper-Petkovšek representation given by Algorithm 9.7 is unique. Use* $t_k = \frac{(k+1)(k+2)}{k}$ *to show that it is essential to take the minimal* $j \in \mathbb{N}$ *in each rewrite step to obtain the Gosper-Petkovšek representation, since otherwise one of the gcd conditions (b) or (c) may be invalid* [Stölting96, p. 30].

Exercise 9.16 *Show that for two hypergeometric terms* s_n *and* t_n *with local data* (c_1, v_1, d_1) *and* (c_2, v_2, d_2) *at infinity, the product* $s_n \cdot t_n$ *has the data* $(c_1 \cdot c_2, v_1 + v_2, d_1 + d_2)$.

Exercise 9.17 *Generate a recurrence equation for which the computation time using Petkovšek's algorithm is at least 1,000 seconds by using the ideas of Sessions 9.14 and 9.17 by adding three sufficiently complicated hypergeometric terms. Use the Maple procedure* `LocalInfiniteType(RE,s(n))` *contained in the* `hsum` *package to recover the infinite types of the three solutions as well as van Hoeij's algorithm to compute its solution basis.*

Exercise 9.18 (Symmetrizing) *Prove the following identity of Rogers*

$$\sum_{k \in \mathbb{Z}} F(n, k) = \sum_{k \in \mathbb{Z}} \frac{(-1)^k \, q^{\frac{k(3k-1)}{2}}}{(q; q)_{n+k} \, (q; q)_{n-k}} = \frac{1}{(q; q)_n}$$

by an application of the q-analogues of Zeilberger's and Petkovšek's algorithms. Note that the q-Zeilberger algorithm returns a recurrence equation of order 3 rather than 1.

By creative symmetrizing [PR97], the q-Zeilberger algorithm can be taught to do better: Note that $F(n, -k) = q^k \, F(n, k)$, hence

$$\sum_{k \in \mathbb{Z}} F(n, k) = \sum_{k \in \mathbb{Z}} q^k \, F(n, k).$$

Show that the application of the q-Zeilberger algorithm to $G(n, k) = \frac{1+q^k}{2} \, F(n, k)$ yields the resulting first order recurrence equation in one step.

Exercise 9.19 [PR97] *Prove the identity*

$$\sum_{k \in \mathbb{Z}} \frac{q^{k^2+k}}{(q; q)_k \, (q; q)_{n-k}} = \sum_{k \in \mathbb{Z}} \frac{(-1)^k \, q^{\frac{k(5k+3)}{2}}}{(q; q)_{n+k+1} \, (q; q)_{n-k}}$$

using the q-Zeilberger algorithm and

(a) the procedure `recursion/compare` of the qsum package,[11]
(b) creative symmetrizing, see Exercise 9.18.

Exercise 9.20 *Prove Jackson's terminating q-analogue of Dixon's identity*

$$\sum_{k\in\mathbb{Z}}(-1)^k \begin{bmatrix} n+b \\ n+k \end{bmatrix}_q \begin{bmatrix} n+c \\ c+k \end{bmatrix}_q \begin{bmatrix} b+c \\ b+k \end{bmatrix}_q q^{\frac{k(3k-1)}{2}} = \begin{bmatrix} n+b+c \\ n,b,c \end{bmatrix}_q \tag{9.38}$$

where

$$\begin{bmatrix} n+b+c \\ n,b,c \end{bmatrix}_q = \frac{[n+b+c]_q!}{[n]_q!\,[b]_q!\,[c]_q!}$$

using the q-Zeilberger algorithm and

(a) the q-Petkovšek algorithm,
(b) creative symmetrizing, see Exercise 9.18.

Show that (9.38) *is equivalent to* [GR90, Appendix (II.15)]:

$$_3\phi_2\left(\begin{matrix} q^{-2n}, b, c \\ q^{1-2n}/b, q^{1-2n}/c \end{matrix}\;\middle|\; q, \frac{q^{2-n}}{bc}\right) = \frac{(b,c;q)_n\,(q,bc;q)_{2n}}{(q,bc;q)_n\,(b,c;q)_{2n}}.$$

References

ABP95. Abramov, S.A., Bronstein, M., Petkovšek, M.: On polynomial solutions of linear operator equations. In: Proceedings of ISSAC 95, pp. 290–296. ACM Press, New York (1995)

APP98. Abramow, S.A., Paule, P., Petkovšek, M.: q-hypergeometric solutions of q-difference equations. Discrete Math. **180**, 3–22 (1998)

Birkhoff30. Birkhoff, G.D.: Formal theory of irregular linear difference equations. Acta Math. **54**, 205–246 (1930)

BK99. Böing, H., Koepf, W.: Algorithms for q-hypergeometric summation in computer algebra. J. Symbolic Comput. **28**, 777–799 (1999)

Bronstein94. Bronstein, M.: An improved algorithm for factoring linear ordinary differential operators. In: Proceedings of ISSAC 94, pp. 336–340. ACM Press, New York (1994)

CvH06. Cluzeau, T., van Hoeij, M.: Computing hypergeometric solutions of linear recurrence equations. Appl. Algebra Eng. Commun. Comput. **17**, 83–115 (2006)

FKTN12. Foupouagnigni, M., Koepf, W., Tcheutia, D.D., Sadjang, P.N.: Representations of q-orthogonal polynomials. J. Symbolic Comput. **47**, 1347–1371 (2012)

GR90. Gasper, G., Rahman, M.: Basic hypergeometric series. In: Encyclopedia of Mathematics and its Applications, vol. 35, 2nd edn. Cambridge University Press, London (1990)

---

[11] `recursion/compare`(rec1,rec2,s(n)) decides whether the two recurrence equations rec1 and rec2 in terms of s_n are equivalent, i.e. are (commutative) multiples of each other, or whether they are compatible, i.e. have different order and the solutions of the recurrence equation of smaller order are also solutions of the second recurrence equation, hence, as operators, are (noncommutative) multiples of each other.

GK95. Gruntz, D., Koepf, W.: Maple package on formal power series. Maple Tech. Newsl. **2**(2), 22–28 (1995)

Hearn95. Hearn, A.C.: REDUCE user's manual, Version 3.6. RAND Co., Santa Monica (1995)

Horn08. Horn, P.: Faktorisierung in Schief-Polynomringen. PhD thesis, Universität Kassel (2008)

HKS12. Horn, P., Koepf, W., Sprenger, T.: m-fold hypergeometric solutions of linear recurrence equations revisited. Math. Comput. Sci. **6**, 61–77 (2012)

KS12. Koepf, W., Sprenger, T.: Algorithmic determination of q-power series for q-holonomic functions. J. Symbolic Comput. **47**, 519–535 (2012)

MA94. Melenk, H., Apel, J.: REDUCE package NCPOLY: Computation in non-commutative polynomial ideals. Konrad-Zuse-Zentrum Berlin (ZIB) (1994)

Nörlund24. Nörlund, N.E.: Vorlesungen über Differenzenrechnung. Springer, Berlin (1924)

PR97. Paule, P., Riese, A.: A Mathematica q-analogue of Zeilberger's algorithm based on an algebraically motivated approach to q-hypergeometric telescoping. In: Ismail, M.E.H., et al. (eds.) Fields Institute Communications, vol. 14, pp. 179–210. American Mathematical Society, Rhode Island (1997)

PS95. Paule, P., Schorn, M.: A Mathematica version of Zeilberger's algorithm for proving binomial coefficient identities. J. Symbolic Comput. **20**, 673–698 (1995)

Petkovšek92. Petkovšek, M.: Hypergeometric solutions of linear recurrences with polynomial coefficients. J. Symbolic Comput. **14**, 243–264 (1992)

PWZ96. Petkovšek, M., Wilf, H., Zeilberger, D.: $A = B$. A. K. Peters, Wellesley (1996)

PBM90. Prudnikov, A.P., Brychkov, Y.A., Marichev, O.I.: Integrals and series. Vol. 3: More Special Functions. Gordon and Breach Publ., New York (1990)

SZ94. Salvy, B., Zimmermann, P.: Gfun: a maple package for the manipulation of generating and holonomic functions in one variable. ACM Trans. Math. Softw. **20**, 163–177 (1994)

Sprenger09. Sprenger, T.: Algorithmen für q-holonome Funktionen und q-hypergeometrische Reihen. PhD thesis, Universität Kassel (2009)

Stanley80. Stanley, R.P.: Differentiably finite power series. Eur. J. Comb. **1**, 175–188 (1980)

Stölting96. Stölting, G.: Algorithmische Berechnung von Summen. Diploma thesis, Freie Universität Berlin (1996)

vdPS03. van der Put, M., Singer, M.F.: Galois theory of linear differential equations. Grundlehren der mathematischen Wissenschaften, vol. 328. Springer, Berlin (2003)

vanHoeij98. van Hoeij, M.: Finite singularities and hypergeometric solutions of linear recurrence equations. J. Pure Appl. Algebra **139**, 109–131 (1998)

Chapter 10
Differential Equations for Sums

In this chapter, as an interesting variation of Zeilberger's method, we present an algorithm which generates *holonomic differential equations* rather than recurrence equations for definite sums of a certain type. Again, we call a differential equation holonomic, if it is homogeneous, linear and has polynomial coefficients. In this book, we deal exclusively with *ordinary differential equations*, in which only derivatives with respect to a single variable, denoted here by x, occur. A function that satisfies a holonomic differential equation is also called holonomic.

A first order holonomic recurrence equation

$$Q_n \, s_{n+1} - P_n \, s_n = 0 \qquad (P, Q \in \mathbb{Q}[n])$$

is equivalent to a rational term ratio

$$\frac{s_{n+1}}{s_n} = \frac{P_n}{Q_n} \in \mathbb{Q}(n),$$

hence to the fact that s_n is a hypergeometric term.

Similarly, a first order holonomic differential equation

$$Q(x) \, S'(x) - P(x) \, S(x) = 0 \qquad (P, Q \in \mathbb{Q}[x])$$

is equivalent to a rational logarithmic derivative

$$\frac{S'(x)}{S(x)} = \frac{P(x)}{Q(x)} \in \mathbb{Q}(x), \tag{10.1}$$

and we call a (differentiable) function of the variable x a *hyperexponential term* if it satisfies (10.1).

It is possible to characterize hyperexponential terms by elementary functions.

W. Koepf, *Hypergeometric Summation*, Universitext,
DOI: 10.1007/978-1-4471-6464-7_10, © Springer-Verlag London 2014

Lemma 10.1 *A function $S(x)$ is a complex hyperexponential term, i.e. $S'(x)/S(x) \in \mathbb{C}(x)$, if and only if it has a representation*

$$S(x) = e^{R(x)} \cdot \prod_{j=1}^{J} (x - x_j)^{\alpha_j} \tag{10.2}$$

for $R(x) \in \mathbb{C}(x)$, and $\alpha_j, x_j \in \mathbb{C}$ ($j = 1, \ldots, J$). Moreover, if $S(x)$ is a rational hyperexponential term, i.e. $S'(x)/S(x) \in \mathbb{Q}(x)$, and if the roots of $Q(x)$ are rational, then we have

$$S(x) = C \cdot e^{R(x)} \cdot \sqrt[m]{T(x)} \tag{10.3}$$

for $C \in \mathbb{C}$, $R, T \in \mathbb{Q}(x)$ and $m \in \mathbb{N}$.

Proof Taking the antiderivative of the defining Eq. (10.1), we get

$$\ln S(x) = \int (\ln S(x))' \, dx = \int \frac{S'(x)}{S(x)} \, dx = \int \frac{P(x)}{Q(x)} \, dx.$$

Expanding the integrand into partial fractions (over $\mathbb{C}$), one can see that the anti-derivative of a rational function is the sum of a rational function and of logarithmic terms of the form

$$\alpha_j \ln(x - x_j) \qquad (\alpha_j, x_j \in \mathbb{C} \ (j = 1, \ldots, J)),$$

x_j ($j = 1, \ldots, J$) being the complex roots of $Q(x)$. Therefore we get

$$\ln S(x) = R(x) + \sum_{j=1}^{J} \alpha_j \ln(x - x_j)$$

for rational $R \in \mathbb{C}(x)$. Exponentiation gives us (10.2). If $S'(x)/S(x) \in \mathbb{Q}(x)$ and if the roots x_j of $Q(x)$ are rational, then it turns out that so are $\alpha_j \in \mathbb{Q}$ ($j = 1, \ldots, J$). Finally (10.3) follows if we take m as the least common multiple of the denominators of α_j ($j = 1, \ldots, J$).

Note that, on the other hand, any function with representation (10.2) or (10.3) is easily seen to be a hyperexponential term. $\qquad\qquad\square$

Hyperexponential terms that have a representation (10.3) with $m = 1$ are of special interest. We call them *strictly hyperexponential terms*.

We call the roots x_j ($j = 1, \ldots, J$) of the denominator polynomial $Q(x)$ of (10.1) the *critical points* of the hyperexponential term $S(x)$ since these are the points where $S'(x)/S(x)$ has a pole. This may, or may not, imply that $S(x)$ is ill-defined at x_j depending on the corresponding value α_j. If $\alpha_j \in \mathbb{N}$ then by (10.2) $S(x)$ is well-defined in a neighborhood of x_j. For example, any zero of an arbitrary polynomial $S(x)$ is a critical point in which case α_j is the order of the zero since

$$S(x) = c(x - x_j)^N + \text{higher order terms}$$

implies

$$\frac{S'(x)}{S(x)} = \frac{N}{x - x_j} + \text{higher order terms.}$$

Next, we state the main algorithm which generates differential equations for definite sums.

Algorithm 10.2 Given $F(x, k)$, this algorithm searches for a homogeneous, linear differential equation for $S(x) = \sum_{k=-\infty}^{\infty} F(x, k)$:

1. Input: $F(x, k) \not\equiv 0$, a hypergeometric term with respect to k, which is hyperexponential with respect to x such that $F'(x, k)/F(x, k) \in \mathbb{Q}(x, k)$. Assume further that

$$S(x) = \sum_{k=-\infty}^{\infty} F(x, k)$$

   converges absolutely and uniformly in a certain region $D \subset \mathbb{C}$ (or $D \subset \mathbb{R}$).
2. Set $J := 1$.
3. Set $a_k := F^{(J)}(x, k) + \sum_{j=0}^{J-1} \sigma_j(x) F^{(j)}(x, k)$ with undetermined variables σ_j depending on x, but independent of k.
4. Apply the adaptation of Gosper's algorithm described in Chap. 6 to a_k: In the last step, solve the linear system for the coefficients of f_k, and at the same time for the unknowns σ_j $(j = 0, \ldots, J - 1)$. In the affirmative case, Gosper's algorithm finds $G(x, k)$ with

$$a_k = G(x, k + 1) - G(x, k). \tag{10.4}$$

   Moreover, the calculation determines the functions $\sigma_j(x)$. If $F(x, k)$ is strictly hyperexponential with respect to x, then $\sigma_j \in \mathbb{Q}(x)$ $(j = 0, \ldots, J - 1)$.
   If the procedure is not successful then increase J by one and continue with Step 3.
5. Output: By summation, we have

$$S^{(J)}(x) + \sum_{j=0}^{J-1} \sigma_j(x) S^{(j)}(x) = 0$$

   for $S(x)$, valid in D. Multiplication by the common denominator results in the differential equation sought. If $F(x, k)$ is strictly hyperexponential with respect to x, then this differential equation is holonomic.

Proof The proof of this algorithm is a straightforward adaptation of the proof of Algorithm 7.5. The main difference is that, in the current case, we cannot restrict ourselves to terms with finite support because the simplest examples do not have this property; see, e.g., Example 10.3 below. Since we assume that

$$\sum_{k=-\infty}^{\infty} F(x,k)$$

converges absolutely and uniformly in D, in particular $F(x,k) \to 0$ for $k \to \pm\infty$. Summation of (10.4) then yields a telescoping right-hand side which by a limit consideration tends to zero. $\square$
We give some examples.

Example 10.3 (Elementary Functions) As a first example, we consider the exponential function, given by its power series representation

$$e^x = S(x) = \sum_{k=0}^{\infty} \frac{1}{k!} x^k.$$

Note that since the given sum is a power series, the following general method is applicable: The holonomic recurrence equation for the coefficient $A_k = 1/k!$,

$$(k+1)\, A_{k+1} - A_k = 0, \tag{10.5}$$

can be converted to the differential equation

$$S'(x) - S(x) = 0 \tag{10.6}$$

by summing (10.5) from $k = 0, \ldots, \infty$. This method of generating a holonomic differential equation can always be applied to functions given by power (or Laurent) series with holonomic coefficients; see, e.g. [Koepf92, Koepf06, GKP94, GK95].

Let us nevertheless use instead Algorithm 10.2, which is more flexible since it is not restricted to power or Laurent series. The summand $F(x,k) = x^k/k!$ does not have finite support (for $k \to +\infty$), but the desired convergence property is valid in the neighborhood of any point $x \in \mathbb{C}$. We define

$$a_k := F(x,k) + \sigma_1\, F'(x,k) = \frac{x^k}{k!} + \sigma_1 \frac{x^{k-1}}{(k-1)!}$$

and get

$$\frac{a_{k+1}}{a_k} = \frac{x\,(x + \sigma_1 k + \sigma_1)}{(k+1)\,(x + \sigma_1 k)},$$

so Gosper's algorithm is applicable. We find that $p_k = x + \sigma_1 k$, $q_k = x$ and $r_k = k$. The degree bound for f_k equals 0, and if we substitute the generic polynomial $f_k = b_0$ into the main equation for f_k, and equate the coefficients, we get the solution

$$\{b_0 = 1, \sigma_1 = -1\}.$$

In particular this establishes the differential equation (5.6), once again.

Similarly, for

$$\sin x = S(x) = \sum_{k=0}^{\infty} \frac{(-1)^k}{(2k+1)!} x^{2k+1},$$

we use $J = 2$, set $F(x,k) := \dfrac{(-1)^k}{(2k+1)!} x^{2k+1}$ and

$$a_k := F(x,k) + \sigma_1 F'(x,k) + \sigma_2 F''(x,k)$$
$$= \frac{(-1)^k}{(2k+1)!} x^{2k+1} + \sigma_1 \frac{(-1)^k}{(2k)!} x^{2k} + \sigma_2 \frac{(-1)^k}{(2k-1)!} x^{2k-1}.$$

This leads to the term ratio

$$\frac{a_{k+1}}{a_k} = -\frac{x^2 \left(x^2 + 2\sigma_1 kx + 3 x\sigma_1 + 4\sigma_2 k^2 + 10\sigma_2 k + 6\sigma_2\right)}{(4k+6)(k+1)\left(x^2 + 2\sigma_1 kx + x\sigma_1 + 4\sigma_2 k^2 + 2\sigma_2 k\right)},$$

and to $p_k = x^2 + 2\sigma_1 xk + \sigma_1 x + 4\sigma_2 k^2 + 2\sigma_2 k$, $q_k = -x^2$ and $r_k = 2(2k+1)k$. The degree bound of f_k is 0 again, and with $f_k = b_0$, we finally find

$$\{\sigma_1 = 0, \sigma_2 = 1, b_0 = -1\}.$$

In particular, we have deduced the differential equation

$$S''(x) + S(x) = 0$$

for the power series $S(x)$.

Example 10.4 (*Legendre Polynomials*) As a more sophisticated and probably more interesting example, we consider the Legendre polynomials once more, given by their representation

$$P_n(x) = S(x) = \sum_{k=-\infty}^{\infty} \binom{n}{k}\binom{-n-1}{k}\left(\frac{1-x}{2}\right)^k.$$

The holonomic differential equation of $P_n(x)$ could be deduced from the hypergeometric representation (2.19) of $P_n(x)$ and the hypergeometric differential equation;

see Exercise 2.3. We use Algorithm 10.2, instead, and therefore set

$$F(x,k) := \binom{n}{k}\binom{-n-1}{k}\left(\frac{1-x}{2}\right)^k.$$

Once again, the choice $J = 1$ is not successful, hence we use $J = 2$. For

$$a_k := F(x,k) + \sigma_1\,F'(x,k) + \sigma_2\,F''(x,k),$$

we get the term ratio

$$\frac{a_{k+1}}{a_k} = \frac{(x-1)\left(1-2x+x^2-\sigma_1 k+\sigma_1 kx-\sigma_1+x\sigma_1+\sigma_2 k^2+\sigma_2 k\right)(n-k)(n+1+k)}{2(k+1)^2\left(1-2x+x^2-\sigma_1 k+\sigma_1 kx+\sigma_2 k^2-\sigma_2 k\right)}.$$

Now, we have

$$\begin{aligned}
p_k &= 1 - 2x + x^2 - \sigma_1 k + \sigma_1 kx + \sigma_2 k^2 - \sigma_2 k,\\
q_k &= (x-1)(n+k)(n-k+1), \qquad r_k = 2k^2,
\end{aligned}$$

and the degree bound 0 for f_k. For $f_k = b_0$, we finally get

$$\left\{b_0 = \frac{x-1}{(n+1)\,n},\, \sigma_1 = -\frac{2x}{(n+1)\,n},\, \sigma_2 = \frac{1-x^2}{(n+1)\,n}\right\},$$

in particular the differential equation

$$\left(1-x^2\right)P_n''(x) - 2x\,P_n'(x) + (n+1)\,n\,P_n(x) = 0.$$

Session 10.5 The Maple procedure

```
sumdiffeq:=proc(F,k,sx)
local x,S,a,b,sigma,rat,p,q,r,upd,deg,f,j,jj,l,var,deq,sol,
num,den,J,cert;
if type(sx,function) then
  S:=op(0,sx); x:=op(1,sx)
else
  x:=sx
end if;
for J from 1 to MAXORDER do
  a:=add(sigma[j]*diff(F,[x$j]),j=0..J-1)+diff(F,x$J);
  rat:=simpcomb(a/subs(k=k-1,a));
  if not type(rat,ratpoly(anything,k)) then
    error 'Algorithm not applicable'
  end if;
  p:=1: q:=subs(k=k-1,numer(rat)): r:=subs(k=k-1,denom(rat)):
  upd:=update(p,q,r,k);
```

```
p:=op(1,upd):  q:=op(2,upd):  r:=op(3,upd):
deg:=degreebound(p,q,r,k);
if deg>=0 then
   f:=add(b[j]*k^j,j=0..deg);
   var:={seq(sigma[jj],jj=0..J-1),seq(b[jj],jj=0..deg)};
   deq:=collect(subs(k=k+1,q)*f-r*subs(k=k-1,f)-p,k);
   sol:={solve({coeffs(deq,k)},var)};
   if not(sol={} or
      {seq(op(2,op(1,op(1,sol))),l=1..nops(op(1,sol))))}={0}) then
      deq:=add(sigma[j]*diff(S(x),[x$j]),j=0..J-1)+diff(S(x),x$J);
      deq:=normal(subs(op(1,sol),deq));
      deq:=numer(deq);
      deq:=collect(deq,[seq(diff(S(x),x$(J-j)),j=0..J-1),S(x)]);
      return map(factor,deq)=0;
   end if;
 end if;
end do;
error cat('Algorithm finds no differential equation of order <= ',
MAXORDER);
end proc:
```

implements Algorithm 10.2. Note that only a few lines from the code of the Maple
procedure `sumrecursion` of Session 7.3 had to be changed: All occurrences of
shifts are replaced by the corresponding derivatives, and that's it. Note moreover that
we do not need a specific simplification procedure for the ratios $f'(x)/f(x)$ of hyperexponential terms $f(x)$ since the differentiation of a term of the form (10.2) by the
chain rule reproduces the exponential term, which therefore cancels in $f'(x)/f(x)$,
to give a result which is automatically rational.

We use the above procedure to calculate the differential equations of our previous
examples, again:

```
>   sumdiffeq(1/k!*x^k,k,S(x));
```

$$S(x) - \tfrac{d}{dx} S(x) = 0$$

```
>   sumdiffeq((-1)^k/(2*k+1)!*x^(2*k+1),k,S(x));
```

$$S(x) + \tfrac{d^2}{dx^2} S(x) = 0$$

```
>   sumdiffeq(binomial(n,k)*binomial(-n-1,k)*((1-x)/2)^k,k,S(x));
```

$$-(-1+x)(x+1)\tfrac{d^2}{dx^2} S(x) - 2\tfrac{d}{dx} S(x) x + S(x) n (n+1) = 0$$

The coefficient $\frac{(-1)^k x^k}{(2k+1)!}$ of the function $S(x) = \sin\sqrt{x}/\sqrt{x}$ is strictly hyperexponential with respect to x so that the differential equation generated by Algorithm 10.2 is
holonomic:

```
>   sumdiffeq((-1)^k/(2*k+1)!*x^k,k,S(x));
```

$$S(x) + 6\tfrac{d}{dx} S(x) + 4x\tfrac{d^2}{dx^2} S(x) = 0$$

This is not so for $S(x) = e^{\sqrt{x}} = \sum_{k=0}^{\infty} \frac{1}{k!} x^{k/2}$ [here we have $m = 2$ in (10.3)], hence we get a differential equation with nonpolynomial coefficients:

```
>    sumdiffeq(1/k!*x^(k/2),k,S(x));
```

$$S(x) - 2\sqrt{x}\, \tfrac{d}{dx}\, S(x) = 0$$

Similarly, we deduce the differential equation

```
>    sumdiffeq((-1)^k/(2*k+1)!*x^((2*k+1)/m),k,S(x));
```

$$S(x) + (m-1)\,m\,x^{\frac{m-2}{m}}\,\tfrac{d}{dx}\,S(x) + m^2\,x^{\frac{2(m-1)}{m}}\,\tfrac{d^2}{dx^2}\,S(x) = 0$$

for $S(x) = \sin \sqrt[m]{x}$. In Exercise 10.7, a modified version of Algorithm 10.2 is to be used to determine *holonomic* differential equations for $e^{\sqrt{x}}$, $e^{\sqrt[3]{x}}$, $\sin \sqrt{x}$, and $\sin \sqrt[3]{x}$.

The *Bessel function*

$$J_0(x) = \sum_{k=0}^{\infty} \frac{(-1)^k}{4^k\,k!^2}\,x^{2k}$$

satisfies the *Bessel differential equation*

```
>    sumdiffeq((-1)^k/(4^k*k!^2)*x^(2*k),k,S(x));
```

$$S(x)\,x + \tfrac{d}{dx}\,S(x) + x\,\tfrac{d^2}{dx^2}\,S(x) = 0$$

See Exercise 10.9 for the complete family $J_n(x)$ of Bessel functions.

Note however, that one has to be careful that the sum under consideration is indeed supported for $k = -\infty, \ldots, \infty$. For the geometric series, we get for example

```
>    sumdiffeq(x^k,k,S(x));
```

$$-(-1+x)\,(x\,b_0 - x - b_0)\,\tfrac{d}{dx}\,S(x) + S(x)\,x = 0$$

Here a differential equation for the bilateral sum

$$\sum_{k=-\infty}^{\infty} x^k = \sum_{k=-\infty}^{-1} x^k + \sum_{k=0}^{\infty} x^k$$

is generated which converges nowhere however since $\sum_{k=0}^{\infty} x^k$ converges inside and $\sum_{k=-\infty}^{-1} x^k$ converges outside the unit disk $\mathbb{D} := \{z \in \mathbb{C} \mid |z| < 1\}$.

On the other hand, the method easily deduces an *inhomogeneous differential equation* for

$$S(x) = \sum_{k=0}^{\infty} x^k.$$

We calculate the antidifference of $a_k = x^k + \sigma_1 k x^{k-1}$ by (5.6)

$$G(x,k) = \frac{r_k}{p_k} f_{k-1} a_k = -x^{k-1}(-kb_0 + kxb_0 - xb_0 - kx + x),$$

so that summation for $k = 0$ to $k = \infty$ yields

$$S(x) - (x-1)(xb_0 - x - b_0) S'(x) = \sum_{k=0}^{\infty} \Big(G(x,k+1) - G(x,k)\Big)$$
$$= \lim_{k \to \infty} G(x,k) - G(x,0) = 1 - b_0.$$

Hence, we might choose $b_0 = 1$ to obtain the homogeneous differential equation

$$S(x) + (x-1) S'(x) = 0$$

for $S(x)$.

Example 10.6 (Hypergeometric Transformations) Here we apply Algorithm 10.2
to obtain proofs of *hypergeometric transformations* like the ones that we met in
Exercise 3.7.

These were the transformations of Kummer

$$e^x \cdot {}_1F_1\left(\begin{array}{c}a\\b\end{array}\bigg| -x\right) = {}_1F_1\left(\begin{array}{c}b-a\\b\end{array}\bigg| x\right), \tag{10.7}$$

Pfaff

$$\frac{1}{(1-x)^a} \cdot {}_2F_1\left(\begin{array}{c}a,b\\c\end{array}\bigg| -\frac{x}{1-x}\right) = {}_2F_1\left(\begin{array}{c}a,c-b\\c\end{array}\bigg| x\right), \tag{10.8}$$

and of Euler

$$(1-x)^{a+b-c} \cdot {}_2F_1\left(\begin{array}{c}a,b\\c\end{array}\bigg| x\right) = {}_2F_1\left(\begin{array}{c}c-a,c-b\\c\end{array}\bigg| x\right). \tag{10.9}$$

In Exercise 3.7 these identities were deduced by an application of the Chu-Vandermonde and Pfaff-Saalschütz identities, respectively, by equating coefficients in each
identity. We can work with differential equations instead. Algorithm 10.2 is able to
generate proofs for such types of hypergeometric transformations.

For the left- and right-hand sides of (10.7)–(10.9), Maple gives:
for the Kummer transformation

```
>   sumdiffeq(exp(x)*hyperterm([a],[b],-x,k),k,S(x));
```

$$x \frac{d^2}{dx^2} S(x) + (b-x) \frac{d}{dx} S(x) + (-b+a) S(x) = 0$$

```
>    sumdiffeq(hyperterm([b-a],[b],x,k),k,S(x));
```

$$x \frac{d^2}{dx^2} S(x) + (b - x) \frac{d}{dx} S(x) + (-b + a) S(x) = 0$$

for the Pfaff transformation

```
>    sumdiffeq(1/(1-x)^a*hyperterm([a,b],[c],-x/(1-x),k),k,S(x));
```

$$-x(-1 + x) \frac{d^2}{dx^2} S(x) - (-x b + x + x a - c + c x) \frac{d}{dx} S(x) + S(x) a (-c + b) = 0$$

```
>    sumdiffeq(hyperterm([a,c-b],[c],x,k),k,S(x));
```

$$-x(-1 + x) \frac{d^2}{dx^2} S(x) - (-x b + x + x a - c + c x) \frac{d}{dx} S(x) + S(x) a (-c + b) = 0$$

and for the Euler transformation

```
>    sumdiffeq((1-x)^(a+b-c)*hyperterm([a,b],[c],x,k),k,S(x));
```

$$x(-1 + x) \frac{d^2}{dx^2} S(x) - (x b - x - 2 c x + x a + c) \frac{d}{dx} S(x)$$
$$+ S(x)(-c + b)(-c + a) = 0$$

```
>    sumdiffeq(hyperterm([c-a,c-b],[c],x,k),k,S(x));
```

$$x(-1 + x) \frac{d^2}{dx^2} S(x) - (x b - x - 2 c x + x a + c) \frac{d}{dx} S(x)$$
$$+ S(x)(-c + b)(-c + a) = 0$$

Hence we obtained in each case the same second-order differential equation. Note that by standard results on ordinary differential equations, the solution $f(x)$ of a differential equation of order J is uniquely determined by J initial values $f^{(j)}(x_0)$ ($j = 0, \ldots, J - 1$) at a regular point x_0.

Therefore, it only remains to check two initial values to finish the proof of each of our identities. We choose $x_0 = 0$, and get
for the Kummer transformation

```
>    simplify(subs(x=0,exp(x)*hypergeom([a],[b],-x)-
>    hypergeom([b-a],[b],x)));
```

$$0$$

```
>    simplify(subs(x=0,diff(exp(x)*hypergeom([a],[b],-x)-
>    hypergeom([b-a],[b],x),x)));
```

$$0$$

for the Pfaff transformation

```
>    simplify(subs(x=0,1/(1-x)^a*hypergeom([a,b],[c],-x/(1-x))-
>    hypergeom([a,c-b],[c],x)));
```

$$0$$

```
>    simplify(subs(x=0,diff(1/(1-x)^a*hypergeom([a,b],[c],-x/(1-x))-
>    hypergeom([a,c-b],[c],x),x)));
```

$$0$$

and for the Euler transformation

```
>   simplify(subs(x=0,(1-x)^(a+b-c)*hypergeom([a,b],[c],x)-
hypergeom([c-a,c-b],[c],x)));
```

$$0$$

```
>   simplify(subs(x=0,diff((1-x)^(a+b-c)*hypergeom([a,b],[c],x)-
>   hypergeom([c-a,c-b],[c],x),x)));
```

$$0$$

and we are done.

Session 10.7 (Derivative Rules) We can modify Algorithms 7.5 and 10.2 to detect
a *derivative rule* of the form

$$S_n'(x) = \sum_{j=0}^{J} \sigma_j(n, x)\, S_{n+j}(x)$$

with $\sigma_j \in \mathbb{Q}(n, x)$ for

$$S_n(x) = \sum_{k=-\infty}^{\infty} F(n, x, k),$$

$F(n, x, k)$ being hypergeometric with respect to n, k and strictly hyperexponential
with respect to x, rather than a pure recurrence or differential equation. This is done
by the Maple implementation

```
sumdiffrule:=proc(F,k,snx)
local n,x,S,a,b,sigma,rat,p,q,r,upd,deg,f,j,jj,l,var,req,sol,
sol2,num,den,J;
if type(snx,function) then
  S:=op(0,snx); n:=op(1,snx); x:=op(2,snx);
else
  n:=op(1,snx); x:=op(2,snx);
end if;
for J from 1 to MAXORDER do
  a:=diff(F,x)-add(sigma[j]*subs(n=n+j,F),j=0..J);
  rat:=ratio(a,k);
  if not type(rat,ratpoly(anything,k)) then
    error 'Algorithm not applicable';
  end if;
  p:=1: q:=subs(k=k-1,numer(rat)): r:=subs(k=k-1,denom(rat)):
  upd:=update(p,q,r,k);
  p:=op(1,upd): q:=op(2,upd): r:=op(3,upd):
  deg:=degreebound(p,q,r,k);
  if deg>=0 then
    f:=add(b[j]*k^j,j=0..deg);
    var:={seq(sigma[jj],jj=0..J),seq(b[jj],jj=0..deg)};
    req:=collect(subs(k=k+1,q)*f-r*subs(k=k-1,f)-p,k);
```

```
sol:={solve({coeffs(req,k)},var)};
if not(sol={} or
   {seq(op(2,op(1,op(1,sol))),l=1..nops(op(1,sol)))}={0})
   then sol2:=add(sigma[j]*S(n+j,x),j=0..J);
   sol2:=subs(op(1,sol),sol2);
   return diff(S(n,x),x)=map(factor,sol2);
   end if;
  end if;
 end do;
 error cat('Algorithm finds no derivative rule of order <= ',
 MAXORDER);
 end proc:
```

Which derivative rule do the Legendre polynomials satisfy? Here we go:

```
>   sumdiffrule
>   (binomial(n,k)*binomial(-n-1,k)*((1-x)/2)^k,k,
>   P(n,x));
```

$$\frac{\partial}{\partial x} P(n,\ x) = -\frac{x\,(n+1)\,P(n,\ x)}{(-1+x)\,(1+x)} + \frac{(n+1)\,P(n+1,\ x)}{(-1+x)\,(1+x)}$$

More general identities between the shifted derivatives of a family of functions $S_n(x)$ are discussed in [Koepf97a]. They can be treated by a similar approach. One more example of this type is given next.

Session 10.8 (Integration Rules) Here we are interested in obtaining representations for the antiderivative $\int S_n(x)dx$ of a hypergeometric sum

$$S_n(x) = \sum_{k=-\infty}^{\infty} F(n, x, k),$$

$F(n, x, k)$ being hypergeometric with respect to n, k and strictly hyperexponential with respect to x, in terms of the original functions S_n.

Assume that we find one or several representations of $S_n(x)$ of the form

$$S_n(x) = \sum_{j=-J}^{J} \sigma_j(n, x)\, S'_{n+j}(x)$$

with $\sigma_j \in \mathbb{Q}(n, x)$ for $S_n(x)$. If one of these representations has coefficients σ_j that are *independent of* x, then we can integrate and get the representation

$$\int S_n(x)\, dx = \sum_{j=-J}^{J} \sigma_j(n)\, S_{n+j}(x)$$

for the antiderivative of $S_n(x)$, an *integration rule* for $S_n(x)$.

By another modification of the previous algorithms, for $J = 1$ this is done by the Maple implementation[1]

```
sumintrule:=proc(F,k,snx)
local n,x,S,a,b,sigma,rat,p,q,r,upd,deg,f,j,jj,l,var,req,sol,pol,
coefflist,DS,sol2,num,den;
if type(snx,function) then
   S:=op(0,snx); n:=op(1,snx); x:=op(2,snx);
else
   n:=op(1,snx); x:=op(2,snx);
end if;
a:=F-add(sigma[j]*diff(subs(n=n+j,F),x),j=-1..1);
rat:=ratio(a,k);
if not type(rat,ratpoly(anything,k)) then
   error `Algorithm not applicable`
end if;
p:=1: q:=subs(k=k-1,numer(rat)): r:=subs(k=k-1,denom(rat)):
upd:=update(p,q,r,k);
p:=op(1,upd): q:=op(2,upd): r:=op(3,upd):
deg:=degreebound(p,q,r,k);
if deg>=0 then
   f:=add(b[j]*k^j,j=0..deg);
   var:={seq(sigma[jj],jj=-1..1),seq(b[jj],jj=0..deg)};
   req:=collect(subs(k=k+1,q)*f-r*subs(k=k-1,f)-p,k);
   sol:={solve({coeffs(req,k)},var)};
   if not(sol={} or
      {seq(op(2,op(1,op(1,sol))),l=1..nops(op(1,sol)))}={0}) then
      req:=S(n,x)-add(sigma[j]*DS(n+j,x),j=-1..1);
      req:=subs(op(1,sol),req);
      req:=numer(normal(req));
      req:=collect(req,S(n,x));
      pol:=collect(coeff(req,S(n,x)),x);
      coefflist:={coeffs(pol,x)} minus {coeff(pol,x,0)};
      for j from -1 to 1 do
         req:=collect(req,DS(n+j,x));
         pol:=collect(coeff(req,DS(n+j,x)),x);
         coefflist:={op(coefflist),op({coeffs(pol,x)} minus
            {coeff(pol,x,0)})};
      end do;
      sol2:={solve(coefflist,indets(var) minus {x,n})};
      if sol2={} then ERROR(`No such identity exists`) end if;
      req:=add(sigma[j]*S(n+j,x),j=-1..1);
      req:=subs(op(1,sol),req);
      req:=subs(op(1,sol2),req);
      req:=map(factor,req);
      return Int(S(n,x),x)=req;
   end if;
```

---

[1] The implementation can be extended easily to $J > 1$, but this is not necessary for our purposes.

```
   end if;
   error `Algorithm finds no integration rule';
   end proc:
```

As an example, we consider the Legendre polynomials again. We obtain the representation

```
>   sumintrule(
>   binomial(n,k)*binomial(-n-1,k)*((1-x)/2)^k,k,P(n,x));
```

$$\int P(n, x)\, dx = -\frac{P(n-1, x)}{2n+1} + \frac{P(n+1, x)}{2n+1}$$

Note that by the same method one can obtain an integration rule for the *Jacobi polynomials*, which cannot be found in mathematical compilations like [OLBC10, Erdélyi53]; see Exercise 10.14. The Jacobi polynomials generalize the Legendre polynomials.

The algorithms of this chapter permit us to find differential equations for series of hyperexponential terms. In the special case of power series this can be handled by a much simpler approach as shown in Example 10.13 which is implemented in the gfun package [SZ94] as gfun[rectodiffeq]. Summation of the series will be successful if the resulting differential equation can be solved.

For the converse question of finding a power series representation of a holonomic function the FormalPowerSeries package can be used which is implemented in the convert(...,FormalPowerSeries) command (see e.g.[Koepf92, GK95]). As an example, one can compute

```
>   convert(exp(arcsinh(x)),FormalPowerSeries);
```

$$x + \sum_{k=0}^{\infty} -\frac{(-1)^k\, 4^{(-k)}\, (2k)!\, x^{2k}}{(k!)^2\, (2k-1)}$$

This procedure is successful if the resulting holonomic recurrence equation for the power series coefficients can be solved. For this purpose van Hoeij's algorithm is used, see Chap. 9.

Using a similar approach the algorithmic computation of formal Fourier series was dealt with in [Nana10].

q-Differential Equations for Sums

As a q-analogue of the differential operator one can define the *q-derivative operator*[2]

$$\mathcal{D}_q\, f(x) := \frac{f(x) - f(qx)}{(1-q)x}.$$

---

[2] See [GR90, p. 22, Exercise 1.12]. There are other q-analogues as well, see [GR90, Sect. 7].

The q-derivative operator has the property

$$\lim_{q \to 1} \mathscr{D}_q f(x) = f'(x).$$

Note that after replacing x by q^x, a *q-differential equation* can be regarded as a recurrence equation.

A q-analogue of Algorithm 10.2 can be used to generate q-differential equations for q-hypergeometric sums. An implementation of this algorithm is given in the qsum package. This package contains the Maple procedure qsumdiffeq(F,q, k,S(x)) for this purpose. The following are q-analogues of the exponential, the sine and cosine functions (see e.g. [GR90, Exercise 1.14]):

$$e_q(x) := {}_1\phi_0 \left(\begin{array}{c} 0 \\ - \end{array} \middle| q, x \right) = \sum_{k=0}^{\infty} \frac{x^k}{(q; q)_k} = \frac{1}{(x; q)_\infty},$$

$$E_q(x) := {}_0\phi_0 \left(\begin{array}{c} - \\ - \end{array} \middle| q, -x \right) = \sum_{k=0}^{\infty} \frac{q^{\binom{k}{2}}}{(q; q)_k} x^k = (-x; q)_\infty,$$

$$\sin_q(x) := \frac{e_q(ix) - e_q(-ix)}{2i} = \sum_{k=0}^{\infty} \frac{(-1)^k x^{2k+1}}{(q; q)_{2k+1}},$$

$$\mathrm{Sin}_q(x) := \frac{E_q(ix) - E_q(-ix)}{2i},$$

$$\cos_q(x) := \frac{e_q(ix) + e_q(-ix)}{2} = \sum_{k=0}^{\infty} \frac{(-1)^k x^{2k}}{(q; q)_{2k}},$$

$$\mathrm{Cos}_q(x) := \frac{E_q(ix) + E_q(-ix)}{2}.$$

The requests

```
> qsumdiffeq(qphihyperterm([0],[],q,x,k),q,k,eq(x));
```

$$eq(x) + (-1 + q) Dq_x(eq(x), x, q) = 0$$

and

```
> qsumdiffeq(qphihyperterm([],[],q,-x,k),q,k,Eq(x));
```

$$Eq(x) + (1 + x)(-1 + q) Dq_x(Eq(x), x, q) = 0$$

e.g., generate the q-differential equations for the two q-exponentials $e_q(x)$ and $E_q(x)$, respectively, and

```
>   qsumdiffeq(
>   (-1)^k*x^(2*k+1)/qpochhammer(q,q,2*k+1),q,k,sinq(x));
```

$$sinq(x) + (-1 + q)^2 \, Dq_x(sinq(x), x, x, q) = 0$$

gives the q-differential equation for the q-sine function $\sin_q(x)$.

If one uses the option `evalqdiff=true`,[3] then the resulting q-differential equation for $S(x)$ is written in terms of $S(x)$, $S(qx)$, $S(q^2x)$, ...:

```
>   qsumdiffeq(
>   (-1)^k*x^(2*k+1)/qpochhammer(q,q,2*k+1),q,k,sinq(x),
>   evalqdiff=true);
```

$$(x^2 + 1) \, q \, sinq(x) - (q + 1) sinq(x \, q) + sinq(x \, q^2) = 0$$

More q-differential equations will be considered in the exercises. Note that the algorithm considered here uses the q-series representation to find q-differential equations. For algorithms using the *algebra of q-holonomic functions* see ([KRM07, KK09, KS12, Sprenger09]).

Exercises

Exercise 10.1 *Show that if $S(x)$ is a hyperexponential term, then so is $S'(x)$. Show further that $S'(x)/S(x) = R(x) \in \mathbb{Q}(x)$ implies that $S^{(n)}/S(x) \in \mathbb{Q}(x)$ for each $n \in \mathbb{N}$. Use the definition of a hyperexponential term directly, and do not use the explicit representation of Lemma 10.1.*

Exercise 10.2 *Show that any rational function $S(x)$ is a hyperexponential term, having critical points at its zeros and at its poles. Determine the corresponding exponents α_j.*

Exercise 10.3 *Show, again, that the three different representations of the Legendre polynomials*

$$P_n(x) = \sum_{k=-\infty}^{\infty} \binom{n}{k} \binom{-n-1}{k} \left(\frac{1-x}{2}\right)^k,$$

$$P_n(x) = \frac{1}{2^n} \sum_{k=0}^{n} \binom{n}{k}^2 (x-1)^{n-k} (x+1)^k$$

and

$$P_n(x) = \frac{1}{2^n} \sum_{k=0}^{\lfloor n/2 \rfloor} (-1)^k \binom{n}{k} \binom{2n-2k}{n} x^{n-2k},$$

---

[3] Or sets the global variable `_qsumdiffeq_evalqdiff:=true;`

define the same family of functions, this time by showing that all three functions
satisfy the same differential equation and initial values. Do they also satisfy the same
derivative rules?

Exercise 10.4 *Find recurrence and differential equations of order two for the consecutive differences* $P_{n+1}(x) - P_n(x)$ *of the Legendre polynomials.*
   Show that the differential equation of order two of $S(x) := P_{n+2}(x) - P_n(x)$ *has
the nice property that the coefficient of* $S'(x)$ *vanishes.*

Exercise 10.5 *In* [KS96], *an article connected with de Branges' proof of the Bieberbach conjecture, the functions*

$$B_n(y) = \frac{1}{2y}\left(C_{n+1}^{-1/2}(1 - 2y) - C_n^{-1/2}(1 - 2y)\right)$$

play a central role. Here,

$$C_n^{-1/2}(x) = (1 - x)\, {}_2F_1\left(\begin{array}{c} 1 - n, n \\ 2 \end{array}\middle|\ \frac{1 - x}{2}\right)$$

denotes the family of Gegenbauer polynomials with upper index $v = -1/2$. *For*
$v > -1/2\ (v \neq 0)$ *the Gegenbauer polynomials are defined by*

$$C_n^v(x) = \frac{(2v)_n}{n!}\, {}_2F_1\left(\begin{array}{c} -n, n + 2v \\ v + 1/2 \end{array}\middle|\ \frac{1 - x}{2}\right).$$

Show that $C_n^{-1/2}(x)$ *are the limiting functions for* $v \to -1/2$ *of* $C_n^v(x)$.
   Find recurrence and differential equations of order two for $B_n(y)$. *Furthermore,
give a hypergeometric representation of* $B_n(y)$.

Exercise 10.6 *In* [KS94], *the Bateman functions* [Bateman31]

$$F_n(t) = \frac{e^{-t}}{n} \sum_{k=1}^{n} \frac{(-1)^k}{(k-1)!}\binom{n}{k}(2t)^k \tag{10.10}$$

were considered. Give a hypergeometric representation of $F_n(t)$.
   Find recurrence and differential equations of order two for $F_n(t)$. *Note how simple
the differential equation looks! Generate a derivative rule for* $F_n(t)$.

Furthermore, show the identities

$$F_n(t) = e^{-t}\Big(L_n(2t) - L_{n-1}(2t)\Big),$$
$$F_n(t) = -e^{-t}\tfrac{2t}{n}\, L^{(1)}_{n-1}(2t),$$
$$(n-1)\Big(F_n(t) - F_{n-1}(t)\Big) + (n+1)\Big(F_n(t) - F_{n+1}(t)\Big) = 2t\, F_n(t),$$
$$F'_n(t) - F'_{n+1}(t) = F_n(t) + F_{n+1}(t),$$

all developed in [Bateman31], *where*

$$L_n^{(\alpha)}(x) = \sum_{k=0}^{n} \frac{(-1)^k}{k!}\binom{n+\alpha}{n-k} x^k$$

denote the (generalized) Laguerre polynomials $L_n(x) = L_n^{(x)}$.

Exercise 10.7 *Find holonomic differential equations for* $e^{\sqrt{x}}$, $e^{\sqrt{3}x}$, $\sin\sqrt{x}$, *and* $\sin\sqrt{3}x$ *using Algorithm 10.2 with an appropriately chosen* J.[4] *Hint: If* J *is large enough, the resulting differential equation is not unique. Choose the open parameters such that the coefficients give polynomials.*

Exercise 10.8 *Find a differential equation with respect to* x *and recurrence equations with respect to* n *and* m *for the associated Legendre functions* [OLBC10, (14.3.6)]

$$P_n^m(x) = \frac{1}{\Gamma(1-m)}\left(\frac{x+1}{x-1}\right)^{\frac{m}{2}} \cdot {}_2F_1\left(\begin{array}{c}-n, n+1 \\ 1-m\end{array}\bigg|\,\frac{1-x}{2}\right).$$

Exercise 10.9 *Find holonomic differential and recurrence equations and a derivative rule for the Bessel functions*

$$J_n(x) = \left(\frac{x}{2}\right)^n \sum_{k=0}^{\infty} \frac{(-1)^k}{4^k\, k!\,\Gamma(k+1+n)}\, x^{2k}.$$

Exercise 10.10 *Find derivative rules for the generalized Laguerre polynomials (see Exercise 10.6) and for the Hermite polynomials*

$$H_n(x) := n! \sum_{k=0}^{\lfloor n/2 \rfloor} \frac{(-1)^k}{(n-2k)!\,k!}(2x)^{n-2k}.$$

---

[4] For another approach you can use the `FormalPowerSeries` package ([Koepf92, Koepf06]) which is implemented in `convert(...,FormalPowerSeries)` and can be invoked by `bind(FormalPowerSeries);` One of our examples is then given by the command `HolonomicDE(exp(sqrt(x)),F(x));`

Exercise 10.11 (Quadratic Transformations) *Show the following* quadratic trans-
formations of the Gauss hypergeometric function:

(a) $\;{}_2F_1\left(\begin{matrix} a, b \\ 2b \end{matrix}\;\middle|\; \dfrac{4x}{(1+x)^2}\right) = (1+x)^{2a} \cdot {}_2F_1\left(\begin{matrix} a, a-b+1/2 \\ b+1/2 \end{matrix}\;\middle|\; x^2\right),$

(b) $\;{}_2F_1\left(\begin{matrix} a, b \\ a+b+1/2 \end{matrix}\;\middle|\; 4x(1-x)\right) = {}_2F_1\left(\begin{matrix} 2a, 2b \\ a+b+1/2 \end{matrix}\;\middle|\; x\right),$

(c) $\;{}_2F_1\left(\begin{matrix} a, 1-a \\ c \end{matrix}\;\middle|\; x\right) = (1-x)^{c-1} \cdot {}_2F_1\left(\begin{matrix} \frac{c-a}{2}, \frac{c+a-1}{2} \\ c \end{matrix}\;\middle|\; 4x(1-x)\right).$

Exercise 10.12 (Whipple Transformation) *Show that*

$${}_3F_2\left(\begin{matrix} a/2, a/2+1/2, 1+a-b-c \\ 1+a-b, 1+a-c \end{matrix}\;\middle|\; -\frac{4x}{(1-x)^2}\right) = (1-x)^a \cdot {}_3F_2\left(\begin{matrix} a, b, c \\ 1+a-b, 1+a-c \end{matrix}\;\middle|\; x\right).$$

Exercise 10.13 (Kummer's Second Identity) *Show the identity*

$$e^{-x}\,{}_1F_1\left(\begin{matrix} a \\ 2a \end{matrix}\;\middle|\; 2x\right) = {}_0F_1\left(\begin{matrix} - \\ a+1/2 \end{matrix}\;\middle|\; \frac{x^2}{4}\right).$$

Exercise 10.14 (Jacobi Polynomials) *The Jacobi polynomials $P_n^{(\alpha,\beta)}(x)$ were
defined in Exercise 7.30 by the hypergeometric representation*

$$P_n^{(\alpha,\beta)}(x) = \binom{n+\alpha}{n}\,{}_2F_1\left(\begin{matrix} -n, n+\alpha+\beta+1 \\ \alpha+1 \end{matrix}\;\middle|\; \frac{1-x}{2}\right).$$

Find an integration rule for the Jacobi polynomials (see [Koepf97a]).

Exercise 10.15 (Bessel Polynomials) *The Bessel polynomials $B_n^{(\alpha)}(x)$ are given
by*

$$B_n^{(\alpha)}(x) = {}_2F_0\left(\begin{matrix} -n, n+\alpha+1 \\ - \end{matrix}\;\middle|\; -\frac{x}{2}\right) = \frac{(n+\alpha+1)_n}{2^n}\, x^n\,{}_1F_1\left(\begin{matrix} -n \\ -2n-\alpha \end{matrix}\;\middle|\; \frac{2}{x}\right).$$

*Show, by changing the order of summation, that the two representations agree. Find
a derivative rule and an integration rule for these polynomials; see [KS98b].*

Exercise 10.16 (Classical Orthogonal Polynomials) *The Jacobi polynomials (see
Exercise 10.14), the Laguerre polynomials (see Exercise 10.6), the Hermite polynomials (see Exercise 10.10), and the Bessel polynomials (see Exercise 10.15), are
called the classical orthogonal polynomials.*
Show that these satisfy a differential equation of the form

$$\sigma(x)\, p_n''(x) + \tau(x)\, p_n'(x) + \lambda_n\, p_n(x) = 0,$$

*where $\sigma(x)$ and $\tau(x)$ are polynomials of degree at most 2 and 1, respectively; compare
Exercise 7.11.*

Exercise 10.17 (Ramanujan's Notebook) *In Ramanujan's second notebook* [Ramanujan57] *on p. 258 one finds the identity*

$$_2F_1\left(\begin{array}{c}\frac{1}{3},\frac{2}{3}\\1\end{array}\Bigg|\,1-\left(\frac{1-x}{1+2x}\right)^3\right)=(1+2x)\,_2F_1\left(\begin{array}{c}\frac{1}{3},\frac{2}{3}\\1\end{array}\Bigg|\,x^3\right).$$

Following Garvan [Garvan95] *we might ask to generalize this result: For which* A, B, C, a, b, c, d *is*

$$_2F_1\left(\begin{array}{c}A,B\\C\end{array}\Bigg|\,1-\left(\frac{1-x}{1+2x}\right)^3\right)=(1+2x)^d\,_2F_1\left(\begin{array}{c}a,b\\c\end{array}\Bigg|\,x^3\right)?\qquad(10.11)$$

Hint: *Find the differential equations for both sides of* (10.11), *and arbitrary* A, B, C, a, b, c *and* d. *These differential equations must be multiples of each other. Multiplying by suitable polynomials, and equating coefficients, one gets a nonlinear system of equations for the unknowns* A, B, C, a, b, c *and* d *which (e.g., by computing a Gröbner basis) can be solved to obtain the (essentially) unique solution.*

Exercise 10.18 *Investigate the behavior of the q-analogues of the exponential and trigonometric functions in the limiting case $q \to 1$.*

Exercise 10.19 *Compute q-differential equations for*

(a) $S(x) := \mathrm{Sin}_q(x),$
(b) $S(x) := \cos_q(x),$
(c) $S(x) := \mathrm{Cos}_q(x),$
(d) $S(x) := \cos_q(x) + i\,\sin_q(x),$
(e) $S(x) := \mathrm{Cos}_q(x) + i\,\mathrm{Sin}_q(x).$

Show that

$$e_q(ix) = \cos_q(x) + i\,\sin_q(x) \quad\text{and}\quad E_q(ix) = \mathrm{Cos}_q(x) + i\,\mathrm{Sin}_q(x).$$

Rewrite these q-differential equations for $S(x)$ in terms of $S(x)$, $S(qx)$ and $S(q^2x)$.

Exercise 10.20 *Find q-differential equations for*

(a) $S(x) := {}_2\phi_1\left(\begin{array}{c}a,b\\c\end{array}\Bigg|\,q,x\right),$

(b) $S(x) := {}_3\phi_2\left(\begin{array}{c}a,b,c\\d,e\end{array}\Bigg|\,q,x\right).$

What is the order of the q-differential equation for $_p\phi_q(x)$?

References

Bateman31. Bateman, H.: The k-function, a particular case of the confluent hypergeometric function. Trans. Am. Math. Soc. **33**, 817–831 (1931)

Erdélyi53. Erdélyi, A., et al.: Higher Transcendental Functions, vol. 1. McGraw-Hill, New York (1953)

Garvan95. Garvan, F.G.: Ramanujan's theories of elliptic functions to alternative bases-a symbolic excursion. J. Symbolic Comput. **20**, 517–536 (1995)

GR90. Gasper, G., Rahman, M.: Basic hypergeometric series. In: Encyclopedia of Mathematics and its Applications, vol. 35, 2nd edn in 2004. Cambridge University Press, London (1990)

GKP94. Graham, R.L., Knuth, D.E., Patashnik, O.: Concrete Mathematics, 2nd edn. A Foundation for Computer Science. Addison-Wesley, Reading (1994)

GK95. Gruntz, D., Koepf, W.: Maple package on formal power series. Maple Tech. Newsl. **2**(2), 22–28 (1995)

KK09. Kauers, M., Koutschan, C.: A Mathematica package for q-holonomic sequences and power series. Ramanujan J. **19**, 137–150 (2009)

Koepf92. Koepf, W.: Power series in computer algebra. J. Symbolic Comput. **13**, 581–603 (1992)

Koepf97a. Koepf, W.: Identities for families of orthogonal polynomials and special functions. Integr. Transforms Special Functions **5**, 69–102 (1997)

Koepf06. Koepf, W.: Computeralgebra. Springer, Berlin (2006)

KS94. Koepf, W., Schmersau, D.: Bounded nonvanishing functions and Bateman functions. Complex Variables **25**, 237–259 (1994)

KS96. Koepf, W., Schmersau, D.: On the de Branges Theorem. Complex Variables **31**, 213–230 (1996)

KS98b. Koepf, W., Schmersau, D.: Representations of orthogonal polynomials. J. Comput. Appl. Math. **90**, 57–94 (1998)

KRM07. Koepf, W., Rajkovic, P.M., Marinkovic, S.D.: Properties of q-holonomic functions. J. Difference Equ. Appl. **13**, 621–638 (2007)

KS12. Koepf, W., Sprenger, T.: Algorithmic determination of q-power series for q-holonomic functions. J. Symbolic Comput. **47**, 519–535 (2012)

Nana10. Nana Chiadjeu, E.: Algorithmic Computation of Formal Fourier Series. PhD thesis, Universität Kassel (2010)

OLBC10. Olver, F.W.J., Lozier, D.W., Boisvert, R.F., Clark, C.W.: NIST Handbook of Mathematical Functions. Cambridge University Press, New York (2010). http://dlmf.nist.gov

Ramanujan57. Ramanujan, S.: Notebooks (2 vols). Tata Institute of Fundamental Research, Bombay (1957)

SZ94. Salvy, B., Zimmermann, P.: Gfun: a Maple package for the manipulation of generating and holonomic functions in one variable. ACM Trans. Math. Softw. **20**, 163–177 (1994)

Sprenger09. Sprenger, T.: Algorithmen für q-holonome Funktionen und q-hypergeometrische Reihen. PhD thesis, Universität Kassel (2009)

Chapter 11
Hyperexponential Antiderivatives

In this chapter, we consider a continuous counterpart of Gosper's algorithm. The appropriate question is to find a hyperexponential term antiderivative $G(x)$ of a given $f(x)$ whenever one exists.

In the affirmative case, as in the discrete case, the input function $f(x)$ itself must be a hyperexponential term since $G'(x) = R(x)G(x)\,(R \in \mathbb{Q}(x))$ implies

$$G''(x) = R'(x)G(x) + R(x)G'(x) = \left(\frac{R'(x)}{R(x)} + R(x) \right) G'(x),$$

and therefore

$$\frac{f'(x)}{f(x)} = \frac{G''(x)}{G'(x)} = \frac{R'(x)}{R(x)} + R(x)$$

is rational.

Note that the *Risch-Bronstein algorithm* ([Risch69, Risch70, Bronstein92, Bronstein96, GCL92, Chaps. 11–12]) is much more powerful so that from this perspective not much seems to be gained by a continuous version of Gosper's algorithm. On the other hand, we are mainly interested in applying the resulting algorithm, this time to *definite integration* rather than summation. This will be done in Chap. 12.

The following continuous counterpart of Gosper's algorithm for *indefinite integration* in hyperexponential terms is due to Almkvist and Zeilberger ([AZ90], Sect. 5) and its proof is along the lines of that of Gosper's algorithm.

Algorithm 11.1 (*Continuous Gosper Algorithm*) Given $f(x)$, the following algorithm decides whether there is a hyperexponential term antiderivative $G(x)$, and returns it if there is one:

1. Input: $f(x) \not\equiv 0$, a hyperexponential term.
2. Calculate the logarithmic derivative $f'(x)/f(x)$. Cancellation of exponential terms yields $b(x), c(x) \in \mathbb{Q}[x]$ for which

W. Koepf, *Hypergeometric Summation*, Universitext,
DOI: 10.1007/978-1-4471-6464-7_11, © Springer-Verlag London 2014

$$\frac{f'(x)}{f(x)} = \frac{b(x)}{c(x)}$$

with $\gcd(b(x), c(x)) = 1$.

3. Calculate $p(x), q(x)$ and $r(x) \in \mathbb{Q}[x]$ satisfying

$$\frac{f'(x)}{f(x)} = \frac{p'(x)}{p(x)} + \frac{q(x)}{r(x)} \tag{11.1}$$

with the property

$$\gcd\left(r(x), q(x) - jr'(x)\right) = 1 \qquad \text{for all } j \in \mathbb{N}_{\geq 0} \tag{11.2}$$

by a rewriting process starting with $p(x) = 1, q(x) = b(x)$ and $r(x) = c(x)$, and applying the rewrite rules

$$\tilde{p}(x) = p(x) g(x)^j ,$$
$$\tilde{q}(x) = j\frac{d}{dx}\left(\frac{r(x)}{g(x)}\right) + \frac{q(x) - jr'(x)}{g(x)} \quad \text{and}$$
$$\tilde{r}(x) = \frac{r(x)}{g(x)} \tag{11.3}$$

whenever the resultant of $q(x) - jr'(x)$ and $r(x)$ has a nonnegative integer root j (see e.g. [DST88], Appendix, [GCL92], Chap. 7), and therefore

$$g(x) = \gcd\left(r(x), q(x) - jr'(x)\right) \neq 1 .$$

4. Use an adapted version of Algorithm 5.5 (see Session 11.3 and Exercise 11.6) to determine the degree bound $M \in \mathbb{N}$ for the polynomial $\tilde{f}(x) \in \mathbb{Q}[x]$ for which

$$G(x) = \frac{r(x)\tilde{f}(x)}{p(x)} f(x) . \tag{11.4}$$

If $M < 0$, then return "no hyperexponential term solution exists"; exit.

5. Substitute the generic polynomial

$$\tilde{f}(x) = b_0 + b_1 x + b_2 x^2 + \ldots + b_M x^M$$

into the functional equation

$$p(x) = \left(q(x) + r'(x)\right)\tilde{f}(x) + r(x)\,\tilde{f}'(x) \tag{11.5}$$

for $\tilde{f}(x)$. Equate coefficients, and solve the resulting linear system for the unknowns $b_l (l = 0, \ldots, M)$.

6. If there is no solution, then return "no hyperexponential term solution exists"; exit.

7. Output: $G(x)$ according to (11.4).

Proof First, we have to show that the rewriting of step 3 leaves the term

$$\frac{f'(x)}{f(x)} = \frac{p'(x)}{p(x)} + \frac{q(x)}{r(x)}$$

invariant. This is established by the calculation

$$\frac{\tilde{p}'(x)}{\tilde{p}(x)} + \frac{\tilde{q}(x)}{\tilde{r}(x)} = \frac{p'(x)}{p(x)} + j\frac{g'(x)}{g(x)} + \left(j\frac{d}{dx}\left(\frac{r(x)}{g(x)}\right) + \frac{q(x) - j\,r'(x)}{g(x)} \right)\frac{g(x)}{r(x)}$$

$$= \frac{p'(x)}{p(x)} + j\frac{g'(x)}{g(x)} + j\frac{r'(x)}{r(x)} - j\frac{g'(x)}{g(x)} + \frac{q(x)}{r(x)} - j\frac{r'(x)}{r(x)}$$

$$= \frac{p'(x)}{p(x)} + \frac{q(x)}{r(x)}.$$

Next we define $\tilde{f}(x)$ by (11.4), take the derivative of this equation and substitute the resulting equation into the identity $f(x) = G'(x)$ to get

$$f(x) = \frac{d}{dx}\left(\frac{r(x)\tilde{f}(x)}{p(x)} f(x) \right)$$

$$= r(x)\tilde{f}(x)\frac{d}{dx}\left(\frac{f(x)}{p(x)}\right) + r(x)\tilde{f}'(x)\frac{f(x)}{p(x)} + r'(x)\tilde{f}(x)\frac{f(x)}{p(x)}$$

$$= r(x)\tilde{f}(x)\left(\frac{f'(x)}{f(x)} - \frac{p'(x)}{p(x)}\right)\frac{f(x)}{p(x)} + r(x)\tilde{f}'(x)\frac{f(x)}{p(x)} + r'(x)\tilde{f}(x)\frac{f(x)}{p(x)}$$

$$= r(x)\tilde{f}(x)\frac{q(x)}{r(x)}\frac{f(x)}{p(x)} + r(x)\tilde{f}'(x)\frac{f(x)}{p(x)} + r'(x)\tilde{f}(x)\frac{f(x)}{p(x)},$$

using (11.1). Multiplying by $p(x)/f(x)$ yields (11.5).

Note that, again, the main single fact to be proved is that $\tilde{f}(x)$, defined by (11.4), is a polynomial. Since $\tilde{f}(x)$ is clearly rational, we may write

$$\tilde{f}(x) = \frac{u(x)}{v(x)} \tag{11.6}$$

with polynomials $u(x), v(x) \in \mathbb{Q}[x]$ satisfying

$$\gcd(u(x), v(x)) = 1. \tag{11.7}$$

Assume now that the degree of $v(x)$ is positive. Then substitution of (11.6) in (11.5) yields

$$p(x) = q(x) \frac{u(x)}{v(x)} + r'(x) \frac{u(x)}{v(x)} + r(x) \left(\frac{u'(x)v(x) - v'(x)u(x)}{v^2(x)} \right),$$

and after multiplication by $v^2(x)$, we have

$$p(x)v^2(x) = q(x)u(x)v(x) + r'(x)u(x)v(x) + r(x)u'(x)v(x) - r(x)v'(x)u(x). \tag{11.8}$$

Let $w(x)$ denote any nonconstant prime factor over $\mathbb{Q}$ of $v(x)$. Such a factor exists since by assumption $v(x)$ is not constant. Then we have, for some $k \in \mathbb{N}$, the representation[1]

$$v(x) = w^k(x)\, \tilde{v}(x) \tag{11.9}$$

for which the conditions

$$\gcd\,(w(x), w'(x)) = 1 \quad \text{and} \quad \gcd\,(w(x), \tilde{v}(x)) = 1 \tag{11.10}$$

are satisfied. After division by $w^k(x)$, substitution of (11.9) into (11.8) gives the identity

$$p(x)w^k(x)\tilde{v}^2(x) = q(x)u(x)\tilde{v}(x) + r'(x)u(x)\tilde{v}(x) + r(x)u'(x)\tilde{v}(x)$$
$$- r(x)u(x)\tilde{v}'(x) - \frac{kr(x)u(x)w'(x)\tilde{v}(x)}{w(x)}. \tag{11.11}$$

Therefore, $r(x)u(x)w'(x)\tilde{v}(x)$ is a constant multiple of $w(x)$, and, by conditions (11.7) and (11.10), we conclude that $w(x)$ must be a divisor of $r(x)$. Hence, we can write

$$r(x) = w(x)\,\tilde{r}(x) \tag{11.12}$$

for some $\tilde{r}(x) \in \mathbb{Q}[x]$. Note that differentiation yields

$$w'(x)\tilde{r}(x) = r'(x) - w(x)\tilde{r}'(x). \tag{11.13}$$

Substitution of (11.12) in (11.11) then gives

$$p(x)w^k(x)\tilde{v}^2(x) = q(x)u(x)\tilde{v}(x) + r'(x)u(x)\tilde{v}(x) + w(x)\tilde{r}(x)u'(x)\tilde{v}(x)$$
$$- w(x)\tilde{r}(x)u(x)\tilde{v}'(x) - ku(x)w'(x)\tilde{r}(x)\tilde{v}(x)$$
$$= q(x)u(x)\tilde{v}(x) + r'(x)u(x)\tilde{v}(x) + w(x)\tilde{r}(x)u'(x)\tilde{v}(x)$$
$$- w(x)\tilde{r}(x)u(x)\tilde{v}'(x) - ku(x)\tilde{v}(x)\left(r'(x) - w(x)\tilde{r}'(x) \right)$$

---

[1] For details see, e.g., [DST88], Appendix.

where we have used (11.13). Dividing by $w(x)$ again, we get finally

$$p(x)w^{k-1}(x)\tilde{v}^2(x) = \tilde{r}(x)u'(x)\tilde{v}(x) - \tilde{r}(x)u(x)\tilde{v}'(x) + ku(x)\tilde{v}(x)\tilde{r}'(x)$$
$$+ \frac{q(x) - (k-1)r'(x)}{w(x)}u(x)\tilde{v}(x).$$

The conditions (11.7) and (11.10) imply that neither $u(x)$ nor $\tilde{v}(x)$ have a common divisor with $w(x)$ so that from the above identity we see that $w(x)$ is a divisor of $q(x) - (k-1)r'(x)$. This, however, contradicts the main gcd condition (11.2) so that our assumption about the degree of $v(x)$ cannot be valid. Hence $\tilde{f}(x)$ is a polynomial.

The degree bound algorithm for $\tilde{f}(x)$ is very similar to the discrete case, and its complete description is given in the Maple procedure `contdegreebound` (p,q,r,x) in Session 11.3 and its proof is left to the reader (Exercise 11.6). $\square$

Example 11.2 Let us consider

$$f(x) = e^{-x^2}\left(1 - 2x^2\right)$$

and check whether the antiderivative

$$G(x) = \int e^{-x^2}\left(1 - 2x^2\right) dx$$

is a hyperexponential term. We have

$$\frac{f'(x)}{f(x)} = -\frac{2x\left(2x^2 - 3\right)}{2x^2 - 1},$$

so that we set initially

$$p(x) = 1, \quad q(x) = -2x\left(2x^2 - 3\right) \quad \text{and} \quad r(x) = 2x^2 - 1.$$

The dispersion condition (11.2) yields

$$J := \{j \in \mathbb{N}_{\geq 0} \mid \gcd\left(r(x), q(x) - jr'(x)\right) \neq 1\} = \{1\}$$

and the rewrite procedure (11.3) generates the final choice

$$p(x) = 2x^2 - 1, \quad q(x) = -2x \quad \text{and} \quad r(x) = 1.$$

The degree bound for $\tilde{f}(x)$ is 1, and we substitute the generic first order polynomial into the main Eq. (11.5). Equating coefficients gives $\tilde{f}(x) = -x$, and by (11.4) we have finally

$$G(x) = \frac{r(x)\tilde{f}(x)}{p(x)} \quad f(x) = xe^{-x^2}.$$

For the similar example function

$$f(x) = e^{-x^2}\left(1 - x^2\right)$$

the same procedure yields $p(x) = x^2 - 1$, $q(x) = -2x$ and $r(x) = 1$, and the degree bound for $\tilde{f}(x)$ turns out to be equal to 1 again, but in this case no first order polynomial $\tilde{f}(x)$ satisfies (11.5). Hence we conclude that $f(x)$ does not have a hyperexponential antiderivative. Indeed, its antiderivative

$$G(x) = \int e^{-x^2}\left(1 - x^2\right)dx = \frac{\sqrt{\pi}}{4}\mathrm{erf}(x) + \frac{x}{2}e^{-x^2},$$

is in terms of the *error function*

$$\mathrm{erf}(x): = \frac{2}{\sqrt{\pi}}\int_0^x e^{-t^2}dt.$$

Note that one can prove that the error function is not representable by elementary functions. This result cannot be obtained by the current method. Algorithm 11.1 proves instead the weaker statement that $G(x)$ as well as erf x are not hyperexponential terms.   △

Session 11.3 The following is a complete implementation of Algorithm 11.1.

```
contratio:=proc(f,x)
   simpcomb(diff(f,x)/f);
end proc:


contdispersionset:=proc(q,r,x)
# finds the nonnegative integer dispersion values j
local j,res,s,l;
res:=frontend(resultant,[r,q-j*diff(r,x),x]); # (11.2)
s:=simplify({solve(res,j)});
l:={};
for j in s do
   if type(j,nonnegint) then l:=l union {j} end if;
end do;
return convert(l,set);
end proc:


contupdate:=proc(p,q,r,x)
# updates the triple [p,q,r] according to gcd-condition
local dis,g,pnew,qnew,rnew,j;
g:=frontend(gcd,[r,q-diff(r,x)]); # (11.2), j=1
if has(g,x) then
   pnew:=normal(p*g); # (11.3), j=1
```

```
      qnew:=normal(diff(r/g,x)+(q-diff(r,x))/g); # (11.3), j=1
      rnew:=normal(r/g); # (11.3), j=1
   else
      pnew:=p; qnew:=q; rnew:=r;
   end if;
   dis:=contdispersionset(qnew,rnew,x);
   for j in dis do
      g:=frontend(gcd,[rnew,qnew-j*diff(rnew,x)]); # (11.2)
      if has(g,x) then
         pnew:=normal(pnew*g^j); # (11.3)
         qnew:=normal(j*diff(rnew/g,x)+(qnew-j*diff(rnew,x))/g); # (11.3)
         rnew:=normal(rnew/g); # (11.3)
      end if;
   end do;
   return [pnew,qnew,rnew];
end proc:


contdegreebound:=proc(p,q,r,x)
# calculates the degree bound for f
local pol1,pol2,deg1,deg2,a,b;
pol1:=collect(r,x);
pol2:=collect(q+diff(r,x),x);
if pol1=0 then deg1:=-1 else deg1:=degree(pol1,x) end if;
if pol2=0 then deg2:=-1 else deg2:=degree(pol2,x) end if;
if deg1<=deg2 then return degree(p,x)-deg2 end if;
   a:=coeff(pol1,x,deg1);
   b:=coeff(pol2,x,deg1-1);
   if not(type(-b/a,nonnegint)) then
      return degree(p,x)-deg1+1
   else
      return max(-b/a,degree(p,x)-deg1+1)
   end if;
end proc:


contfindf:=proc(p,q,r,x)
# finds ftilde, given the triple [p,q,r]
local deg,ftilde,a,j,deq,sol,result;
deg:=contdegreebound(p,q,r,x);
if deg<0 then error `No polynomial ftilde exists` end if;
ftilde:=add(a[j]*x^j,j=0..deg);
deq:=collect((q+diff(r,x))*ftilde+r*diff(ftilde,x)-p,x); # (11.5)
sol:={solve({coeffs(deq,x)},{seq(a[j],j=0..deg)})};
if sol={} then
   error `No polynomial ftilde exists`
else
   result:=subs(op(1,sol),ftilde);
end if;
for j from 0 to deg do
   result:=subs(a[j]=0,result);
end do;
return result;
end proc:


contgosper:=proc(f,x)
# implements the continuous version of Gosper's algorithm
```

```
local rat,p,q,r,pqr,ftilde;
rat:=contratio(f,x);
if not type(rat,ratpoly(anything,x)) then
  error 'Algorithm not applicable'
end if;
p:=1; q:=numer(rat); r:=denom(rat);
pqr:=contupdate(p,q,r,x);
p:=op(1,pqr); q:=op(2,pqr); r:=op(3,pqr);
try
  ftilde:=contfindf(p,q,r,x);
catch:
  error 'No hyperexponential antiderivative exists';
end try;
return normal(r*ftilde*f/p); # (11.4)
end proc:
```

The procedure `contgosper(f,x)` invokes all particular subalgorithms and yields the hyperexponential type antiderivative sought, or one of the error messages "algorithm not applicable" or "no hyperexponential antiderivative exists" is issued.

Note that the main difference between our implementations of Gosper's algorithm (see Exercise 5.6) and the current algorithm is the fact that we could not avoid a resultant computation in the latter case; compare with Algorithm 4.2. Since $j = 1$ typically occurs in the rewriting step 3, this case is handled separately by the procedure `contupdate(p,q,r,x)` to avoid unnecessary resultant computations.

The above examples are handled by the statements

```
>   contgosper(exp(-x^2)*(1-2*x^2),x);
```

$$x\,e^{-x^2}$$

```
>   contgosper(exp(-x^2)*(1-x^2),x);
```

```
Error, (in contgosper) No hyperexponential antiderivative exists
```

We give some more examples

```
>   contgosper(exp(-x^2),x);
```

```
Error, (in contgosper) No hyperexponential antiderivative exists
```

```
>   contgosper(x*exp(-x^2),x);
```

$$-\frac{1}{2}\,e^{-x^2}$$

```
>   f:=diff(exp((1+x)/(1-x))*(1+x^2)/(1-x^2),x);
```

$$\frac{\left(\dfrac{1}{1-x}+\dfrac{x+1}{(1-x)^2}\right)e^{\frac{1+x}{1-x}}\left(x^2+1\right)}{1-x^2}+\frac{2\,x\,e^{\frac{1+x}{1-x}}}{1-x^2}+\frac{2\,e^{\frac{1+x}{1-x}}\left(x^2+1\right)x}{(1-x^2)^2}$$

```
>   contgosper(f,x);
```

$$\frac{e^{\frac{1+x}{1-x}}\left(x^2+1\right)}{1-x^2}$$

Example 11.4 (*Rational Functions*) If a rational function $f(x) \in \mathbb{Q}(x)$ has a rational antiderivative, Algorithm 11.1 will find it. In particular, in such a case, Algorithm 11.1 will give the output safely as a rational function $G(x) \in \mathbb{Q}(x)$. This is not the case with the Maple procedure `int` as the following example shows.[2]

```
>   term:=diff((1+x^2)/(1-x^10),x);
```

$$\frac{2x}{1-x^{10}} + \frac{10(x^2+1)x^9}{(1-x^{10})^2}$$

```
>   contgosper(term,x);
```

$$-\frac{(1+x^8)x^2}{(x^6 - x^5 + x - 1)(x^4 + x^3 + x^2 + x + 1)}$$

```
>   integral:=int(term,x);
```

$$-1/5\,(-1+x)^{-1} - 2/5 \arctan\left(\frac{4x+1+\sqrt{5}}{\sqrt{10-2\sqrt{5}}}\right)\frac{1}{\sqrt{10-2\sqrt{5}}} - 2/5 \arctan\left(\frac{4x+1-\sqrt{5}}{\sqrt{10+2\sqrt{5}}}\right)\frac{1}{\sqrt{10+2\sqrt{5}}}$$

$$+1/5\,(x+1)^{-1} + 2/5 \arctan\left(\frac{4x-1-\sqrt{5}}{\sqrt{10-2\sqrt{5}}}\right)\frac{1}{\sqrt{10-2\sqrt{5}}} + 2/5 \arctan\left(\frac{4x-1+\sqrt{5}}{\sqrt{10+2\sqrt{5}}}\right)\frac{1}{\sqrt{10+2\sqrt{5}}}$$

$$+1/5\,\frac{\left(-8\sqrt{5}-\left(-5+\sqrt{5}\right)\left(\sqrt{5}-1\right)\right)x - 2\sqrt{5}\left(\sqrt{5}-1\right)+20-4\sqrt{5}}{\left(10+2\sqrt{5}\right)\left(2x^2-x+\sqrt{5}x+2\right)}$$

$$+1/5\,\frac{\left(8\sqrt{5}-\left(-\sqrt{5}-5\right)\left(-1-\sqrt{5}\right)\right)x + 2\sqrt{5}\left(-1-\sqrt{5}\right)+4\sqrt{5}+20}{\left(10-2\sqrt{5}\right)\left(2x^2-x-\sqrt{5}x+2\right)}$$

$$+1/5\,\frac{\left(-8\sqrt{5}-\left(-\sqrt{5}-5\right)\left(\sqrt{5}+1\right)\right)x - 2\sqrt{5}\left(\sqrt{5}+1\right)+4\sqrt{5}+20}{\left(10-2\sqrt{5}\right)\left(2x^2+x+\sqrt{5}x+2\right)}$$

$$+1/5\,\frac{\left(8\sqrt{5}-\left(-5+\sqrt{5}\right)\left(1-\sqrt{5}\right)\right)x + 2\sqrt{5}\left(1-\sqrt{5}\right)+20-4\sqrt{5}}{\left(10+2\sqrt{5}\right)\left(2x^2+x-\sqrt{5}x+2\right)}$$

$$-4 \arctan\left(\frac{4x-1+\sqrt{5}}{\sqrt{10+2\sqrt{5}}}\right)\left(10+2\sqrt{5}\right)^{-3/2} - 4 \arctan\left(\frac{4x-1-\sqrt{5}}{\sqrt{10-2\sqrt{5}}}\right)\left(10-2\sqrt{5}\right)^{-3/2}$$

$$+4 \arctan\left(\frac{4x+1+\sqrt{5}}{\sqrt{10-2\sqrt{5}}}\right)\left(10-2\sqrt{5}\right)^{-3/2} + 4 \arctan\left(\frac{4x+1-\sqrt{5}}{\sqrt{10+2\sqrt{5}}}\right)\left(10+2\sqrt{5}\right)^{-3/2}$$

$$-4/5 \arctan\left(\frac{4x-1+\sqrt{5}}{\sqrt{10+2\sqrt{5}}}\right)\sqrt{5}\left(10+2\sqrt{5}\right)^{-3/2} + 4/5 \arctan\left(\frac{4x-1-\sqrt{5}}{\sqrt{10-2\sqrt{5}}}\right)\sqrt{5}\left(10-2\sqrt{5}\right)^{-3/2}$$

$$-4/5 \arctan\left(\frac{4x+1+\sqrt{5}}{\sqrt{10-2\sqrt{5}}}\right)\sqrt{5}\left(10-2\sqrt{5}\right)^{-3/2} + 4/5 \arctan\left(\frac{4x+1-\sqrt{5}}{\sqrt{10+2\sqrt{5}}}\right)\sqrt{5}\left(10+2\sqrt{5}\right)^{-3/2}$$

```
>   integral:=normal(integral);
```

$$320\,(x^2+1)\Big/\Big((x+1)(\sqrt{5}-5)(2x^2+x+5^{(1/2)}x+2)(5+\sqrt{5})$$

$$(-2x^2-x+\sqrt{5}x-2)(-1+x)(2x^2-x+\sqrt{5}x+2)(-2x^2+x+\sqrt{5}x-2)\Big)$$

```
>   normal(integral,expanded);
```

---

[2] This result may depend on the version of Maple you use.

$$\frac{-x^2 - 1}{-1 + x^{10}}$$

Note that even with `normal` we could not get rid of the square roots to convert the result to an expression in $\mathbb{Q}(x)$, but we need the option `expanded`.

Maple has an implementation of Risch's algorithm ([Risch69]–[Risch70]) which unfortunately is not invoked in the current example. Risch's algorithm can be directly invoked using `'int/risch'`, which yields correctly

```
>    'int/risch'(term,x);
```

$$\frac{2\left(-\dfrac{1}{2} - \dfrac{1}{2}x^2\right)}{-1 + x^{10}}$$

Further Reading

For further reading on the algorithms of this chapter see [AZ90].

Exercises

Exercise 11.1 *Which of the following functions have a hyperexponential antiderivative?*

(a) $f(x) = e^{1/x}$,

(b) $f(x) = \dfrac{e^{x^2}(4x^2+1)}{\sqrt{x}}$,

(c) $f(x) = \dfrac{(1+x)^\alpha}{(1-x)^{\alpha+1}}$,

(d) $f(x) = \dfrac{(1+x)^\alpha}{(1-x)^{\alpha+2}}$,

(e) $f(x) = e^{x^3+x^2}\left(1 + 2x + 2x^2 + 5x^3 + 3x^4\right)$.

In the affirmative cases, compare the results with those of Maple's `int` *and (if applicable)* `'int/risch'` *commands.*

Exercise 11.2 *Use Maple's differentiation procedure to differentiate* 10 *different hyperexponential terms, integrate them with* `contgosper`, *and check the results.*

Exercise 11.3 *Take the fifth (tenth) derivative of* $f(x) = \dfrac{1+x^2}{1\pm x^{10}}$. *Use* `contgosper` *iteratively to reconstruct* f. *Try the same procedure using* `int` *and* `'int/risch'`.

Exercise 11.4 *Compare the calculation for the antiderivatives of*

$$f(x) = x^{m-1}\frac{(1 + x^m)^\alpha}{(1 - x^m)^{\alpha+n}}$$

for $m = 1, 2, 3$ *and* $n = 2, \ldots, 5$, *using Algorithm 11.1 in form of the implementation* `contgosper` *of Session 11.3, and the* `int` *procedure, respectively.*

Exercise 11.5 *Describe the rational certification mechanism connected with Algorithm* 11.1.

Exercise 11.6 *Give a complete description and proof of the degree bound computation of Algorithm* 11.1.

Exercise 11.7 *Give a hypergeometric representation of the error function.*

References

AZ90. Almkvist, G., Zeilberger, D.: The method of differentiating under the integral sign. J. Symbolic Comput. **10**, 571–591 (1990)

Bronstein92. Bronstein, M.: Integration and differential equations in computer algebra. Programmirovanie **18**(5), 26–44 (1992)

Bronstein96. Bronstein, M.: Introduction to Symbolic Integration. Algorithms and Computation in Mathematics, vol. 1. Springer, Berlin (1996)

DST88. Davenport, J.H., Siret, Y., Tournier, E.: Computer Algebra: Systems and Algorithms for Algebraic Computation. Academic Press, London (1988)

GCL92. Geddes, Ko, Czapor, S.R., Labahn, G.: Algorithms for Computer Algebra. Kluwer Academic Publishers, Boston, Dordrecht, London (1992)

Risch69. Risch, R.: The problem of integration in finite terms. Trans. Amer. Math. Soc. **139**, 167–189 (1969)

Risch70. Risch, R.: The solution of the problem of integration in finite terms. Bull. Amer. Math. Soc. **76**, 605–608 (1970)

Chapter 12
Holonomic Equations for Integrals

Now we are ready to consider *definite integration* of hyperexponential terms. If the corresponding indefinite integral is a hyperexponential term again, then Algorithm 11.1 applies, and definite integration is trivial.

In this chapter we consider definite integrals of the type

$$I_n = \int_a^b F(n, t)\, dt \tag{12.1}$$

where $F(n, t)$ is a hypergeometric term with respect to n and a hyperexponential term with respect to t, or of the type

$$I(x) = \int_a^b F(x, t)\, dt, \tag{12.2}$$

$F(x, t)$ being a hyperexponential term with respect to both x and t.

Generally, we cannot expect that a hyperexponential term antiderivative exists, so the next best thing we can hope for is a holonomic equation for I_n or $I(x)$, in other words a holonomic recurrence equation for I_n or a holonomic differential equation for $I(x)$, respectively. As in the way we derived holonomic equations for definite sums applying Gosper's algorithm to an appropriate auxiliary function, we can apply the continuous version of Gosper's algorithm to the current problem.

This procedure results in the following pair of algorithms [AZ90, AZ91].

Algorithm 12.1 (*Almkvist, Zeilberger*) Given $F(n, t)$ or $F(x, t)$, this algorithm searches for a holonomic recurrence or differential equation for I_n or $I(x)$, defined by (12.1) or (12.2), respectively.

1. Input: $F(n, t) \not\equiv 0$ ($F(x, t) \not\equiv 0$), a hypergeometric term with respect to n, hyperexponential with respect to t (or: a hyperexponential term with respect to both x and t).

W. Koepf, *Hypergeometric Summation*, Universitext,
DOI: 10.1007/978-1-4471-6464-7_12, © Springer-Verlag London 2014

2. Set $J := 1$.
3. Set

$$f(t) := F(n, t) + \sum_{j=1}^{J} \sigma_j(n) \, F(n + j, t)$$

or

$$f(t) := F(x, t) + \sum_{j=1}^{J} \sigma_j(x) \, \frac{\partial^j}{\partial x^j} F(x, t),$$

respectively, with undetermined variables σ_j depending on n (or x), but independent of t.

4. Apply the continuous Gosper algorithm adapted in the following way to $f(t)$: In the last step, solve the linear system for the coefficients of $\tilde{f}(t)$, and at the same time for the unknowns σ_j ($j = 1, \ldots, J$). In the affirmative case, the continuous Gosper algorithm finds $G(n, t)$ with

$$\frac{d}{dt} G(n, t) = f(t)$$

or $G(x, t)$ with

$$\frac{\partial}{\partial t} G(x, t) = f(t).$$

Pay attention to possible nonnegative integer denominator zeros with respect to n (or x) of the rational certificate

$$\tilde{R}(n, t) = \frac{G(n, t)}{f(t)} \qquad \left(\tilde{R}(x, t) = \frac{G(x, t)}{f(t)} \right)$$

where the resulting holonomic equation might not be valid.
The calculation also determines the functions σ_j ($j = 1, \ldots, J$) that are in $\mathbb{Q}(n)$ or $\mathbb{Q}(x)$. If the procedure is not successful then increase J by one and continue with step 3.

5. Output: By integration from $t = a$ to $t = b$, from the fundamental theorem of calculus it follows that

$$I_n + \sum_{j=1}^{J} \sigma_j(n) \, I_{n+j} = G(n, b) - G(n, a) = G(n, t) \Big|_{t=a}^{t=b} \tag{12.3}$$

or

$$I(x) + \sum_{j=1}^{J} \sigma_j(x) \, I^{(j)}(x) = G(x, b) - G(x, a) = G(x, t) \Big|_{t=a}^{t=b} \tag{12.4}$$

for I_n $(I(x))$ where the right-hand sides of (12.3)–(12.4) in actual situations may have to be defined as limits. Multiplication by the common denominator results in the equation sought. $\qquad\square$

Note that generally an inhomogeneous recurrence or differential equation results. In many interesting cases, however, we can choose the integration bounds such that the resulting holonomic equation is homogeneous.

Example 12.2 We come back to the definition of the Γ function. Let

$$I(n) := \int_0^\infty F(n, t)\, dt = \int_0^\infty t^{n-1}\, e^{-t}\, dt.$$

With $J := 1$, for

$$f(t) = F(n, t) + \sigma_1 F(n + 1, t) = t^{n-1}\, e^{-t} + \sigma_1 t^n\, e^{-t}$$

we have

$$\frac{f'(t)}{f(t)} = \frac{-t + n - 1 - \sigma_1 t^2 + \sigma_1 nt}{t\,(1 + \sigma_1 t)},$$

and the usual rewriting step yields the triple

$$p(t) = 1 + \sigma_1 t\,, \quad q(t) = n - 1 - t \quad \text{and} \quad r(t) = t.$$

The degree bound for $\tilde{f}(t)$ is zero, and equating coefficients in the main Eq. (11.5) for $\tilde{f}(t) = b_0$ implies $b_0 = \frac{1}{n}$, $\sigma_1 = -\frac{1}{n}$. Since for $\operatorname{Re} n > 0$ we have

$$G(n, t)\Big|_{t=0}^{t=\infty} = \frac{t}{n - t}\, f(t)\Big|_{t=0}^{t=\infty} = 0$$

(requiring limit computations), we arrive at the holonomic recurrence equation

$$n\, I_n - I_{n+1} = 0 \qquad (\operatorname{Re} n > 0).$$

Note that this equation holds for all $n \in \mathbb{C}$ with $\operatorname{Re} n > 0$ since no further assumption was needed in the derivation. A hypergeometric term solution, however, can only be given using a suitable initial value, $I_1 = 1$, say, so that for integer $n \in \mathbb{N}$ we get $I_n = (n - 1)!$

It is typical that we could also have deduced the result by an intelligent application of integration by parts as we did in Chap. 1. The main advantage of the technique given here is that the result is derived completely automatically.

Example 12.3 Next, we would like to find a holonomic differential equation for

$$I(x) := \int_0^\infty F(x, t)\, dt = \int_0^\infty e^{-\frac{x^2}{t^2} - t^2}\, dt$$

(see [AZ90]). For $J = 1$, the procedure does not return a result, hence we set $J := 2$. After some complicated calculations, we get

$$p(t) = t^4 - 2\sigma_1 x t^2 - 2\sigma_2 t^2 + 4\sigma_2 x^2,$$

$$q(t) = 2x^2 - 4t^2 - 2t^4 \quad \text{and} \quad r(t) = t^3.$$

The polynomial $\tilde{f}(t)$ is of degree zero, $\tilde{f}(t) = b_0$ say, and we get $b_0 = -1/2$, $\sigma_1 = 0$ and $\sigma_2 = -1/4$. Hence we are led to the differential equation

$$I''(x) - 4I(x) = 0.$$

Note that in this particular case, the resulting differential equation can be used easily to find a hyperexponential term representing the integral under consideration.

Since the functions e^{2x} and e^{-2x} are linearly independent solutions of the given differential equation, its general solution is a linear combination

$$I(x) = \alpha e^{2x} + \beta e^{-2x}.$$

Since $\lim\limits_{x \to \infty} I(x) = 0$, we must have $\alpha = 0$.

Furthermore, using the initial value (see 1.15)

$$I(0) = \int_0^\infty e^{-t^2}\, dt = \frac{\sqrt{\pi}}{2},$$

we have finally $\beta = \sqrt{\pi}/2$ and

$$I(x) = \frac{\sqrt{\pi}}{2} e^{-2x},$$

a rather surprising outcome, indeed!

Session 12.4 The procedures `intrecursion(F,t,S(n))` and `intdiffeq` `(F,t,S(x))` below implement Algorithm 12.1. Their output is a homogeneous recurrence/differential equation for the definite integral under consideration assuming that the integration bounds are chosen such that the right-hand sides of (12.3) and (12.4) vanish.

```
intrecursion:=proc(F,t,sn)
local S,n,f,b,sigma,rat,p,q,r,upd,deg,ftilde,j,jj,l,var,req,sol,
num,den,J,cert;
if type(sn,function) then
  S:=op(0,sn); n:=op(1,sn) else n:=sn
end if;
for J from 1 to MAXORDER do
  f:=F+add(sigma[j]*subs(n=n+j,F),j=1..J);
  rat:=contratio(f,t);
  if not type(rat,ratpoly(anything,t)) then
    error 'Algorithm not applicable';
  end if;
  p:=1: q:=numer(rat): r:=denom(rat):
  upd:=contupdate(p,q,r,t);
  p:=op(1,upd): q:=op(2,upd): r:=op(3,upd):
  deg:=contdegreebound(p,q,r,t);
  if deg>=0 then
    ftilde:=add(b[j]*t^j,j=0..deg);
    var:={seq(sigma[jj],jj=1..J),seq(b[jj],jj=0..deg)};
    req:=collect((q+diff(r,t))*ftilde+r*diff(ftilde,t)-p,t);
    sol:={solve({coeffs(req,t)},var)};
    if not(sol={} or
    {seq(op(2,op(1,op(1,sol))),l=1..nops(op(1,sol)))}={0}) then
      req:=S(n)+add(sigma[j]*S(n+j),j=1..J);
      req:=normal(subs(op(1,sol),req));
      req:=collect(numer(req),[seq(S(n+J-j),j=0..J)]);
      return map(factor,req)=0;
    end if;
  end if;
end do;
error cat('Algorithm finds no recurrence equation of order <= ',
MAXORDER);
end proc:


intdiffeq:=proc(F,t,sx)
local x,S,f,b,sigma,rat,p,q,r,upd,deg,ftilde,j,jj,l,var,deq,sol,
num,den,J,cert;
if type(sx,function) then
  S:=op(0,sx); x:=op(1,sx) else x:=sx
end if;
for J from 1 to MAXORDER do
  f:=F+add(sigma[j]*diff(F,x$j),j=1..J);
  rat:=contratio(f,t);
  if not type(rat,ratpoly(anything,t)) then
    error 'Algorithm not applicable';
  end if;
  p:=1: q:=numer(rat): r:=denom(rat):
  upd:=contupdate(p,q,r,t);
  p:=op(1,upd): q:=op(2,upd): r:=op(3,upd):
  deg:=contdegreebound(p,q,r,t);
  if deg>=0 then
    ftilde:=add(b[j]*t^j,j=0..deg);
    var:={seq(sigma[jj],jj=1..J),seq(b[jj],jj=0..deg)};
```

```
deq:=collect((q+diff(r,t))*ftilde+r*diff(ftilde,t)-p,t);
sol:={solve({coeffs(deq,t)},var)};
if not(sol={} or
{seq(op(2,op(1,op(1,sol))),l=1..nops(op(1,sol))))}={0}) then
   deq:=S(x)+add(sigma[j]*diff(S(x),x$j),j=1..J);
   deq:=normal(subs(op(1,sol),deq));
   deq:=numer(deq);
   deq:=collect(deq,[seq(diff(S(x),x$(J-j)),j=0..J-1),S(x)]);
   deq:=numer(normal(deq));
   deq:=collect(deq,[seq(diff(S(x),x$(J-j)),j=0..J-1),S(x)]);
   return map(factor,deq)=0;
   end if;
  end if;
 end do;
 error cat('Algorithm finds no differential equation of order <= ',
 MAXORDER);
 end proc:
```

Note that these procedures are only slightly adapted versions of the previous
procedures `sumrecursion(F,k,S(n))` and `sumdiffeq(F,k,S(x))`.
We repeat the above examples with Maple[1]:

```
>   intrecursion(exp(-t)*t^(z-1),t,S(z));
```
$$S(z)\, z - S(z+1) = 0$$
```
>   intdiffeq(exp(-x^2/t^2-t^2),t,S(x));
```
$$4\,S(x) - \frac{d^2}{dx^2}\,S(x) = 0$$

Let us rediscover the connection of the Beta function, defined by

$$B(z,w) = \int_0^1 t^{z-1}\,(1-t)^{w-1}\,dt$$

with the Γ function, namely

$$\int_0^1 t^{z-1}\,(1-t)^{w-1}\,dt = \frac{\Gamma(z)\Gamma(w)}{\Gamma(z+w)} \tag{12.5}$$

(see Theorem 1.1) for integer $z, w \in \mathbb{N}_{\geq 1}$:

```
>   intrecursion(t^(z-1)*(1-t)^(w-1),t,S(z));
```
$$-(z+w)\,S(z+1) + S(z)\,z = 0$$
```
>   intrecursion(t^(z-1)*(1-t)^(w-1),t,S(w));
```
$$-(z+w)\,S(w+1) + S(w)\,w = 0$$

The initial values for $w = 1$ and $z = 1$ are easily established.

---

[1] Unfortunately, one cannot use a function name like `I(z)` since `I` is Maple's complex unit.

Example 12.5 Again, we might interpret the resulting holonomic equations in an appropriate noncommutative polynomial ring. Let's have another look at Example 12.3 from this point of view. Here the resulting differential equation can be written as

$$\left(D^2 - 4\right) I(x) = 0$$

using the differential operator D. The product rule shows that $D(xf(x)) = f(x) + xf'(x) = (1 + xD)f(x)$, i.e. the commutator rule $Dx - xD = 1$ is valid. The operator polynomial $D^2 - 4$ has the factorizations

$$D^2 - 4 = (D - 2)(D + 2) = (D + 2)(D - 2)$$

with the two different right factors $D \pm 2$ corresponding to the particular hyperexponential solutions $e^{\pm 2x}$ that we met in Example 12.3. In particular, we realize that Algorithm 12.1 did not discover the differential equation of lowest order valid for $I(x)$.

This was a very simple example. Let's do a more complicated one! We consider

$$I(x) := \int_{-\infty}^{\infty} \frac{x^2}{\left(x^4 + t^2\right)\left(1 + t^2\right)} \, dt. \tag{12.6}$$

Algorithm 12.1 yields the differential equation *DE2* of second order

```
>   DE2:=intdiffeq(x^2/((x^4+t^2)*(1+t^2)),t,S(x));
```

$$DE2 := (x - 1)(x + 1)(x^2 + 1)x \tfrac{d^2}{dx^2} S(x) + (7x^4 + 1)\tfrac{d}{dx} S(x) + 8x^3 S(x) = 0$$

for $I(x)$. This corresponds to the operator equation $P(D, x)I(x) = 0$ with the operator polynomial

$$P(D, x) := (x^4 - 1)x D^2 + \left(7x^4 + 1\right) D + 8x^3. \tag{12.7}$$

The REDUCE implementation `ncpoly` [MA94] mentioned earlier yields 60 different noncommutative polynomial factorizations of $P(D, x)$ (wow!), one of which is given by[2]

$$P(D, x) = \left((x^2 - 1)x D + (3x^2 + 1)\right)\left((x^2 + 1)D + 2x\right). \tag{12.8}$$

The right factor $(x^2 + 1)D + 2x$ corresponds to the differential equation

$$(x^2 + 1)I'(x) + 2xI(x) = 0, \tag{12.9}$$

---

[2] Singular's [GLMS10] `FirstWeyl` command of the `ncfactor` library can also find this factorization.

that turns out to be valid for $I(x)$. This can be seen by the calculations

```
>   integral:=contgosper((x^2+1)*
>   diff(x^2/((x^4+t^2)*(1+t^2)),x)+2*x*x^2/((x^4+t^2)*(1+t^2)),x);
```

$$-\frac{-x^2 + t^2}{(x^4 + t^2)(1 + t^2)}$$

```
>   limit(integral,t=infinity);
```

$$0$$

Solving (12.9) we get

$$\ln I(x) = \int \frac{I'(x)}{I(x)}\,dx = -\int \frac{2x}{1+x^2}\,dx = -\ln(1 + x^2) + C_1,$$

hence

$$I(x) = \frac{C}{1 + x^2}.$$

By the calculation

$$I(1) = \int_{-\infty}^{\infty} \frac{1}{(1+t^2)^2}\,dt = \frac{\pi}{2}$$

we have $C = \pi$, and finally for $x \neq 0$ the result

$$I(x) = \int_{-\infty}^{\infty} \frac{x^2}{(x^4 + t^2)(1 + t^2)}\,dt = \frac{\pi}{1 + x^2}.$$

Now, having resolved the particular question of computing the definite integral (12.6), we would like to consider the question of factoring differential operators from a more general perspective. Maple can factor differential operators over $\mathbb{Q}(x)$ (and not only over the polynomial ring $\mathbb{Q}[x]$!) by an implementation of Mark van Hoeij ([vanHoeij96, vanHoeij97], see also [Bronstein94]). This implementation is part of the DEtools package. Note that this implementation is used by the dsolve command to solve holonomic differential equations.

Considering our above example, again, as a first step, we select the names of the operators and convert the differential equation to a differential operator:

```
>   _Envdiffopdomain:=[Dx,x]:
>   P:=DEtools[de2diffop](DE2,S(x));
```

$$P := (x^5 - x)\,Dx^2 + (7\,x^4 + 1)\,Dx + 8\,x^3$$

according to (12.7). The DEtools[DFactor] command can factorize differential operators over $\mathbb{Q}(x)$. In our example, we get

```
>   fac:=DEtools[DFactor](P);
```

$$fac := \left[(x^5 - x)\, Dx + 3\, x^4 + 1, \; Dx + \frac{4\, x^3}{(x-1)\,(x+1)\,(x^2+1)} \right]$$

resulting in a left and a right factor of P. With `DEtools[mult]` one can multiply the two differential operators resulting in P again[3]:

```
>   DEtools[mult](op(1,fac),op(2,fac));
```

$$(x^5 - x)\, Dx^2 + (7\, x^4 + 1)\, Dx + 8\, x^3$$

Of course the given first order right factor corresponds to a hyperexponential solution. We deduce the corresponding first order differential equation

```
>   DE1:=DEtools[diffop2de](op(2,fac),S(x));
```

$$DE1 := \frac{4\, x^3\, S(x)}{(x-1)\,(x+1)\,(x^2+1)} + \left(\tfrac{d}{dx}\, S(x) \right)$$

and get the logarithmic derivative of this solution as

```
>   logder:=solve(DE1,diff(S(x),x))/S(x);
```

$$logder := -\frac{4\, x^3}{(x-1)\,(x+1)\,(x^2+1)}$$

so that the corresponding solution is given as

```
>   exp(int(logder,x));
```

$$\frac{1}{(x-1)\,(x+1)\,(x^2+1)}$$

which therefore is also given by `dsolve`

```
>   dsolve(DE1,S(x));
```

$$S(x) = \frac{\_C1}{(x^2+1)\,(x^2-1)}$$

Having computed this solution, one can find the second linearly independent solution. This explains `dsolve`'s output for our starting differential equation $DE2$

```
>   dsolve(DE2,S(x));
```

$$S(x) = \frac{\_C1}{x^4-1} + \frac{\_C2\, x^2}{x^4-1}$$

which—of course—covers (with $C_1 = -\pi$ and $C_2 = \pi$) the obtained solution of our particular problem.

Example 12.6 (Euler Integral Representation)   Euler gave the following integral representation

$$_2F_1\left(\begin{matrix} a,\, b \\ c \end{matrix} \,\middle|\, x \right) = \frac{\Gamma(c)}{\Gamma(b)\Gamma(c-b)} \int_0^1 t^{b-1}(1-t)^{c-b-1}(1-tx)^{-a}\, dt \qquad (12.10)$$

---

[3] Similarly the two differential operators given in (12.8) can be treated.

for the Gauss hypergeometric function. Can we prove it? We note that (12.10) is valid if its left- and right-hand sides both satisfy the same holonomic differential equation and initial values. Let's first generate their common differential equation using Algorithms 10.2 and 12.1, respectively.

```
>   sumdiffeq(hyperterm([a,b],[c],x,k),k,S(x));
```

$$x (x-1) \tfrac{d^2}{dx^2} S(x) + (x + x\,a + x\,b - c) \tfrac{d}{dx} S(x) + S(x)\,a\,b = 0$$

```
>   intdiffeq(GAMMA(c)/(GAMMA(b)*GAMMA(c-b))*
>   t^(b-1)*(1-t)^(c-b-1)*(1-t*x)^(-a),t,S(x));
```

$$x (x-1) \tfrac{d^2}{dx^2} S(x) + (x + x\,a + x\,b - c) \tfrac{d}{dx} S(x) + S(x)\,a\,b = 0$$

For $x = 0$, the initial value statement

$$1 = \frac{\Gamma(c)}{\Gamma(b)\Gamma(c-b)} \int_0^1 t^{b-1}(1-t)^{c-b-1}\, dt$$

is a particular case of the Beta function identity (12.5). The equality of the corresponding derivatives at $x = 0$ yields

$$\frac{b}{c} = \frac{\Gamma(c)}{\Gamma(b)\Gamma(c-b)} \int_0^1 t^b (1-t)^{c-b-1}\, dt$$

another case of the Beta function identity. These results are also suggested by the computations

```
>   intrecursion(
>   GAMMA(c)/(GAMMA(b)*GAMMA(c-b))*t^(b-1)*(1-t)^(c-b-1),t,S(b));
```

$$S(b) - S(b+1) = 0$$

```
>   intrecursion(c/b*GAMMA(c)/(GAMMA(b)*GAMMA(c-b))*
>   subs(x=0,diff(t^(b-1)*(1-t)^(c-b-1)*(1-t*x)^(-a),x)),t,S(b));
```

$$S(b) - S(b+1) = 0$$

Note that (12.10) turns out to be valid whenever $|x| < 1$ and $\mathrm{Re}\,(c) > \mathrm{Re}\,(b) > 0$.

Example 12.7 (Bateman Integral Representation) Bateman discovered a hypergeometric representation for

$$\int_0^1 t^{c-1} (1-t)^{d-1}\, {}_2F_1 \left(\begin{matrix} a, b \\ c \end{matrix} \,\middle|\, tx \right) dt.$$

How can we rediscover his representation? In Example 7.8 for a similar discrete example we changed the order of summation, and were successful. Hence we change the order between integration and summation and interpret the above integral as

$$\sum_{k=-\infty}^{\infty} S_k = \sum_{k=-\infty}^{\infty} \int_0^1 t^{c-1} (1-t)^{d-1} \frac{(a)_k\,(b)_k}{(c)_k\,k!} (tx)^k \, dt.$$

The calculation

```
>   intrecursion(
>   t^(c-1)*(1-t)^(d-1)*hyperterm([a,b],[c],t*x,k),t,S(k));
```

$$-(k+1)\,(d+k+c)\,S(k+1) + S(k)\,x\,(b+k)\,(a+k) = 0$$

shows that the resulting function is indeed a hypergeometric sum. Evaluating S_k using the initial value

$$S_0 = \int_0^1 t^{c-1} (1-t)^{d-1} \, dt = \frac{\Gamma(c)\Gamma(d)}{\Gamma(c+d)},$$

we have finally deduced *Bateman's identity* ([Bateman09, Erdélyi53], p. 78)

$$\int_0^1 t^{c-1} (1-t)^{d-1} \, {}_2F_1\left(\begin{array}{c} a,b \\ c \end{array} \Bigg| \, tx \right) dt = \frac{\Gamma(c)\Gamma(d)}{\Gamma(c+d)} \, {}_2F_1\left(\begin{array}{c} a,b \\ c+d \end{array} \Bigg| \, x \right).$$

Note that this method can be used to prove and discover many hypergeometric representations of integrals of sums, in particular for integrals of orthogonal polynomials, see e.g. [Feldheim43, AF69], or [AG71]. Some more examples of this type are given in the exercises.

Further Reading

For further reading on the algorithms of this chapter see [AZ90].

Exercises

Exercise 12.1 *Find both a holonomic recurrence and differential equation for the* Abramowitz functions (*see* [AS64], 27.5)

$$A(n, x) := \int_0^\infty t^n \, e^{-t^2 - \frac{x}{t}} \, dt. \tag{12.11}$$

Exercise 12.2 *Prove the following integral representations for the Kummer hypergeometric function*

$$\frac{\Gamma(a)\Gamma(b-a)}{\Gamma(b)}\; {}_1F_1\left(\begin{matrix} a \\ b \end{matrix}\;\middle|\; x\right) = \int_0^1 e^{tx}\, t^{a-1}\,(1-t)^{b-a-1}\,dt$$

$$= 2^{1-b}\, e^{x/2} \int_{-1}^1 e^{-\frac{tx}{2}}\,(1+t)^{b-a-1}\,(1-t)^{a-1}\,dt.$$

Exercise 12.3 *Evaluate the integral*

$$\int_0^1 e^{-1/t}\, t^{-3-n}\,(1-t)^n\,dt.$$

Exercise 12.4 *Describe a continuous version of the WZ method for the proof of a hypergeometric term integration identity. Apply the method to the appropriate examples of this chapter, in particular to Exercise 12.3.*

Exercise 12.5 *Deduce the identities*

(a) $\;{}_3F_2\left(\begin{matrix} a,b,c \\ d,e \end{matrix}\;\middle|\; x\right) = \frac{\Gamma(d)}{\Gamma(a)\Gamma(d-a)} \int_0^1 t^{a-1}\,(1-t)^{d-a-1}\; {}_2F_1\left(\begin{matrix} b,c \\ e \end{matrix}\;\middle|\; xt\right)dt,$

(b) $\;{}_1F_1\left(\begin{matrix} a \\ b \end{matrix}\;\middle|\; x\right) = \frac{1}{\Gamma(a)} \int_0^\infty e^{-t}\, t^{a-1}\; {}_0F_1\left(\begin{matrix} - \\ b \end{matrix}\;\middle|\; xt\right)dt,$

(c) $\;{}_2F_1\left(\begin{matrix} a,b-d \\ c \end{matrix}\;\middle|\; x\right) = \frac{\Gamma(b)}{\Gamma(d)\Gamma(b-d)} \int_0^1 t^{b-d-1}\,(1-t)^{d-1}\; {}_2F_1\left(\begin{matrix} a,b \\ c \end{matrix}\;\middle|\; xt\right)dt\,.$

◇ **Exercise 12.6 [Derivative Rules]** *Implement* `intdiffrule(F,t,S(n,x))` *as a modified version of Algorithm 10.2 which detects a derivative rule (see Example 10.7) of the form*

$$I_n'(x) = \sum_{j=0}^J \sigma_j(n,x)\, I_{n+j}(x)$$

with $\sigma_j \in \mathbb{Q}(n,x)$ for

$$I_n(x) = \int_a^b F(n,x,t)\,dt,$$

$F(n,x,t)$ *being hypergeometric with respect to n, hyperexponential with respect to t, and strictly hyperexponential with respect to x, under the hypothesis that the integration bounds are chosen such that the resulting equation is homogeneous.*

Apply the procedure to the Abramowitz functions (12.11).

Exercise 12.7 (Gauss Identity) *In Exercise 7.6 the Gauss identity (3.1)*

$$_2F_1\left(\begin{matrix} a\,,b \\ c \end{matrix}\,\middle|\,1\right) = \frac{\Gamma(c)\Gamma(c-a-b)}{\Gamma(c-a)\Gamma(c-b)} \tag{12.12}$$

of our hypergeometric database in Chap. 3 was proved for arbitrary a, b and c. Give an alternative proof of (12.12) by using Euler's integral representation of Example 12.6.

Exercise 12.8 *Find a homogeneous differential equation for (see [AS64], (27.6))*

$$I(x) := \int_0^\infty \frac{e^{-t^2}}{t+x}\,dt.$$

Exercise 12.9 (Schläfli's Integral) Show that the contour integral

$$P_n(x) = \frac{1}{2\pi i}\int_\gamma \frac{(t^2-1)^n}{2^n\,(t-x)^{n+1}}\,dt$$

where γ denotes a closed curve surrounding the point $x \in \mathbb{C}$ once in the counter-clockwise direction, is a representation of the Legendre polynomials.

Exercise 12.10 (Airy Integral) *Assume γ is a curve that is not closed but chosen such that the integrand of*

$$A(x) := \frac{1}{2\pi i}\int_\gamma e^{tx-\frac{t^3}{3}}\,dt$$

vanishes at its boundary points. What does γ look like? Can one assume that γ lies entirely on the real axis? Derive a holonomic differential equation for $A(x)$. Note that the resulting differential equation is one of the simplest differential equations whose solutions Ai(x) and Bi(x), called Airy functions (see e.g. [OLBC10], (9.2.1)), are not elementary.

Exercise 12.11 *Show that the Bessel functions*

$$J_n(x) = \left(\frac{x}{2}\right)^n \sum_{k=0}^\infty \frac{(-1)^k}{4^k\,k!\,\Gamma(k+1+n)}\,x^{2k},$$

which we defined in Exercise 10.9, have the integral representation

$$J_n(x) = \frac{(x/2)^n}{\sqrt{\pi}\,\Gamma(n+1/2)}\int_{-1}^1 e^{ixt}\,(1-t^2)^{n-1/2}\,dt.$$

Exercise 12.12 *Give a hypergeometric representation of the integral*

$$\int_0^\infty e^{-a^2 x^2}\, x^{m-1}\, J_n(bx)\, dx,$$

where $J_n(x)$ denote the Bessel functions.

Exercise 12.13 [KS94] *Show that the Bateman functions* (10.10)

$$F_n(t) = \frac{e^{-t}}{n} \sum_{k=1}^n \frac{(-1)^k}{(k-1)!} \binom{n}{k} (2t)^k,$$

which we defined in Exercise 10.6, have the representation

$$F_n(t) = \frac{1}{\pi} \int_{-\infty}^\infty e^{it\tau} \frac{(\tau+i)^{n-1}}{(\tau-i)^{n+1}}\, d\tau.$$

Exercise 12.14 *Calculate*

$$\int_0^\infty t^n e^{-t} L_n\left(t\frac{1-x}{2}\right) dt,$$

$L_n(x) = L_n^{(0)}(x)$ *denoting the Laguerre polynomials.*

Exercise 12.15 *Algorithm 12.1 may yield a differential equation with nonpolynomial coefficients if*

$$\frac{\frac{\partial}{\partial t} F(x,t)}{F(x,t)} \in \mathbb{Q}(t) \quad \text{but} \quad \notin \mathbb{Q}(x)\,.$$

Show that the algorithm generates the differential equation

$$-2\left(-x^2 + \ln(x)\right) x \left(I'(x)\right) + \left(-1 + 2x^2\right) I(x) = 0$$

for

$$I(x) := \int_{-\infty}^\infty x^{t^2} e^{-x^2 t^2}\, dt.$$

Use dsolve *to solve this differential equation, and hence find the explicit formula*

$$I(x) = \frac{\sqrt{\pi}}{\sqrt{x^2 - \ln(x)}}.$$

for $I(x)$. Check this result numerically for $x = 1, \ldots, 5$.

Use the same procedure to evaluate the definite integrals

(a) $\displaystyle\int_{-\infty}^{\infty} x^t\, e^{-x^2 t^2}\, dt$,

(b) $\displaystyle\int_{-\infty}^{\infty} x^{a+bt-ct^2}\, e^{-x^2 t^2}\, dt$.

Exercise 12.16 *Find a differential equation for*

$$G(t) = \int_0^{\infty} e^{-\frac{t}{x}} e^{-\frac{x^2}{2}}\, dx \; ;$$

see SIAM Review 37, 1995, Problem 95-16 [Glasser95].

Exercise 12.17 *Prove the integral representation*

$$\sum_{k=0}^{n} \frac{1}{k!} x^k = \frac{x^{n+1} e^x}{n!} \int_1^{\infty} t^n e^{-xt}\, dt$$

for the partial sums of the exponential series (see e.g. [AS64], (5.2.8)). Hint: Use Exercise 2.19 to find a common differential equation. Note that the limits of integration are not the natural ones.

References

AS64. Abramowitz, M., Stegun, I.A.: Handbook of Mathematical Functions. Dover Publications, New York (1964)

AZ90. Almkvist, G., Zeilberger, D.: The method of differentiating under the integral sign. J. Symb. Comput. **10**, 571–591 (1990)

AZ91. Almkvist, G., Zeilberger, D.: A Maple program that finds, and proves, recurrences and differential equations satisfied by hyperexponential integrals. SIGSAM Bull. **25**, 14–17 (1991)

AG71. Askey, R., Gasper, G.: Jacobi polynomial expansions of Jacobi polynomials with non-negative coefficients. Proc. Camb. Phil. Soc. **70**, 243–255 (1971)

AF69. Askey, R., Fitch, J.: Integral representations for Jacobi polynomials and some applications. J. Math. Anal. Appl. **26**, 411–437 (1969)

Bateman09. Bateman, H.: The solution of linear differential equations by means of definite integrals. Trans. Cambridge Philos. Soc. **21**, 171–196 (1909)

Bronstein94. Bronstein, M.: An improved algorithm for factoring linear ordinary differential operators. In: Proceedings of ISSAC 94, ACM Press, New York, 1994, pp. 336–340

Erdélyi53. Erdélyi, A., et al.: Higher Transcendental Functions, vol. 1. McGraw-Hill, New York (1953)

Feldheim43. Feldheim, E.: Relations entre les polynomes de Jacobi. Laguerre et Hermite. Acta Math. 75, 117–138 (1943)

Glasser95. Glasser, M.L.: Problem 95–16. SIAM Rev. **37**, 608 (1995)
GLMS10. Greuel, G.-M., Levandovskyy, V., Motsak, A., Schönemann, H.: Plural. A Singular 3.1 Subsystem for Computations with Non-commutative Polynomial Algebras. Centre for Computer Algebra, TU Kaiserslautern. http://www.singular.uni-kl.de (2010)
KS94. Koepf, W., Schmersau, D.: Bounded nonvanishing functions and Bateman functions. Complex Variables **25**, 237–259 (1994)
OLBC10. Olver, F.W.J., Lozier, D.W., Boisvert, R. F., Clark, C.W.: NIST Handbook of Mathematical Functions. Cambridge University Press, New York (2010) http://dlmf.nist.gov
MA94. Melenk, H., Apel, J.: REDUCE package NCPOLY: Computation in Non-commutative Polynomial Ideals. Konrad-Zuse-Zentrum, Berlin (ZIB) (1994)
vanHoeij96. van Hoeij, M.: Factorization of linear differential operators. Ph.D. thesis, Universiteit Nijmegen (1996)
vanHoeij97. van Hoeij, M.: Factorization of differential operators with rational function coefficients. J. Symb. Comput. **24**, 537–561 (1997)

Chapter 13
Rodrigues Formulas and Generating Functions

In this chapter we use the algorithms of the preceding chapter to obtain holonomic equations for function families given by Rodrigues type formulas and generating functions [AZ90]. For this purpose we must assume that the reader is familiar with complex *contour integration* and the *Cauchy integral formula*, see e.g. [Ahlfors53].

Let $f(x)$ denote a function of a complex variable x that is analytic in a simply-connected domain $D \subset \mathbb{C}$ (for example in a disk). Recall that the Cauchy integral formula states that such an $f(x)$ is infinitely differentiable, and its derivatives can be expressed by the integrals

$$f^{(n)}(x) = \frac{n!}{2\pi i} \int_\gamma \frac{f(t)}{(t-x)^{n+1}} \, dt \tag{13.1}$$

where γ denotes a closed curve lying completely in D and winding around the point $x \in \mathbb{C}$ once in the counterclockwise direction. This will be assumed throughout the current chapter.

Assume now that a family of functions $f_n(x)$ is given by a *Rodrigues type formula*

$$f_n(x) = g_n(x) \frac{d^n}{dx^n} h_n(x) \tag{13.2}$$

in terms of the nth derivative of another function $h_n(x)$. Applying (13.1) to $f = h_n$, we can write $f_n(x)$ as the integral

$$f_n(x) = g_n(x) \frac{n!}{2\pi i} \int_\gamma \frac{h_n(t)}{(t-x)^{n+1}} \, dt. \tag{13.3}$$

W. Koepf, *Hypergeometric Summation*, Universitext,
DOI: 10.1007/978-1-4471-6464-7_13, © Springer-Verlag London 2014

Example 13.1 (*Legendre Polynomials*) As an example, we consider the family $f_n(x)$ given by the Rodrigues formula

$$f_n(x) = \frac{(-1)^n}{2^n\,n!}\,\frac{d^n}{dx^n}(1-x^2)^n. \tag{13.4}$$

An application of (13.3) shows that $f_n(x)$ has the integral representation

$$f_n(x) = \frac{(-1)^n}{2^n}\,\frac{1}{2\pi i}\int_\gamma \frac{(1-t^2)^n}{(t-x)^{n+1}}\,dt.$$

On the other hand, in Exercise 12.9 it was shown that the Legendre polynomials are given by Schläfli's integral

$$P_n(x) = \frac{1}{2\pi i}\int_\gamma \frac{(t^2-1)^n}{2^n\,(t-x)^{n+1}}\,dt.$$

Hence we have verified that the Rodrigues formula (13.4) represents the Legendre polynomials

$$P_n(x) = \frac{(-1)^n}{2^n\,n!}\,\frac{d^n}{dx^n}(1-x^2)^n.$$

Session 13.2 Representation (13.3) makes it possible to utilize Algorithm 12.1 for functions given by a Rodrigues type formula (13.2).
The Maple procedures

```
rodriguesrecursion:=proc(g,h,x,sn)
local S,n,t,result;
if type(sn,function) then
  S:=op(0,sn); n:=op(1,sn);
else
  n:=op(1,sn);
end if;
result:=intrecursion(n!*g*subs(x=t,h)/(t-x)^(n+1),t,S(n));
end proc:


rodriguesdiffeq:=proc(g,h,n,sx)
local S,x,t,result;
if type(sx,function) then
  S:=op(0,sx); x:=op(1,sx);
else
  x:=op(1,sx);
end if;
result:=intdiffeq(g*subs(x=t,h)/(t-x)^(n+1),t,S(x));
end proc:
```

use the integral representation (13.3) to find recurrence and differential equations for $f_n(x)$, given by (13.2).

Note that the resulting holonomic equations are always homogeneous since γ is closed.

We do some examples. The calculations

```
>   rodriguesrecursion((-1)^n/(2^n*n!),(1-x^2)^n,x,P(n));
```

$$(n+2)\,P(n+2) - x\,(3+2n)\,P(n+1) + (n+1)\,P(n) = 0$$

```
>   rodriguesdiffeq((-1)^n/(2^n*n!),(1-x^2)^n,n,P(x));
```

$$-(-1+x)\,(1+x)\,\tfrac{d^2}{dx^2}\,P(x) - 2\,x\,\tfrac{d}{dx}\,P(x) + P(x)\,n\,(n+1) = 0$$

yield Example 13.1 again, namely the recurrence and differential equations of the Legendre polynomials. These equations might be compared with those calculated from the series representation of the Legendre polynomials

```
>   sumrecursion(binomial(n,k)*binomial(-n-1,k)*((1-x)/2)^k,k,P(n));
```

$$(n+2)\,P(n+2) - x\,(3+2n)\,P(n+1) + (n+1)\,P(n) = 0$$

```
>   sumdiffeq(binomial(n,k)*binomial(-n-1,k)*((1-x)/2)^k,k,P(x));
```

$$-(-1+x)\,(1+x)\,\tfrac{d^2}{dx^2}\,P(x) - 2\,x\,\tfrac{d}{dx}\,P(x) + P(x)\,n\,(n+1) = 0$$

Having therefore proved the above, the family

$$f_n(x) = \frac{(-1)^n}{2^n\,n!}\,\frac{d^n}{dx^n}(1-x^2)^n$$

now satisfies the same recurrence equation as $P_n(x)$ does. To show that $f_n(x) = P_n(x)$ for all $n \in \mathbb{N}_{\geq 0}$ we have only to check the two initial values

$$f_0(x) = \frac{(-1)^n}{2^n\,n!}\,\frac{d^n}{dx^n}(1-x^2)^n\bigg|_{n=0} = 1 = P_0(x)$$

and

$$f_1(x) = \frac{(-1)^n}{2^n\,n!}\,\frac{d^n}{dx^n}(1-x^2)^n\bigg|_{n=1} = -\frac{1}{2}\,(-2x) = x = P_1(x).$$

The following are the recurrence and differential equations of the generalized Laguerre polynomials

$$L_n^{(\alpha)}(x) := \sum_{k=0}^{n} \frac{(-1)^k}{k!}\binom{n+\alpha}{n-k}x^k$$

```
>    sumrecursion((-1)^k/k!*binomial(n+alpha,n-k)*x^k,k,L(n));
```

$$(n+2)\,L(n+2) - (\alpha+3-x+2n)\,L(n+1) + (n+\alpha+1)\,L(n) = 0$$

```
>    sumdiffeq((-1)^k/k!*binomial(n+alpha,n-k)*x^k,k,L(x));
```

$$x\,\frac{d^2}{dx^2}\,L(x) + (-x+\alpha+1)\,\frac{d}{dx}\,L(x) + L(x)\,n = 0$$

stemming from their series representation. These calculations can be compared with

```
>    rodriguesrecursion(
>    exp(x)/(n!*x^alpha),exp(-x)*x^(alpha+n),x,L(n));
```

$$(n+2)\,L(n+2) - (\alpha+3-x+2n)\,L(n+1) + (n+\alpha+1)\,L(n) = 0$$

```
>    rodriguesdiffeq(exp(x)/(n!*x^alpha),exp(-x)*x^(alpha+n),n,L(x));
```

$$x\,\frac{d^2}{dx^2}\,L(x) + (-x+\alpha+1)\,\frac{d}{dx}\,L(x) + L(x)\,n = 0$$

After checking the initial values, these calculations prove the valid Rodrigues formula

$$L_n^{(\alpha)}(x) = \frac{e^x}{n!\,x^\alpha}\,\frac{d^n}{dx^n}\left(e^{-x}\,x^{\alpha+n}\right)$$

for the generalized Laguerre polynomials.

Next, we would like to find recurrence equations for families given by *discrete Rodrigues formulas*. These are in terms of the nth power of one of the difference operators Δ or ∇ instead of the differential operator $\frac{d}{dx}$. Note that this research has been done by Fischer [Fischer13].

There are two difference operators, the *backward difference operator*

$$\nabla f(x) = f(x) - f(x-1)$$

and the *forward difference operator*

$$\Delta f(x) = f(x+1) - f(x).$$

Since the theories with ∇ and Δ are very similar, we consider here only the ∇ case.

To be able to apply Zeilberger's algorithm in a suitable way it is essential to have a representation of $\nabla^n f(x)$ as a series, similarly as Cauchy's theorem yields an integral representation for $\frac{d^n}{dx^n} f(x)$. However, such a representation is easy to find. Since

$$\nabla^2 f(x) = f(x) - 2\,f(x-1) + f(x-2),$$
$$\nabla^3 f(x) = f(x) - 3\,f(x-1) + 3\,f(x-2) - f(x-3),$$

by induction we get for $n \in \mathbb{N}_{\geq 0}$

$$\nabla^n f(x) = \sum_{k=0}^{n} (-1)^k \binom{n}{k} f(x - k). \qquad (13.5)$$

Therefore, using (13.5), we can apply Zeilberger's algorithm to every family of functions $f_n(x)$ of the form

$$f_n(x) = g_n(x)\, \nabla^n h_n(x) \qquad (13.6)$$

in terms of the nth power of ∇ of $h_n(x)$.

Session 13.3 Representation (13.5) makes it possible to utilize Algorithm 8.5 for functions given by a discrete Rodrigues type formula (13.6).
   The Maple procedure

```
nablarodriguesrec:=proc(g,h,x,sn)
local S,n,k,result;
if type(sn,function) then
  S:=op(0,sn); n:=op(1,sn);
else
  n:=op(1,sn);
end if;
result:=sumrecursion(g*(-1)^k*binomial(n,k)*subs(x=x-k,h),k,S(n));
end proc:
```

uses the series representation (13.5) to find a recurrence equation w.r.t. n for $f_n(x)$, given by (13.6). Note that the resulting holonomic equation is always homogeneous since the binomial coefficient in (13.5) makes the bounds natural.
   Assume the family $f_n(x)$ is given by the formula

$$f_n(x) = \frac{x!}{a^x}\, \nabla^n \frac{a^x}{x!}.$$

Then we can compute the following recurrence equation w.r.t. n

```
>   nablarodriguesrec(x!/a^x,a^x/x!,x,C(n));
```

$$a\, C(n+2) - (a + n - x + 1)\, C(n+1) + (n+1)\, C(n) = 0$$

for $f_n(x)$. Comparing this recurrence equation with that for the *Charlier polynomials* (see Exercise 7.10) given by the hypergeometric representation

$$C_n(x, a) = {}_2F_0\left(\begin{array}{c} -n, -x \\ - \end{array} \middle| -\frac{1}{a} \right),$$

through the computation

```
>   sumrecursion(hyperterm([-n,-x],[],-1/a,k),k,C(n));
```

$$a\, C(n+2) - (a + n - x + 1)\, C(n+1) + (n+1)\, C(n) = 0$$

we see that $f_n(x) = C_n(x, a)$ by just checking two initial values again. This yields the representation

$$C_n(x, a) = \frac{x!}{a^x} \nabla^n \frac{a^x}{x!}$$

for the Charlier polynomials given in ([KLS10], (9.14.10)).

Note, however, that the discrete Rodrigues formula can even be translated *directly* towards a series representation by (13.5). In the Charlier case, we get for example

$$f_n(x) = \frac{x!}{a^x} \nabla^n \frac{a^x}{x!}$$

$$= \frac{x!}{a^x} \sum_{k=0}^{n} (-1)^k \binom{n}{k} \frac{a^{x-k}}{(x-k)!} = {}_2F_0 \left(\begin{matrix} -n, -x \\ - \end{matrix} \,\middle|\, -\frac{1}{a} \right)$$

as the computation

```
>    Sumtohyper(x!/a^x*(-1)^k*binomial(n,k)*subs(x=x-k,a^x/x!),k);
```

$$\mathrm{Hypergeom}\left([-n, -x], [], -\frac{1}{a} \right)$$

shows.

Similarly the code

```
nablarodriguesdiffeq:=proc(g,h,n,sx)
local S,x,k,result;
if type(sx,function) then
  S:=op(0,sx); x:=op(1,sx);
else
  x:=op(1,sx);
end if;
result:=sumrecursion(g*(-1)^k*binomial(n,k)*subs(x=x-k,h),k,S(x));
end proc:
```

generates the two identical difference equations (which are also recurrence equations) for the two different representations:

```
>    nablarodriguesdiffeq(x!/a^x,a^x/x!,n,C(x));
```

$$a\,C(x+2) - (a - n + x + 1)\,C(x+1) + (x+1)\,C(x) = 0$$

```
>    sumrecursion(hyperterm([-n,-x],[],-1/a,k),k,C(x));
```

$$a\,C(x+2) - (a - n + x + 1)\,C(x+1) + (x+1)\,C(x) = 0$$

Other families can be treated in a similar way, see Exercise 13.7.

Next, we examine *generating functions* of the family of functions $f_n(x)$. Note that the generating function of $f_n(x)$ is given by the power series

$$F(z) = \sum_{n=0}^{\infty} f_n(x)\, z^n.$$

Often generating functions are not elementary functions (see e.g. (6) and Exercise 13.11) or the formal power series is nowhere convergent. Therefore we deal with the more general generating function

$$F(z) = \sum_{n=0}^{\infty} a_n\, f_n(x)\, z^n, \tag{13.7}$$

a_n being a given sequence. In particular, for $a_n = 1/n!$, this is called the *exponential generating function* of $f_n(x)$.

Applying the Cauchy integral formula (13.1) to $F(z)$ and using Taylor's theorem we get the integral representation

$$f_n(x) = \frac{1}{a_n}\frac{F^{(n)}(0)}{n!} = \frac{1}{a_n}\frac{1}{2\pi i}\int_\gamma \frac{F(t)}{t^{n+1}}\, dt \tag{13.8}$$

for $f_n(x)$, which, in combination with Algorithm 12.1, can be used to obtain recurrence and differential equations for functions whose generating function $F(z)$ (given by (13.7)) is a hyperexponential term.

Session 13.4 The Maple procedures

```
GFrecursion:=proc(F,a,z,sn)
local S,n;
if type(sn,function) then
   S:=op(0,sn); n:=op(1,sn);
else
   n:=op(1,sn);
end if;
return intrecursion(F/a/z^(n+1),z,S(n));
end proc:


GFdiffeq:=proc(F,a,z,n,sx)
local S,x;
if type(sx,function) then
   S:=op(0,sx); x:=op(1,sx);
else
   x:=op(1,sx);
end if;
return intdiffeq(F/a/z^(n+1),z,S(x));
end proc:
```

use Algorithm 12.1 to find recurrence and differential equations for $f_n(x)$, given by (13.7), according to (13.8).

For example, the calculations

```
>   GFrecursion((1-z)^(-alpha-1)*exp(x*z/(z-1)),1,z,L(n));
```

$$(n+2)\,L(n+2) - (\alpha+3-x+2n)\,L(n+1) + (n+\alpha+1)\,L(n) = 0$$

```
>   GFdiffeq((1-z)^(-alpha-1)*exp(x*z/(z-1)),1,z,n,L(x));
```

$$x\,\tfrac{d^2}{dx^2}\,L(x) + (-x+\alpha+1)\,\tfrac{d}{dx}\,L(x) + L(x)\,n = 0$$

show that

$$(1-z)^{-\alpha-1}\,e^{\frac{xz}{z-1}} = \sum_{n=0}^{\infty} L_n^{(\alpha)}(x)\,z^n$$

is the generating function of the generalized Laguerre polynomials. The initial values can be checked by using Taylor's theorem.

Whereas `GFrecursion` works for hyperexponential input, a more general procedure can find the recurrence equation for the power series coefficients for every holonomic input. This is implemented in the `FormalPowerSeries` package ([Koepf92, Koepf06]) and can be invoked by the `convert` command.

For the same example as above we get

```
>   convert((1-z)^(-alpha-1)*exp(x*z/(z-1)),
>   FormalPowerSeries,z,L(n),recurrence);
```

$$(n+2)\,L(n+2) + (-2n+x-\alpha-3)\,L(n+1) + (n+1+\alpha)\,L(n) = 0$$

which is of course the same recurrence equation.

q-Rodrigues Formulas

In this section we consider the computation of q-recurrence equations for families given by a q-Rodrigues formula. The research of this section was again developed by Fischer [Fischer13].

In ([KLS10], (14.21.12)), the following formula is given for the q-Laguerre polynomials

$$L_n^{(\alpha)}(x;q) = \frac{(1-q)^n}{(q,q)_n\,w(x;\alpha;q)}\,\mathscr{D}_q^n w(x;\alpha+n;q) \qquad (13.9)$$

where

$$w(x;\alpha;q) = \frac{x^\alpha}{(-x,q)_\infty}.$$

To continue, we need a series representation for the nth power of the q-derivative operator $\mathscr{D}_q$. Such a representation was given by ([KRM07], (4))

$$\mathscr{D}_q^n f(x) = \frac{1}{(1-q)^n\, x^n} \sum_{k=0}^n (-1)^k \begin{bmatrix} n \\ k \end{bmatrix}_q q^{\binom{k}{2}-(n-1)k} f(q^k x). \tag{13.10}$$

For a proof see also ([Sprenger09], 1.12). Note that another formula for $\mathscr{D}_q^n f(x)$ appeared in ([AM08], 2.13), which contains a misprint, though.[1]

Session 13.5 To get a recurrence equation for

$$f_n(x) = g_n(x)\, \mathscr{D}_q^n h_n(x),$$

we apply the q-Zeilberger algorithm to (13.10) and therefore get the implementation

```
qrodriguesrec:=proc(g,h,q,x,sn)
local S,n,k,result;
if type(sn,function) then
   S:=op(0,sn); n:=op(1,sn);
else
   n:=op(1,sn);
end if;
result:=qsumrecursion(g/((1-q)^n*x^n)*(-1)^k*qbinomial(n,k,q)*
   q^(binomial(k,2)-(n-1)*k)*subs(x=x*q^k,h),q,k,S(n));
end proc:
```

Applying this procedure to (13.9) we get

```
>   w:=x^alpha/qpochhammer(-x,q,infinity);
```

$$w := \frac{x^\alpha}{\text{qpochhammer}(-x,\, q,\, \infty)}$$

```
>   qrodriguesrec((1-q)^n/(qpochhammer(q,q,n)*w),
>   subs(alpha=alpha+n,w),q,x,L(n));
```

$$(-1+q^n)\, q\, L(n) + (-q^{(2n+\alpha)} x - q^{(\alpha+n+1)} - q^{(n+1)} + q^2 + q)\, L(n-1)$$
$$+ (q^{(\alpha+n)} - q)\, q\, L(-2+n) = 0$$

Of course the series representation

---

[1] Their corrected formula $\mathscr{D}_q^n f(x) = \frac{(-1)^n q^{-\binom{n}{2}}}{(1-q)^n\, x^n} \sum_{r=0}^n (-1)^r \begin{bmatrix} n \\ r \end{bmatrix}_q q^{\binom{r}{2}} f(q^{n-r} x)$ follows from
(13.10) by changing the order of summation $r = n - k$.

$$L_n^{(\alpha)}(x; q) = \frac{(q^{\alpha+1}; q)_n}{(q; q)_n} \, {}_1\phi_1\left(\begin{array}{c} q^{-n} \\ q^{\alpha+1} \end{array}\bigg| q, -xq^{n+\alpha+1}\right)$$

yields the same recurrence equation for $L_n^{(\alpha)}(x; q)$:

```
>    qsumrecursion(qpochhammer(q^(alpha+1),q,n)/qpochhammer(q,q,n)*
>    qphihyperterm([q^(-n)],[q^(alpha+1)],q,-x*q^(n+alpha+1),k),
>    q,k,L(n));
```

$$(-1 + q^n) q \, L(n) + (-q^{(2n+\alpha)} x - q^{(\alpha+n+1)} - q^{(n+1)} + q^2 + q) L(n-1)$$
$$+ (q^{(\alpha+n)} - q) q \, L(-2+n) = 0$$

As in the discrete case, we can use (13.10) directly to rewrite the q-Rodrigues formula (13.9) as a series. This can be done by the computation

```
>    sum2qhyper((1-q)^n/
>    (qpochhammer(q,q,n)*w)/((1-q)^n*x^(n))*(-1)^k*
>    qbinomial(n,k,q)*q^(binomial(k,2)-(n-1)*k)*
>    subs({x=x*q^k,alpha=alpha+n},w),q,k);
```

$$\frac{\phi([q^{(-n)}, -x], [0], q, q^\alpha q^n q)}{\mathrm{qpochhammer}(q, q, n)}$$

We have therefore found a second independent hypergeometric representation

$$L_n^{(\alpha)}(x; q) = \frac{1}{(q; q)_n} \, {}_2\phi_1\left(\begin{array}{c} q^{-n}, -x \\ 0 \end{array}\bigg| q, q^{n+\alpha+1}\right).$$

Compare ([KLS10], (14.21.1)).

Further Reading

For further reading on the algorithms of this chapter see [AZ90].

Exercises

Exercise 13.1 (Bateman Functions) *Prove that the Bateman functions* [Bateman31, KS94]

$$F_n(t) = \frac{e^{-t}}{n} \sum_{k=1}^{n} \frac{(-1)^k}{(k-1)!} \binom{n}{k} (2t)^k$$

(see Exercise 10.6) satisfy the Rodrigues formula

$$F_n(t) = \frac{t\, e^t}{n!} \frac{d^n}{dt^n} \left(e^{-2t}\, t^{n-1} \right)$$

and have the generating function

$$e^{-t\frac{1+z}{1-z}} = \sum_{n=0}^{\infty} F_n(t)\, z^n.$$

Exercise 13.2 (Hermite Polynomials) *Prove the Rodrigues formula*

$$H_n(x) = (-1)^n\, e^{x^2} \frac{d^n}{dx^n} e^{-x^2}$$

of the Hermite polynomials

$$H_n(x) := n! \sum_{k=0}^{\lfloor n/2 \rfloor} \frac{(-1)^k}{(n-2k)!\,k!} (2x)^{n-2k}.$$

Furthermore use the method of Example 7.8 to deduce the exponential generating function

$$e^{2xz - z^2} = \sum_{n=0}^{\infty} \frac{1}{n!} H_n(x)\, z^n.$$

Verify the result using the method of the present chapter.

Exercise 13.3 (Jacobi Polynomials) *Prove the Rodrigues formula*

$$P_n^{(\alpha,\beta)}(x) = \frac{(-1)^n}{2^n\, n!} (1-x)^{-\alpha}(1+x)^{-\beta} \frac{d^n}{dx^n} \left((1-x)^\alpha (1+x)^\beta\, (1-x^2)^n \right)$$

for the Jacobi polynomials $P_n^{(\alpha,\beta)}(x)$ which are given by the representation

$$P_n^{(\alpha,\beta)}(x) = \binom{n+\alpha}{n}\, {}_2F_1 \left(\begin{matrix} -n,\, n+\alpha+\beta+1 \\ \alpha+1 \end{matrix} \,\middle|\, \frac{1-x}{2} \right).$$

Exercise 13.4 (Gegenbauer Polynomials) *The Gegenbauer polynomials $C_n^\nu(x)$ are for $\nu \neq 0$ given by*

$$C_n^\nu(x) = \frac{(2\nu)_n}{(\nu+1/2)_n}\, P_n^{(\nu-1/2,\,\nu-1/2)}(x).$$

Show that this definition is compatible with the one given in Exercise 10.5:

$$C_n^\nu(x) = \frac{(2\nu)_n}{n!} \, {}_2F_1 \left(\begin{matrix} -n, n+2\nu \\ \nu+1/2 \end{matrix} \,\middle|\, \frac{1-x}{2} \right).$$

Show also that the Gegenbauer polynomials are generated by the simple function

$$\sum_{n=0}^{\infty} C_n^\nu(x) \, z^n = \frac{1}{\left(1 - 2xz + z^2\right)^\nu}.$$

In particular, one has for the Legendre polynomials

$$\sum_{n=0}^{\infty} P_n(x) \, z^n = \frac{1}{\sqrt{1 - 2xz + z^2}}.$$

Exercise 13.5 *Show that*

$$F(z) = \frac{1}{1 - z - z^2}$$

constitutes the generating function of the Fibonacci numbers.

Exercise 13.6 (Laguerre Polynomials) *Prove the identity*

$$e^{-xz} \, (1+z)^\alpha = \sum_{n=0}^{\infty} L_n^{(\alpha-n)}(x) \, z^n.$$

Exercise 13.7 (Discrete Orthogonal Polynomials: Rodrigues Representations) *The Krawtchouk, Meixner and Hahn polynomials were given in Exercise 7.10 by the following hypergeometric representations:*

$$K_n(x; p, N) = {}_2F_1 \left(\begin{matrix} -n \, , -x \\ -N \end{matrix} \,\middle|\, \frac{1}{p} \right),$$

$$M_n(x; \beta, c) = {}_2F_1 \left(\begin{matrix} -n \, , -x \\ \beta \end{matrix} \,\middle|\, 1 - \frac{1}{c} \right),$$

and

$$Q_n(x; \alpha, \beta, N) = {}_3F_2 \left(\begin{matrix} -n \, , -x \, , \alpha + \beta + n + 1 \\ \alpha + 1 \, , -N \end{matrix} \,\middle|\, 1 \right).$$

Show that these polynomial families have the following discrete Rodrigues representations (see [KLS10], Chap. 9)

$$K_n(x; p, N) = \frac{1}{\binom{N}{x} \left(\frac{p}{1-p}\right)^x} \, \nabla^n \left(\binom{N-n}{x} \left(\frac{p}{1-p}\right)^x \right),$$

$$M_n(x; \beta, c) = \frac{x!}{(\beta)_x \, c^x} \, \nabla^n \left(\frac{(\beta + n)_x \, c^x}{x!} \right),$$

and

$$Q_n(x; \alpha, \beta, N) = \frac{(-1)^n \, (\beta + 1)_n}{\omega(x; \alpha, \beta, N) \, (-N)_n} \, \nabla^n \omega(x; \alpha + n, \beta + n, N - n),$$

where

$$\omega(x; \alpha, \beta, N) = \binom{\alpha + x}{x} \binom{\beta + N - x}{N - x}.$$

Exercise 13.8 (Discrete Orthogonal Polynomials: Generating Functions) *Show that the Charlier, Krawtchouk and Meixner polynomials (see Exercises 7.10 and 13.7) have the following generating functions (see* [KLS10], *Chap. 9):*

$$\sum_{n=0}^{\infty} C_n(x, a) \frac{z^n}{n!} = e^z \left(1 - \frac{z}{a} \right)^x,$$

$$\sum_{n=0}^{N} \binom{N}{n} K_n(x; p, N) \, z^n = \left(1 - \frac{(1 - p)}{p} z \right)^x (1 + z)^{N-x},$$

$$\sum_{n=0}^{\infty} \frac{(\beta)_n}{n!} M_n(x; \beta, c) \, z^n = \left(1 - \frac{z}{c} \right)^x (1 - z)^{-x-\beta}.$$

$\diamond$ **Exercise 13.9** *Find and prove a summation formula for $\Delta^n f(x)$ similar to* (13.5). *Then write corresponding Maple procedures* `deltarodriguesrec` *and* `deltarodriguesdiffeq` *to compute recurrence and difference equations for* $f_n(x) = g_n(x) \, \Delta^n h_n(x)$. *Use your programs to compute recurrence equations w.r.t.* n *and* x *for*

$$f_n(x) = \binom{x}{n} \Delta^n \frac{a^n \, b^x}{x!}.$$

$\diamond$ **Exercise 13.10** *If $F(z)$ is a Laurent polynomial*

$$F(z) = \sum_{k=-n}^{n} a_k \, z^k,$$

then the coefficient a_0 is called the constant term of $F(z)$ and is denoted by

$$a_0 = \mathrm{CT}_z \, F(z).$$

Write two Maple procedures `CTrecursion(F,z,s(n))` and `CTdiffeq` `(F,z,s(x))` that calculate a holonomic recurrence and differential equation, respectively, for the constant term of $F(z)$.

Determine a holonomic recurrence equation with respect to n and a holonomic differential equation with respect to x for the function

$$S_n(x) := \mathrm{CT}_z \left(z + x + \frac{1}{z} \right)^n.$$

Show further that the functions

$$P_n(x) = \mathrm{CT}_z \left(\frac{(x+z)^2 - 1}{2z} \right)^n$$

are the Legendre polynomials.

Exercise 13.11 (Legendre Polynomials) *Deduce the exponential generating function (6) of the Legendre polynomials*

$$\sum_{n=0}^{\infty} \frac{1}{n!} P_n(x)\, z^n = e^{xz} J_0 \left(z \sqrt{1 - x^2} \right).$$

Hint: Prove the hypergeometric representation

$$P_n(x) = x^n \, {}_2F_1 \left(\begin{matrix} -n/2,\, (1-n)/2 \\ 1 \end{matrix} \,\middle|\, 1 - \frac{1}{x^2} \right)$$

for the Legendre polynomials, and use it to represent the left-hand side. Then change the order of summation.

Exercise 13.12 (Big q-Jacobi Polynomials) *Prove the following identity for the so-called big q-Jacobi polynomials ([KLS10], Sect. 14.5)*

$$P_n(x; a, b, c; q) = {}_3\phi_2 \left(\begin{matrix} q^{-n},\, abq^{n+1},\, x \\ aq,\, cq \end{matrix} \,\middle|\, q, q \right)$$

$$= \frac{a^n c^n q^{n(n+1)} (1-q)^n}{(aq,q)_n\, (cq,q)_n\, w(x; a, b, c; q)}\, \mathscr{D}_q^n w(x; aq^n, bq^n, cq^n; q)$$

where

$$w(x; a, b, c; q) = \frac{(a^{-1}x, q)_\infty\, (c^{-1}x, q)_\infty}{(x, q)_\infty\, (bc^{-1}x, q)_\infty}.$$

◇ **Exercise 13.13** *Implement a Maple procedure* `qrodriguesdiffeq(g,h,q,` `n,s(x))` *using* `qsumdiffeq`, *and apply it to the q-Laguerre and to the Big q-Jacobi polynomials.*

References

Ahlfors53. Ahlfors, L.: Complex Analysis. McGraw-Hill Book Co., New York (1953)

AZ90. Almkvist, G., Zeilberger, D.: The method of differentiating under the integral sign. J. Symb. Comput. **10**, 571–591 (1990)

AM08. Annaby, M.H., Mansour, Z.S.: q-Taylor and interpolation series for Jackson q-difference operators. J. Math. Anal. Appl. **344**, 472–483 (2008)

Bateman31. Bateman, H.: The k-function, a particular case of the confluent hypergeometric function. Trans. Amer. Math. Soc. **33**, 817–831 (1931)

Fischer13. Fischer, K.: Identifikation spezieller Funktionen, die durch Rodrigues-Formeln gegeben sind. Universität Kassel, Private communication (2013)

KLS10. Koekoek, R., Lesky, P., Swarttouw, R.F.: Hypergeometric Orthogonal Polynomials and their q-Analogues. Springer Monographs in Mathematics. Springer, Berlin (2010)

Koepf92. Koepf, W.: Power series in computer algebra. J. Symb. Comput. **13**, 581–603 (1992)

KS94. Koepf, W., Schmersau, D.: Bounded nonvanishing functions and Bateman functions. Complex Variables **25**, 237–259 (1994)

Koepf06. Koepf, W.: Computeralgebra. Springer, Berlin (2006)

KRM07. Koepf, W., Rajkovic, P.M., Marinkovic, S.D.: Properties of q-holonomic functions. J. Differ. Equat. Appl. **13**, 621–638 (2007)

Sprenger09. Sprenger, T.: Algorithmen für q-holonome Funktionen und q-hypergeometrische Reihen. PhD thesis, Universität Kassel (2009)

Index

Symbols

$P_n(x)$ (Legendre polynomials), 1

$:=$ (definition), 1

$\Gamma(z)$ (Gamma function), 1

$\mathbb{C}$ (field of complex numbers), 1

$\mathrm{Re}\, z$ (Real part of z), 1

$F(t)\Big|_{t=a}^{t=b} = F(b) - F(a)$ (difference), 1

$k!$ (factorial), 1

$\mathbb{N}_{\geq 0}$ (set of nonnegative integers), 1

$\mathbb{Z}$ (ring of integers), 2

$\mathbb{R}$ (field of real numbers), 2

$\mathbb{N}$ (set of positive integers), 2

$(z)_k$ (shifted factorial, Pochhammer symbol), 3

$\mathrm{Res}_{z=z_0} f(z)$ (residue of f at z_0), 3

$\binom{z}{k}$ (z choose k, binomial coefficient), 5

$B(z, w)$ (Beta function), 5

$\mathbb{K}$ (field of characteristic zero), 12

$\mathbb{Q} = \mathbb{Q}(x_1, x_2, \ldots, x_m)$ (extension of field of rational numbers), 12

$_pF_q \left(\begin{array}{c} \text{upper} \\ \text{lower} \end{array} \Big| x \right)$ (generalized hypergeometric function), 12

$_2F_1 \left(\begin{array}{c} a, b \\ c \end{array} \Big| x \right)$ (Gauss hypergeometric function), 14

$_1F_1 \left(\begin{array}{c} a \\ b \end{array} \Big| x \right)$ (Kummer's confluent hypergeometric function), 14

$\mathbb{K}(k)$ (field of rational functions over $\mathbb{K}$), 14

$\mathbb{K}[k]$ (ring of polynomials over $\mathbb{K}$), 14

$a_k|_{k=n}$ (substitution), 20

$_r\phi_s \left(\begin{array}{c} \text{upper} \\ \text{lower} \end{array} \Big| q, x \right)$ (q-hypergeometric series), 26

$(a_1, a_2, \ldots, a_r; q)_k$ (q-Pochhammer notation), 27

$(a; q)_k$ (q-Pochhammer symbol), 27

$[k]_q$ (q-brackets), 28

$[k]_q!$ (q-factorial), 28

$\begin{bmatrix} n \\ k \end{bmatrix}_q$ (q-binomial coefficient), 28

$\Gamma_q(z)$ (q-Gamma function), 28

$\theta f(x) = x f'(x)$ (*differential operator*), 29

$\lfloor x \rfloor$ (floor function), 31

$L_n(x)$ (Laguerre polynomials), 57

Δ (forward difference operator), 64, 141

$p_n(x|q)$ (little q-Legendre polynomials), 71

$P_n(x; c; q)$ (big q-Legendre polynomials), 71

$P_n(x|q)$ (continuous q-Legendre polynomials), 71

$P_n(x; q)$ (continuous q-Legendre polynomials), 71

$L_n^{(\alpha)}(x)$ (generalized Laguerre polynomials), 71

$L_n^{(\alpha)}(x; q)$ (q-Laguerre polynomials), 72

F_n (Fibonacci numbers), 74

$f_n(x)$ (Fasenmyer polynomials), 75

$Z_n(x)$ (Bateman polynomials), 75

$H_n(x)$ (Hermite polynomials), 76

$\gcd(f, g)$ (greatest common divisor of polynomials), 80

$\deg(q_k)$ (degree of polynomial), 82, 87

$\mathrm{disp}(q_k, r_k)$ (dispersion of polynomials), 82

$\max S$ (maximum of set S), 82

H_k (harmonic numbers), 89

$k^{\underline{m}}$ (falling factorial), 95

W. Koepf, *Hypergeometric Summation*, Universitext,
DOI: 10.1007/978-1-4471-6464-7, © Springer-Verlag London 2014

If you have any concerns about our products,
you can contact us on
ProductSafety@springernature.com

In case Publisher is established outside the EU,
the EU authorized representative is:
Springer Nature Customer Service Center GmbH
Europaplatz 3, 69115 Heidelberg, Germany

Printed by Libri Plureos GmbH
in Hamburg, Germany